GENERAL SYSTEMS THEORY

A MATHEMATICAL APPROACH

International Federation for Systems Research
International Series on Systems Science and Engineering

Volume 8 *THE ALTERNATIVE MATHEMATICAL MODEL OF LINGUISTIC SEMANTICS AND PRAGMATICS*
Vilém Novák

Volume 9 *CHAOTIC LOGIC: Language, Thought, and Reality from the Perspective of Complex Systems Science*
Ben Goertzel

Volume 10 *THE FOUNDATIONS OF FUZZY CONTROL*
Harold W. Lewis, III

Volume 11 *FROM COMPLEXITY TO CREATIVITY: Explorations in Evolutionary, Autopoietic, and Cognitive Dynamics*
Ben Goertzel

Volume 12 *GENERAL SYSTEMS THEORY: A Mathematical Approach*
Yi Lin

Volume 13 *PRINCIPLES OF QUANTITATIVE LIVING SYSTEMS SCIENCE*
James R. Simms

Volume 14 *INTELLIGENT ROBOTIC SYSTEMS: Design, Planning, and Control*
Witold Jacak

IFSR was established "to stimulate all activities associated with the scientific study of systems and to coordinate such activities at international level." The aim of this series is to stimulate publication of high-quality monographs and textbooks on various topics of systems science and engineering. This series complements the Federation's other publications.

Volumes 1–6 were published by Pergamon Press.

GENERAL SYSTEMS THEORY

A MATHEMATICAL APPROACH

YI LIN

Slippery Rock University
Slippery Rock, Pennsylvania

SPRINGER SCIENCE+BUSINESS MEDIA, LLC

Library of Congress Cataloging-in-Publication Data

Lin, Yi, 1959-
General systems theory : a mathematical approach / Yi Lin.
p. cm. -- (International Federation for Systems Research international series on systems science and engineering ; v. 12)
Includes bibliographical references and index.
ISBN 978-1-4757-7192-3 ISBN 978-0-306-46962-6 (eBook)
DOI 10.1007/978-0-306-46962-6
1. System analysis. I. Title. II. Series: IFSR international series on systems science and engineering ; v. 12.
QA402.L538 1998
003--dc21 98-42068
CIP

ISBN 978-1-4757-7192-3

Originally published by Kluwer Academic / Plenum Publishers, New York in 1999
Softcover reprint of the hardcover 1st edition 1999

10 9 8 7 6 5 4 3 2 1

A C.I.P. record for this book is available from the Library of Congress.

To my parents—without their comprehensive home teaching
I would not have become who I am

To my wife and children—Kimberly, Dillon, Alyssa, and
Bailey for their love, support, and understanding

Preface

As suggested by the title of this book, I will present a collection of coherently related applications and a theoretical development of a general systems theory. Hopefully, this book will invite all readers to sample an exciting and challenging (even fun!) piece of interdisciplinary research, that has characterized the scientific and technological achievements of the twentieth century. And, I hope that many of them will be motivated to do additional reading and to contribute to topics along the lines described in the following pages.

Since the applications in this volume range through many scientific disciplines, from sociology to atomic physics, from Einstein's relativity theory to Dirac's quantum mechanics, from optimization theory to unreasonable effectiveness of mathematics to foundations of mathematical modeling, from general systems theory to Schwartz's distributions, special care has been given to write each application in a language appropriate to that field. That is, mathematical symbols and abstractions are used at different levels so that readers in various fields will find it possible to read. Also, because of the wide range of applications, each chapter has been written so that, in general, there is no need to reference a different chapter in order to understand a specific application. At the same time, if a reader has the desire to go through the entire book without skipping any chapter, it is strongly suggested to refer back to Chapters 2 and 3 as often as possible.

The motivation to write this book came from the strong influence of historical works by L. von Bertalanffy, George Klir, and M. D. Mesarovic, and the book *On Systems Analysis*, by David Berlinski (MIT Press, Cambridge, Massachusetts, 1976). Berlinski's book and challenges from several scholars really made me decide to write such a book with strong applications in different scientific fields in order to justify the very meaning of existence for a general systems theory. At the same time, one of the important lessons we have learned from the several-decade-old global systems movement, started and supported by many of the most powerful minds of our modern time, is that senseless transfer of statements (more specifically, theoretical conclusions or results) from one discipline to another makes people feel that general systems theory is a doubtful subject. To keep such unnecessary situations from occurring, we develop each application with rigorous logical reasoning. Whenever a bold conclusion is deduced, some relevant gaps in the reasoning process will be pointed out right on the spot or in the final chapter ("Some Unsolved Problems in General Systems Theory"). On the other hand, doubtful people will be as doubtful as they can no matter what facts or evidence are out there to show them their doubt is unfounded. For example, more than 100 years ago, when naive set theory was first introduced and studied, many first-class mathematicians did not treat it as a serious theory at all. Furthermore, Cantor, the founder, was personally attacked by these scholars. As a consequence, he was hospitalized and eventually died in a psychiatric hospital. Today, set theory has succeeded in a great many areas of modern science, including the entire spectrum of mathematics, when the central idea of infinity is employed in systems science, we can still hear doubters saying things like: Infinity? One can be sure that in an infinitely long period of time, a monkey will produce the great Beethoven's music! (A note: according to results in set theory, this statement is not true!)

The structure of my theoretical development in this book is the "top-down" — formalization — approach, launched in 1960 by Mesarovic. This approach is characterized by the following: (1) All concepts are introduced with minimal mathematical structures. (2) Additional mathematical conditions are added when necessary to display the richness of systems properties. At the same time, applicability is always used to test the mathematical conditions added.

Calculus is all that is needed to comprehend this book, since all other mathematical techniques are presented at appropriate levels.

Finally, I would to express my sincere appreciation to many individuals, too many to list. My thanks go to President Robert Aebersold and Vice President and Provost Charles Foust, Deans Charles Zuzak and Jay Harper of Slippery Rock University, Pennsylvania, whose academic support for the past several years was essential to finishing this book. I thank Dr. Ben Fitzpatrick, my Ph.D. supervisor, for his years' teaching and academic influence, Professor Lotfi Zadeh, the father of fuzzy mathematics, for his keen encouragement, Professor Xavier J. R. Avula, President of the International Association for Mathematical and Computer Modeling, for his personal influence and education on professional perfection for the past several years.

I hope you enjoy using and referencing this book, and your comments and suggestions are welcome! Please let me hear from you — my e-mail address is `jeffrey.forrest@sru.edu`.

Yi Lin

Acknowledgments

This book contains many research results previously published in various sources, and I am grateful to the copyright owners for permitting me to use the material. They are International Association for Cybernetics (Namur, Belgium), Gordon and Breach Science Publishers (Yverdon, Switzerland, and New York), Hemisphere (New York), International Federation for Systems Research (Vienna, Austria), Kluwer Academic Publishers (Dordrecht, Netherlands), MCB University Press (Bradford, U.K.), Pergamon Journals, Ltd. (Oxford, England), Principia Scientia (St. Louis), Springer-Verlag (London), Taylor and Francis, Ltd. (London), World Scientific Press (Singapore and New Jersey), and Wroclaw Technical University Press (Wroclaw, Poland).

Contents

CHAPTER 1

Introduction

Although people often talk about "modern science" and "modern technology," what really are their characteristics? The most important and most obvious characteristic is that the forest of specialized and interdisciplinary disciplines can be easily seen, yet the boundaries of disciplines become blurred. That is, the "overall" trend of modern science and technology is to synthesize all areas of knowledge into a few major blocks, as evidenced by the survey of an American national committee on scientific research in 1985 (Mathematical Sciences, 1985). Another characteristic, which is not obvious but dominates the development of modern science and technology, is that the esthetic standard of scientific workers has been changing constantly. Because of the rapid progress of science and technology, researchers and administrators of scientific research have been faced with unprecedented problems: How can they equip themselves with the newest knowledge? How can they handle new knowledge, the amount of which has been increasing in a geometric series? If traditional research methods and esthetic standards are employed without modification, then today's unpublished scientific results could become outdated tomorrow. In this new environment, a new scientific theory — science of sciences — was born (Tian and Wang, 1980). In particular, administrators of scientific research are interested in whether science itself can be studied as a social phenomenon.

Parallel with the science of sciences some natural questions arise. Can the concept of wholeness or connection be used to study similar problems in different scientific fields? Can the point of view of interconnections be employed to observe the world in which we live? In the second decade of this century, von Bertalanffy (1934) wrote: since the fundamental character of living things is its organization, the customary investigation of the single parts and processes cannot provide a complete explanation of the vital phenomena. This investigation gives us no information about the coordination of parts and processes. Thus the chief task of biology must be to discover the laws of biological systems (at all levels of organization). We believe that the attempts to find a foundation for the theoretical point at a fundamental change in the world picture. This view, considered as a method of investigation, we call "organismic biology" and, as an attempt at an explanation, "the system theory of the organism." From this statement it can be seen that parallel to the challenge of modern science and technology, a new concept of "systems" had been proposed formally. Tested in the last seventy some years, this concept has been widely accepted by the entire spectrum of science and technology [for details, see (Blauberg et al., 1977)].

In this chapter we analyze the historical background for systems theory and several important research directions. For a detailed treatment on the history of systems theory, please refer to (von Bertalanffy, 1972; von Bertalanffy, 1968).

1.1. Historical Background

Before discussing the historical background of systems theory, let us first talk about what systems methodology is. Scholars in the field understand it in different ways. However, their fundamental points are the same, loosely speaking. In the following, let us first look at typical opinions of the two important figures in modern science: H. Quastler and L. Zadeh.

Quastler (1965) wrote: generally speaking, systems methodology is essentially the establishment of a structural foundation for four kinds of theories of organization: cybernetics, game theory, decision theory, and information theory. It employs the concepts of a black box and a white box (a white box is a known object, which is formed in certain way, reflecting the efficiency of the system's given input and output), showing that research problems, appearing in the aforementioned theories on organizations, can be represented as white boxes, and their environments as black boxes. The objects of systems are classified into several categories: motivators, objects needed by the system to produce, sensors, and effectors. Sensors are the elements of the system that receive information, and effectors are the elements of the system that produce real reactions. Through a set of rules, policies, and regulations, sensors and effectors do what they are supposed to do. By using these objects, Quastler proved that the common structure of the four theories of organization can be described by the following laws: (1) Interactions are between systems and between systems and their environments. (2) A system's efficiency is stimulated by its internal movements and the reception of information about its environment.

Zadeh (1962) listed some important problems in systems theory as follows: systems characteristics, systems classification, systems identification, signal representation, signal classification, systems analysis, systems synthesis, systems control and programming, systems optimization, learning and adaptation, systems liability, stability, and controllability. The characteristics of his opinion are that the main task of systems theory is the study of general properties of systems without considering their physical specifics. Systems methodology in Zadeh's viewpoint is that systems theory is an independent scientific discipline whose job is to develop an abstract foundation with concepts and frames in order to study various behaviors of different kinds of systems. Therefore, systems theory is a theory of mathematical structures of systems with the purpose of studying the foundation of organizations and systems structures.

Even though the concept of systems has been a hot spot for discussion in almost all areas of modern science and technology, and was first introduced formally in the second decade of this century in biology (von Bertalanffy, 1924), as all new concepts in science, the ideas and thinking logic of systems have a long history. For example, Chinese traditional medicine, treating each human body as a whole, can be traced to the time of Yellow Emperor about 2800 years ago; and Aristotle's statement "the whole is greater than the sum of its parts" has been a fundamental problem in systems theory. That is, over the centuries, mankind has been studying and exploring nature by using the thinking logic of systems. Only in modern times have some new contents been added to the ancient systems thinking. The methodology of studying systems as wholes adequately agrees with the development trend of modern science, namely to divide the object of consideration into parts as small as possible and studying all of them, seek interactions and connections between phenomena, and to observe and comprehend more and bigger pictures of nature.

In the history of science, although the word "system" was never emphasized, we can still find many explanatory terms concerning the concept of systems. For example, Nicholas of Cusa, that profound thinker of the fifteenth century, linking Medieval mysticism with the first beginning of modern science, introduced the notion of *coincidentia oppositorum*, the opposition or indeed fight among the parts within a whole which nevertheless forms a unity of higher order. Leibniz's hierarchy of monads looks quite like that of modern systems; his *mathesis universalis* presages an expanded mathematics which is not limited to quantitative or numerical expressions and is able to formulate much conceptual thought. Hegel and Marx emphasized the dialectic structure of thought and of the universe it produces: the deep insight that no proposition can exhaust reality but only approaches its coincidence of opposites by the dialectic process of thesis, antithesis, and synthesis. Gustav Fechner, known as the author of the psychophysical law, elaborated, in the way of the natural philosophers of the nineteenth century, supraindividual organizations of higher order than the usual objects of observation — for example, life communities and the entire earth, thus romantically anticipating the ecosystems of modern parlance. Here, only a few names are listed. For a more comprehensive study, see (von Bertalanffy, 1972).

Even though Aristotelian teleology was eliminated in the development of modern science, problems contained in it, such as "the whole is greater than the sum of its parts," the order and goal directedness of living things, etc., are still among the problems of today's systems theory research. For example, what is a "whole"? What does "the sum of its parts" mean? All these problems have not been studied in all classical branches of science, since these branches have been established on Descartes' second principle and Galileo's method, where Descartes' second principle says: divide each problem into parts as small as possible, and

Galileo's method implies simplifying a complicated process as basic portions and processes [for details see (Kuhn, 1962)].

From this superficial discussion, it can be seen that the concept of systems we are studying today is not simply a product of yesterday. It is a reappearance of some ancient thought and a modern form of an ancient quest. This quest has been recognized in the human struggle for survival with nature, and has been studied at various points in time by using the languages of different historical moments.

Ackoff (1959) commented that during the past two decades, we witnessed the appearance of the key concept "systems" in scientific research. However, with the appearance of the concept, what changes have occurred in modern science? Under the name "systems research", many branches of modern science have shown the trend of synthetic development; research methods and results in various disciplines have been intertwined to influence overall research progress, so one feels the tendency of synthetic development in scientific activities. This synthetic development requires the introduction of new concepts and new thoughts in the entire spectrum of science. In a certain sense, all of this can be considered as the center of the concept of "systems." One Soviet expert described the progress of modern science as follows (Hahn, 1967, p. 185): refining specific methods of systems research is a widespread tendency in the exploration of modern scientific knowledge, just as science in the nineteenth century with forming natural theoretical systems and processes of science as its characteristics.

In (von Bertalanffy, 1972), von Bertalanffy described the scientific revolution in the sixteenth century as follows: "(The Scientific Revolution of the sixteenth to seventeenth centuries) replaced the descriptive–metaphysical conception of the universe epitomized in Aristotle's doctrine by the mathematical–positivistic or Galilean conception. That is, the vision of the world as a teleological cosmos was replaced by the description of events in causal, mathematical laws." Based on this description, can we describe the change in today's science and technology as follows? At the same time of continuously using Descartes' second principle and Galileo's method, systems methodology was introduced in order to deal with problems of order or organization.

Should we continue to use Descartes' second principle and Galileo's method? Yes, for two reasons. First, they have been extremely effective in scientific research and administration, where all problems and phenomena could be decomposed into causal chains, which could be treated individually. That has been the foundation for all basic theoretical research and modern laboratory activities. In addition, they won victories for physics and led to several technological revolutions. Second, modern science and technology are not Utopian projects as described by Popper (1945), reknitting every corner for a new world, but based on the known knowledge base, they are progressing in all directions with more depth, more applicability, and a higher level of difficulty.

On the other hand, the world is not a pile of infinitely many isolated objects, where not every problem or phenomenon can be simply described by a single causal relation. The fundamental characteristics of this world are its organizational structure and connections of interior and exterior relations of different matters. The study of either an isolated part or a single causality of problems can hardly explain completely or relatively globally our surrounding world. At this junction, the research progress of the three-body problem in mechanics is an adequate example. So, as human race advances, studying problems with multicausality or multirelation will become more and more significant.

In the history of scientific development, the exploration of nature has always moved back and forth between specific matters or phenomena and generalities. Scientific theories need foundations rooted deep inside practice while theories are used to explain natural phenomena so that our understanding can be greatly enhanced. In the following, we discuss the technological background for systems theory — that is, the need which arose in the development of technology and requirement for higher level production.

There have had been many advances in technology: energies produced by various devices, such as steam engines, motors, computers, and automatic controllers; self-controlled equipment from domestic temperature controllers to self-directed missiles; and the information highway, which has resulted in increased communication of new scientific results. On the other hand, increased speed of communication furthers scientific development to a different level. Also, societal changes have brought more pressing demands for new construction materials. From these examples, it can be seen that the development of technology forces mankind to consider not only a single machine or matter or phenomenon, but also "systems" of machines and/or "systems" of matter and phenomena. The design of steam engines, automobiles, cordless equipment, etc., can all be handled by specially trained engineers, but when dealing with the designs of missiles, aircraft, or new construction materials, for example, a collective effort, combining many different aspects of knowledge, has to be in place, which includes the combination of various techniques, machines, electronic technology, chemical reactions, people, etc. Here the relation between people and machines becomes more obvious, and uncountable financial, economic, social, and political problems are intertwined to form a giant, complicated system, consisting of men, machines, and many other components. It was the great political, technical and personnel arrangement success of the American Apollo project, landing humans on the moon, that hinted that history has reached a point where all aspects of science and technology have been maturely developed so that each rational combination of information or knowledge could result in unexpected consumable products.

A great many problems in production require locating the optimal point of the maximum economic effect and minimum cost in an extremely complicated network. This kind problem not only appears in industry, agriculture, military affairs, and business, but politicians are also using similar (systems) methods to seek answers to problems like air and water pollution, transportation blockages, decline of inner cities, and crimes committed by teenage gangs.

In production, there exists a tendency that bigger or more accurate products with more profits are designed and produced. In fact, under different interpretations, all areas of learning have been faced with complexity, totality, and "systemality." This tendency betokens a sharp change in scientific thinking. By comparing Descartes' second principle and Galileo's method with systems methodology and considering the development tendency, as described previously, appearing in the world of learning and production, it is not hard to see that because of systems concepts, another new scientific and technological revolution will soon arrive. To this end, for example, each application of systems theory points out the fact that the relevant classical theory needs to be modified somehow [see (Klir, 1970; Berlinski, 1976; Lilienfeld, 1978)]. However, not all scientific workers have this kind of optimistic opinion. Some scholars believe that it is an omen that systems theory itself is facing a crisis [see (Wood-Harper and Fitzgerald, 1982) for details]. Only time and history will have the right to tell who is right and who is wrong.

1.2. Several Aspects of Systems Theory

In the past half century or so, since the concept of systems was first introduced by von Bertalanffy, systems thinking and methodology have seeped into all corners of science and technology. Because of such a wide range of applications, scholars in systems science have been divided into two big groups. One group follows the direction pointed out by Quastler and Zedah, and the other contains scholars with the desire to discover general principles while solving practical and specific problems. Thus, under the name "systems methodology," many problem-oriented systems approaches have appeared. Here we look at the taxonomy of different approaches, made available by Wood-Harper and Fitzgerald based on the field of application of each approach. For the original work, see (Wood-Harper and Fitzgerald, 1982).

1.2.1. General Systems Methodology

General systems methodology is a theoretical methodology whose purpose is to realize the ideal model of a systems theory described by von Bertalanffy (1968), Quastler (1965) and Zadeh (1962) — that is, establish a general theory applicable to explain phenomena in various specific systems. Based on the idea and logic of establishing such a general systems theory, many scholars have tried to employ known results in general systems theory to solve practical problems (Klir, 1970). However, Berlinski (1976) and Lilienfeld (1978) thought that all these attempts were extremely unsuccessful, since if a conclusion were obtained by applying results from general systems theory, it would require modification of the relevant theory. However, from the previous discussion, it can be seen that it is natural for the application of results from general systems theory to bring about such a phenomenon, because we are using a different methodology than our forefathers used. On the other hand, if an application of results of general systems theory requires a complete rebuilding of the structure and rules of the relevant theory, we surely need to study the correctness of the application of the results from general systems theory. Also, because of this phenomenon, some systems analysts believe that considering applications of general systems theory is not practical. At the same time, many scholars hope to locate the reason why this phenomenon ever appeared and to fix the problem so that the resulting general systems theory will be more like what von Bertalanffy, Quastler, and Zadeh dreamed about, and much easier to use to solve practical problems.

1.2.2. Human Activity Systems Approach

Checkland (1975) tried to establish a systems theory in order to solve problems with unclear meanings and incomplete definition of structures. He believes that this kind of system appears in environmental protection, administration of organizations, and business. He also attempted to look for a specific solution among many candidates of solutions so his approach would be more meaningful in the sense of application, because when practical problems are being solved often the effort is difficult to define and causes controversy.

Checkland established his methodology based on general systems theory and modified some aspects of systems theory so that the results would be more applicable in the real world. This methodology has been applied to many different problems and obtained some obvious consequences. It can be classified into general systems methodology, with its emphasis on problems without definite structures, and complicated environments. Before solving a problem, this method requires effort to explore and test the problem structure.

1.2.3. The Participant Approach

The participant approach emphasizes the importance of the researcher(s) in systems analysis (Mumford et al., 1978). Mumford believes that men must be considered as a part of the system under consideration, either as participants of the activities of the system or as designers of the system.

1.2.4. The Traditional Approach

The traditional approach was introduced by the American National Computer Center. To a certain degree, this method is accepted by many systems analysts.

Its basis is that certain problems are solvable with computers. Each application is considered individually; and problems are usually solved by finding and designing the optimal subsystems. Based on research on the properties of systems of interest, the condition of optimality is determined. According to the desired output, the input is designed. This process is employed to complete the systems design. See (1978–1979) for details.

1.2.5. Data Analysis Approach

Data analysis is established on the principle that the fundamental structure of systems is data [see (1978) for details]. The hypothesis is that if we can, under certain conditions, classify the data, then we will be able to comprehend the essential properties of the system under consideration. For the same set of data, different mathematical models can be established. However, all these models must possess some common properties. This implies that the given set of data can be the initial point for studying properties of the underlying structure, and there is no need to define various models for different applications. Also, if the relations between the data have been established, then the essential structure of the system, represented by the set of data, will appear naturally. Therefore, data analysis can be considered as the "median" steps leading to the ultimate understanding of the structure of organization.

1.2.6. Structured Systems Analysis Approach

The structured systems analysis approach was described individually by de Marco (1980) and Gane and Sarsons (1979). It has been attracting the attention of more and more businesses and organizations. This method was developed to solve problems contained in classical research methodologies. For example, how can the concept of subsystems be used to coordinate a large group of analysts and to solve complicated problems concerning big machines and large systems? This method offers new tools for analysis and literature editing. For instance, concepts such as data flow diagrams, data dictionary, and constructed English have been introduced. These concepts have allowed the literature of known systems to be edited more clearly; consequently discoveries of new structures of the systems can be expected.

From the aforementioned classification of systems approaches due to Wood-Harper and Fitzgerald, it can be seen that the basic assumptions of various systems methodologies can be considered as axioms or as results of general systems theory. In a certain sense, all the systems approaches listed are applications of general systems theory. At the same time, the first three approaches emphasize understanding of systems structures and theoretical analysis of systems, while the other three approaches concentrate on problem solving with the hope that the results will allow us to understand these systems on a theoretical level. Therefore, the first three approaches can be termed theoretical systems methods, while the others can be called applied systems techniques. For detailed discussion on various systems approaches, see (Klir, 1970).

To conclude this chapter, we discuss the foundations of various systems approaches and some research directions in general systems theory.

1.2.7. Some Research Directions in General Systems Theory

Among the main tasks of general systems theory are (1) define the meaning of "systems" and related concepts; (2) classify systems and find their properties in the most general sense; (3) model systems behaviors; (4) study special systems models logically and methodologically. The first three tasks constitute the theoretical basics for the concepts of special systems, such as control systems theory, automata, and information systems theory. That is, the goal of general systems theory is to establish a higher-level abstract theory without touching on any specific physical properties. That was why von Bertalanffy believed that general systems theory deals with formal characteristics of systems structures so that its results can be applied to research topics in various scientific fields, and that it can be used not only in physical systems but in any "whole" consisting of interacting "parts" (von Bertalanffy, 1967, pp. 125–126).

Even though it follows from the previous discussion that systems have a long history and that many problems discussed in modern systems theory have had been considered by many great thinkers from different angles in the past several centuries with different languages, as a theory, especially, as a methodology of general systems, it is still relative new. In (von Bertalanffy, 1972), von Bertalanffy asked how we can define a system. That is, even though the idea of systems has permeated all of modern science, we still have not obtained an ideal definition for the concept of systems!

As a matter of fact, as stated in (Department of Business Administration, 1983), "system" is a generalization of the concept of "structure" in physics but differs from a structure. A system possesses three important features: (i) order or levels; (ii) structure of a set; that is, each system consists of at least two subsystems; (iii) relativity and wholeness; that is, the existence of subsystems is relative and dependent on each other and conditioning each other.

Tarski (1954–1955) defined the concept of a system with relations as a nonempty set, called the domain of the system with relations, and a finite sequence of relations defined on the domain. Hall and Fagen (1956) discussed the definition of systems. They believed that a system consisted of a set of objects, some relations between the objects, and some relations between the attributes of the objects. In (Systems, 1969–1976), Uyomov proved that the attributes connecting the objects of a system can be considered as new objects of the system.

In this way, Hall and Fagen's definition of systems can be simplified as follows: A system is a set of objects and some relations between the objects. Mesarovic (1964) used the language of set theory to define a system as a relation. Therefore, general systems theory becomes a theory of relations. However, ten years later, Mesarovic and Takahara (Mesarovic and Takahara, 1975) pointed out that general systems theory, produced with this definition of systems, does not contain any results. Does this imply that the definition of systems, given by Mesarovic, contains something inappropriate? The authors also pointed out that there is no need to be concerned because the entire research literature of systems theory contains no results. If that is a fact, does it imply that it is still too early to establish such a theory? Yi Lin and Yonghao Ma (Lin, 1987; Lin and Ma, 1987) studied the attempts at defining systems, synthesized all opinions into one with a consideration of future development, and introduced a definition of (general) systems as follows: S is a (general) system if, and only if, S is an ordered pair (M,R) of sets, where M is the set of all objects of S and R is a set of relations defined on M. Here a relation r belongs to R means that there exists an ordinal number n (nonzero) such that r is a subset of the Cartesian product M^n. The referee(s) of (Ma and Lin, 1987) claimed that the definition introduced by Lin and Ma will produce a good quantity of results, which has been evidenced by results contained in (Lin, 1987; Lin and Ma, 1987; Ma and Lin, 1987; Lin, 1989d). In this book, the reader will be led to some of these results. Here, uncertainty, contained in the concept of "sets," makes application of systems theory, developed on the definition of systems by Lin and Ma, contain many uncertain factors. For example, for a given experimental material, is the totality of molecules constituting the material a set? Is the collection of all relations between molecules in the material a set? To deal with these uncertainties, Yi Lin introduced the following epistemological axiom: The existence of an object is determined by the existence of some particles of a certain level and of some relations between the particles, which constitute the object; for details [see (Lin, 1989d)].

We now briefly introduce several research directions in general systems theory.

1.2.7.1. Classical systems theory. Classical systems theory is a mathematical theory, based on calculus, to study the general principles of structures and structures with specific properties, and has been used for research and description. Its results have also been used to solve concrete problems. Because of the generality of the problems studied, even with special features, parts, and relations between the parts of a structure, one can still obtain some formal properties of the structure under concern. For instance, generalized principles of dynamics can be applied to the totality of molecules or biological objects. That is, these principles can find applications in chemistry and biology. There are many references here, one of which is (Wonham, 1979).

1.2.7.2. Catastrophe theory. Catastrophe theory, initiated by Newton and Leibniz three centuries ago, is a different way of thinking and reasoning about "discrete," "jump" or "discontinuous" changes in a course of events, in an object's shape, change in a system's behavior, and change in ideas themselves. As suggested by its name, this study has been applied to research of literal catastrophes (e.g., collapse of a bridge, downfall of an empire) and to quiet changes, (e.g., dancing of sunlight on the bottom of a pool, transition from wakefulness to sleep). For more details, see (Poston and Steward, 1978).

1.2.7.3. Compartment theory. In (Rescigno and Segre, 1966), Rescigno and Segre first studied compartment theory. The basic idea is that the problem or structure under consideration can be described as a whole consisting of parts satisfying certain boundary conditions, between which there appear processes of transportation, such as "chains" or "nipples." That is, the parts can be chained together by transportation, or transportation appears between a central part and its surrounding parts. The mathematical difficulty in dealing with structures with three or more parts becomes obvious. In this area, Laplace transformation and networks and graph theory have been successfully used.

1.2.7.4. Cybernetics. Cybernetics is a theory on systems and their environments, internal information transportation of systems, and impacts on the environment of controlled systems. This theory has found many applications. Because of its usefulness, many scholars do not see it as a part of systems theory. Cybernetics has often been used to describe the formal structure of some "action processes." Even though the structure of the system under concern is not clearly given, and the system is an input–output "black" box, cybernetics can still offer some understanding about that structure. Therefore, cybernetics has been applied in many areas, including hydraulics, electricity, ecology, markets, etc. For more information, see (Bayliss, 1966; Milsum, 1966).

1.2.7.5. Fuzzy mathematics. Fuzzy mathematics is a theory dealing with the rapprochement between the precision of classical mathematics and the pervasive imprecision of the real world. The fundamental concept here is fuzzy sets, which are classes with boundaries that are not sharp in which the transition from membership to nonmembership is gradual rather than abrupt. At present, we are unable to design machines that can compete with humans in the performance of tasks such as recognition of speech, translation of languages, comprehension of meanings, abstraction and generalization, decision making with uncertainty, and, above all, summarization of information. This stems from a fundamental difference between human intelligence, containing fuzziness, and machine intelligence, based on precision. Consisting of a body of new concepts and techniques, this theory accepts fuzziness as an all-pervasive reality of human existence, which has opened many new frontiers in psychology, sociology, political science, philosophy, physiology, economics, linguistics, operations research, management science, and other fields. The literature in this research area is plentiful, and I will not list any here.

1.2.7.6. Game theory. According to Quastler, game theory is a systems theory, since it deals with the behaviors of (n) ideal players with the ability to reason and make decisions, whose goal is to win more than to lose. One goal of this research is how to apply optimal strategies to confront other players (or the environment). That is, game theory is about the study of systems with special reaction forces. A good reference is (Shubik, 1983).

1.2.7.7. Genetic algorithms. Genetic algorithms (GAs) have been applied in many areas, including economics, political science, psychology, linguistics, biology, computer science, etc. GAs are search procedures based on the mechanics of natural selection and natural genetics. They combine survival of the fittest among string structures with a structured yet randomized information exchange to form a search algorithm with some of the innovative flair of human search. The goals of this research are twofold: (1) to abstract and rigorously explain the adaptive processes of natural systems, and (2) to design artificial systems software that retains the important mechanisms of natural systems. This approach has led to important discoveries in both natural and artificial science. The central theme of research here has been robustness, the balance between efficiency and efficacy necessary for survival in many different environments. A good introductory book on this subject matter is (Goldberg, 1989), and the classic is (Holland, 1975).

1.2.7.8. Graph theory. Besides the quantitative study of structures, many problems deal with the organization and topological structures of the systems of interest. There exist many methods useful to this research. Graph analysis, especially directed graph theory, describes in detail the construction of relations between systems. This method has been successfully applied in many different areas, including biology [see, for example, (Rashevsky, 1956)]. Combined with matrix theory in mathematics, the characteristics of the models, developed in graph theory are similar to those of compartment theory, so it establishes relations with the theory of open systems.

1.2.7.9. Information theory. Based on the opinion of Shannon and Weaver (1949), information theory is established on the concept of information, which is defined by an expression similar to one with negative entropy in thermodynamics. Quastler (1955) believes that information can be considered as a measure of the structure of organization. Although information theory is very important in the engineering of communication, research results in other areas of application have not yielded any convincing consequences.

1.2.7.10. Navier–Stokes equation and chaos. Even though the Navier–Stokes equation has been successfully applied in weather forecasting (short term only), its study has recently given some new understanding, which has led to the new science called Chaos. On one hand, the concept of chaos, which has several different meanings, has created a burst of research in almost all areas of science. People are trying to find and understand chaos. On the other hand, some scholars have shown that the original chaos, introduced by Lorenz in the study of simplifying Navier–Stokes equation, has nothing to do with modern chaos. In fact, it is either an illusion of a concept of a higher dimensional space or a misunderstanding in computer-aided calculation. Here I list only two references (Cohen and Steward, 1994; Gleick, 1987).

1.2.7.11. Networks. Network theory can be considered as part of set theory, or graph theory, or compartment theory. The main structure dealt with is net structures of systems. It has been applied to neural networks; for details, see (Rapoport, 1949).

1.2.7.12. Set theory. All the general definitions and properties of structures have been given in terms of set theory. For example, open and closed systems can be described by axioms using the language of set theory. All research along this line was originated by Mesarovic (1964, pp. 1–24). Most of this research, carried out since 1964, is related to the concept of systems and related topics, not to solving practical problems, at least until recently [see (Lin and Ma, 1987; Ma and Lin, 1987; Lin, 1989d; Lin and Qiu, 1987)]. Because of the use of the language of set theory, the language and methodology of category theory can be employed to classify structures. On the other hand, algebraic methods can be introduced so that concepts, such as linear and nonlinear structures, have been introduced. For related references, see (Mesarovic and Takahara, 1975).

1.2.7.13. Simulation. If the structures, described by differential equations, contain nonlinear equations, in general, there is no way to solve the systems. Because of this reason, computer simulation becomes an important method. This method not only saves time and human labor, but also opens up research areas in which, mathematically, we do not know if the solution exists. This kind of system cannot be solved by known mathematics. On the other hand, specific experiments in laboratories can be replaced by computer simulations. For example, computer-aided design uses computer simulation as a substitute for laboratory experiments. At the same time, the mathematical structure or model obtained by computer simulation can be evaluated by experimental data. For example, Hess (1969) employed this method to compute a 14-th step chemical reaction in human cells, which was described by a system of more than 100 nonlinear differential equations. As a matter of fact, this method has often been used in the study of markets and populations.

1.2.7.14. Statistics. Einstein said (1922, p. 60):

> The belief in an external world independent of the perceiving subject is the basis of all natural science. Since, however, sense perception only gives information of this external world or of "physical reality" indirectly, we can only grasp the latter by speculative means. It follows from this that our notion of physical reality can never be final.

Based on this understanding, statistics is a mathematical theory of how to predict and comprehend "reality" according to a small sample, collected with reference to some prior mathematical analysis. This theory is not of what event causes another event (i.e., causality) but about what event will imply that another will occur. Successful applications of statistics can be found in almost all applied areas of human endeavors. One of the best references here is Box, Hunter, and Hunter's classic (Box et al., 1978).

1.2.7.15. Theory of automata. The theory of automata concerns an ideal automaton with input and output, and its ability to be corrected and learn (Minsky, 1967). One such example is the Turing machine (Turing, 1936). In ordinary language, the Turing machine is an ideal machine which can print "1" and "0" on an infinitely long tape. It can be shown that, no matter how complicated a process can be, if it can be expressed by finite steps of logic rules and calculations, then this process can be simulated by a machine. From this fact, the future of this theory can be seen.

1.3. A Few Thoughts

It can be seen that systems theory is very young. Researchers are trying to find more practical examples and general properties, and are testing and rethinking the theory. That is why new ideas have been appearing quickly, and consequent discussions have been intensive. It can be foreseen that the development of modern science and technology will be the effective motivation and method to test the duration and significance of ideas and thoughts in general systems theory. Based on the thoughts of general systems theory, some scientific workers have simplified the question in which administrators of scientific research are interested to the following: Can science be studied as a system? This is an important and practical problem since if this question can be studied deeply with reasonable outcome, the human exploration of nature might be able to go to a higher level (Yablonsky, 1984).

CHAPTER 2

Naive Set Theory

Set theory, created by Georg Cantor (1845–1918), is one of the greatest creations of the human mind. Everyone who studies it is fascinated by it. More important, however, is that the theory has become of the greatest importance for the entire spectrum of mathematics. It has given rise to new branches of mathematics, or at least first rendered the possibility for their further development, such as the theory of point sets, theory of real functions, and topology. Finally, the theory of sets has had influence on the investigation of the foundation of mathematics, as well as through the generality of its concepts, as a connecting link between mathematics and philosophy.

We will present the basic features of Cantor's naive set theory. The basic concepts and methods of proofs will be used in the development of general systems theory.

2.1. A Set and Its Classification

According to Cantor a set M is "a collection into a whole, of finite, well-distinguished objects (called the 'elements' of M) of our perception or of our thoughts." For example, the 38 students in a classroom constitute a set of 38 elements; the totality of even numbers, a set of infinitely many elements; the vertices of a die, a set of 8 elements; the points on a circle, a set of infinitely many elements.

Two sets M and N are equal, in symbols $M = N$, if they contain the same elements. For example, if $M = \{1,2,3\}$ and $N = \{2,1,3\}$, then $M = N$. (Sets, in general, are designated by enclosing their elements in braces or by a symbol like $\{x : \Phi(x)\}$, which means that the set consists of all the elements satisfying the proposition Φ.)

The notation $M \neq N$ means that the sets M and N are different. Further, $m \in M$ means that m is an element of the set M, whereas $m \notin M$ denotes that m is not an element in M. $M \subseteq N$ denotes inclusion; that is, M is a subset of N, or all elements of M also belong to N.

Proposition 2.1.1. *$M = N$ iff $M \subseteq N$ and $N \subseteq M$.*

A mapping (or a function) from a set M into a set N is a rule under which each element in M is assigned to an element in N. Generally, lower case letters, say, f, g, and h, will be used to indicate mappings. Let f be a mapping from the set M into the set N, denoted by $f : M \to N$. Then for each element $m \in M, f(m)$ indicates the element in N which is assigned to the element m.

The mapping $f : M \to N$ is onto (or surjective), if for each element $n \in N$, there exists an element $m \in M$ such that $f(m) = n$. The mapping $f : M \to N$ is 1–1 (or one-to-one, or injective) if for arbitrary distinct elements x and $y \in M$,

$$f(x) \neq f(y) \tag{2.1}$$

The mapping $f : M \to N$ is a bijection (or a bijective mapping) if f is both surjective and injective.

Sets M and N are equipollent if there exists a bijection of M onto N. The equipollence relation between sets classifies the collection of all sets into classes such that each class contains all the equipollent sets. Another coarse classification of sets distinguishes the collection of all sets into finite, denumerable and uncountable sets according to whether the sets contain a finite number of elements, or are equipollent to the set of all natural numbers N, or do not satisfy the previous two conditions; i.e., they contain infinitely many elements and are not equipollent to N.

Proposition 2.1.2. *A set X is denumerable iff it can be written as a sequence $\{x_0, x_1, x_2, x_3, \ldots\}$.*

This fact implies that an infinite set X is denumerable iff to every element x of the set, precisely one natural number corresponds to it and *vice versa*.

Theorem 2.1.1. *The set of all integers is denumerable.*

Proof: Let us write the integers as follows:

$$0, 1, 2, 3, \ldots \tag{2.2}$$

$$-1, -2, -3, \ldots \tag{2.3}$$

Then the set of all integers can be written as the sequence

$$\{0, 1, -1, 2, -2, 3, -3, \ldots, n, -n, \ldots\}$$

■

Theorem 2.1.2. *The set of all rational numbers is denumerable.*

Proof: Let us first deal with the set of all positive rational numbers. We can write all whole numbers in order of magnitude (i.e., all numbers with denominator 1), then all fractions with denominator 2, then all fractions with denominator 3, etc. Thus, we have the rows of numbers

$$\begin{array}{cccccccc}
1 & \rightarrow & 2 & & 3 & \rightarrow & 4 & \cdots \\
 & \swarrow & & \nearrow & & \swarrow & & \\
\frac{1}{2} & & \frac{2}{2} & & \frac{3}{2} & & \frac{4}{2} & \cdots \\
\downarrow & \nearrow & & \swarrow & & & & \\
\frac{1}{3} & & \frac{2}{3} & & \frac{3}{3} & & \frac{4}{3} & \cdots \\
 & \swarrow & & & & & & \\
\frac{1}{4} & & \frac{2}{4} & & \frac{3}{4} & & \frac{4}{4} & \cdots \\
\downarrow & & & & & & & \\
\vdots & & \vdots & & \vdots & & \vdots & \vdots
\end{array} \tag{2.4}$$

If we write the numbers in the order indicated by the arrows (leaving out numbers which have already appeared), then every positive rational number certainly appears, and only once. The collection of these rational numbers is thus written as a sequence

$$1, 2, \frac{1}{2}, \frac{1}{3}, 3, 4, \frac{3}{2}, \frac{2}{3}, \frac{1}{4}, \dots \tag{2.5}$$

If we denote this sequence by $\{r_1, r_2, r_3, \dots\}$, then obviously

$$\{0, r_1, -r_1, r_2, -r_2, r_3, -r_3, \dots\}$$

is the set of all rational numbers, and the denumerability of this set is established.■

The following theorem shows that not every set of real numbers is denumerable.

Theorem 2.1.3. *The set of all points in the closed interval* $[0,1]$ *is uncountable.*

Proof: We prove the theorem by a diagonal method, called the second, or Cantor, diagonal method. Suppose that the closed interval $[0,1]$ is denumerable; then the open internal $(0,1)$ is also denumerable. From Proposition 2.1.2 it follows that the points in $(0,1)$ can be written as the sequence

$$b_0, b_1, b_2, b_3, \dots \tag{2.6}$$

We now write each of them as infinite decimal numbers by putting in enough zeros whenever necessary. Then

$$\begin{aligned} b_0 &= 0.b_{00}b_{01}b_{02}b_{03}\dots \\ b_1 &= 0.b_{10}b_{11}b_{12}b_{13}\dots \\ b_2 &= 0.b_{20}b_{21}b_{22}b_{23}\dots \\ b_3 &= 0.b_{30}b_{31}b_{32}b_{33}\dots \\ &\vdots \end{aligned} \tag{2.7}$$

Now we construct a number $b \in (0,1)$, defined by

$$b = 0.b_0b_1b_2b_3\dots \tag{2.8}$$

such that $b_0 \neq b_{00}, b_1 \neq b_{11}, b_2 \neq b_{22}, b_3 \neq b_{33}, \dots$. Then b is not contained in the sequence $\{b_0, b_1, b_2, \dots\}$, contradiction. This contradiction implies that the set of all points in the interval $(0,1)$ is uncountable. ■

The sets X and Y are of the same cardinality if they are equipollent. Generally, to every set X the cardinality of X is assigned and is denoted by $|X|$. The equality $|X| = |Y|$ holds iff X and Y are equipollent. For a finite set X the cardinality of X is equal to the number of elements of X. The cardinality assigned to each denumerable set is denoted by $\aleph_0$ (read *aleph zero*), and the cardinality assigned to the set of all real numbers is denoted by c (called the *continuum*).

Let m and n be two cardinalities, and let $|X| = m$ and $|Y| = n$. We say that m is not larger than n, or that n is not smaller than m, and we write $m \leq n$ or $n \geq m$, if there exists a one-to-one mapping of X into Y. We say that m is smaller than n, or that n is larger than m, and we write $m < n$ or $n > m$, if $m \leq n$ and X and Y are not equipollent.

Corollary 2.1.1. $\aleph_0 < c$.

The following fact says that for any fixed cardinality m there exists a set whose cardinality is larger than m.

Theorem 2.1.4. *Let X be a set and 2^X the set of all subsets of X, which is called the power set of X. Then*

$$|X| < |2^X| \tag{2.9}$$

Proof: Define a mapping $f : X \to 2^X$ by letting $f(x) = \{x\}$ for each $x \in X$. Then f is one-to-one, so $|X| \leq |2^X|$. We must now show that X and 2^X are not equipollent. We prove this by contradiction. Suppose that X and 2^X are equipollent. There then exists a bijection h from X onto 2^X. Define a subset of X as follows:

$$A = \{x \in X : x \notin h(x)\} \tag{2.10}$$

Since h is onto, it follows that there exists an element $y \in X$ such that

$$h(y) = A \tag{2.11}$$

There now exist two possibilities: (1) $y \in A$ or (2) $y \notin A$. If case (1) holds, then $y \in A = h(y)$. This implies that $y \notin A$, contradiction. If case (2) is true, then $y \notin h(y) = A$. So $y \in A$, another contradiction. Therefore, the bijection h cannot exist, and $|X| < |2^X|$. ∎

2.2. Operations of Sets

Let X be a set. A set Y is called a proper subset of X, denoted $Y \subset X$, if Y is a subset of X and $X \neq Y$. A set Z of those elements of X which do not belong at the same time to the subset Y is called the complement of Y with respect to X; in symbols, $Z = X - Y$. When Y is not a proper subset of X, Z may not contain any element. In this case, we introduce an ideal set, called the empty set, $\emptyset$. The empty set is classed with the finite sets. It is a subset of every set and, in particular, of itself.

Theorem 2.2.1. *Every subset of a denumerable set is at most denumerable.*

Proof: Let $X = \{x_0, x_1, x_2, \dots\}$ be a denumerable set and $Y \subset X$. Then in X there exists a first element x_{k_1} which belongs to Y; let $y_1 = x_{k_1}$. This is followed again by a first element, denoted by x_{k_2}, in X, which belongs to Y; let $y_2 = x_{k_2}$, etc. The process does or does not terminate, according to whether Y is finite. Since X contains all the elements of Y, the possibly terminating sequence $\{k_1, k_2, k_3, \dots\}$ comprises precisely the elements of Y. Therefore, Y is finite or denumerable. ∎

The union of finitely or infinitely many sets is defined to be the set of those elements which belong to at least one of the sets. The union U of an at most denumerable number of sets $X_0, X_1, X_2, \dots$ is written in the form

$$U = X_0 \cup X_1 \cup X_2 \cup \cdots, \quad \text{or} \quad U = \bigcup_{k=0}^{\infty} X_k \tag{2.12}$$

The intersection of arbitrarily many sets is defined to be the set of those elements which belong to each of the aforesaid sets. For the intersection D of two sets X_1 and X_2, we write

$$D = X_1 \cap X_2 \tag{2.13}$$

and for the intersection of at most denumerably many sets,

$$D = \bigcap_{k=1}^{\infty} X_k \tag{2.14}$$

Let I be a collection of sets. The union U and the intersection D of the sets in I is written

$$U = \bigcup_{X \in I} X \text{ or } U = \bigcup \{X : X \in I\} \tag{2.15}$$

and

$$D = \bigcap_{X \in I} X \text{ or } D = \bigcap \{X : X \in I\} \tag{2.16}$$

Theorem 2.2.2 [De Morgan's Laws]. *Suppose that A, B, and C are sets. Then*

$$A - (B \cap C) = (A - B) \cup (A - C) \tag{2.17}$$
$$A - (B \cup C) = (A - B) \cap (A - C) \tag{2.18}$$

Proof: We prove only the first equality. For any element x, $x \in A - (B \cap C)$ iff $x \in A$ and $x \notin B \cap C$, iff $x \in A$ and either $x \notin B$ or $x \notin C$, iff either $x \in A$ and $x \notin B$ or $x \in A$ and $x \notin C$, iff $x \in (A - B) \cup (A - C)$. ■

Proposition 2.2.1 [Commutative Laws]. *Let A and B be two sets. Then*

$$A \cup B = B \cup A \text{ and } A \cap B = B \cap A \tag{2.19}$$

The proof follows immediately from the definitions of union and intersection.

Proposition 2.2.2 [Associative Laws]. *Let A, B, and C be sets. Then*

$$A \cup (B \cup C) = (A \cup B) \cup C \text{ and } A \cap (B \cap C) = (A \cap B) \cap C \tag{2.20}$$

The proof follows immediately from the definitions of union and intersection.

Proposition 2.2.3 [Distributive Laws]. *Let A, B, and C be sets. Then,*

$$A \cap (B \cup C) = (A \cap B) \cup (A \cap C) \tag{2.21}$$
$$A \cup (B \cap C) = (A \cup B) \cap (A \cup C) \tag{2.22}$$

The proof is left to the reader.

Proposition 2.2.4 [Laws of Tautology]. *For any set A,*

$$A \cup A = A \text{ and } A \cap A = A \tag{2.23}$$

The proof is straightforward and is omitted.
The following identities are given without proof.

Proposition 2.2.5. *Let A, B and C be sets. Then*

(1) $A \cup (B - A) = A \cup B$,

(2) $A - B = A - (A \cap B)$,

(3) $A \cap (B - C) = (A \cap B) - C$,

(4) $(A \cup B) - C = (A - C) \cup (B - C)$,

(5) $A - (B - C) = (A - B) \cup (A \cap C)$,

(6) $A - (B \cup C) = (A - B) - C$.

The following formulas illustrate the analogy between the inclusion relation and the "less than or equal to" relation in arithmetic:

Proposition 2.2.6. *For arbitrary sets A, B, C, and D, the following are true:*

(1) $(A \subseteq B)$ *and* $(C \subseteq D) \to (A \cup C \subseteq B \cup D)$,

(2) $(A \subseteq B)$ *and* $(C \subseteq D) \to (A \cap C \subseteq B \cap D)$,

(3) $(A \subseteq B)$ *and* $(C \subseteq D) \to (A - D \subseteq B - C)$,

where $\to$ *means "imply."*

For the union and intersection of a collection I of sets, we have the following results.

Proposition 2.2.7 [De Morgan's Laws]. *Let A be an arbitrary set. Then*

(1) $A - \cap_{X \in I} X = \bigcup_{X \in I}(A - X)$,

(2) $A - \cup_{X \in I} X = \bigcap_{X \in I}(A - X)$.

Proof:

(1) For any element x, $x \in A - \cap_{X \in I} X$ iff $x \in A$ and $x \notin \bigcap_{X \in I} X$, iff $x \in A$ and there exists an $X \in I$ such that $x \notin X$, iff there exists $X \in I$ such that $x \in A - X$, iff $x \in \bigcup_{X \in I}(A - X)$.

(2) For any element x, $x \in A - \bigcup_{X \in I} X$ iff $x \in A$ and for each set $X \in I$, $x \notin X$, iff for each set $X \in I$, $x \in A - X$, iff $x \in \bigcap_{X \in I}(A - X)$. ■

Let x and y be two elements. The ordered pair (x,y) is the set $\{\{x\},\{x,y\}\}$. We call x the first term (or coordinate) of (x,y) and y the second term (or coordinate) of (x,y).

Theorem 2.2.3. *In order that $(a,b) = (c,d)$, it is necessary and sufficient that $a = c$ and $b = d$.*

Proof: The sufficiency part is clear. To prove necessity, suppose that $(a,b) = (c,d)$. From the definition of ordered pairs, it follows that $\{c\} \in (a,b)$ and $\{c,d\} \in (a,b)$; that is, (i) $\{c\} = \{a\}$ or (ii) $\{c\} = \{a,b\}$, and (iii) $\{c,d\} = \{a\}$ or (iv) $\{c,d\} = \{a,b\}$.

Equality (ii) holds if $a = b = c$. Equalities (iii) and (iv) are then equivalent and it follows that $c = d = a$. Hence, we obtain $a = c = b = d$, in which case the theorem holds. Similarly, one can check that the theorem holds for case (iii). It remains to show that the theorem holds for cases (i) and (iv). We then have $c = a$ and either $c = b$ or $d = b$. If $c = b$, then (ii) holds and this case has already been considered. If $d = b$, then $a = c$ and $b = d$, which proves the theorem. ■

The Cartesian product of two sets X and Y is defined to be the set of all ordered pairs (x,y) such that $x \in X$ and $y \in Y$. This product is denoted by $X \times Y$. If $X = \emptyset$ or $Y = \emptyset$, then obviously $X \times Y = \emptyset$. Certain properties of Cartesian products are similar to the properties of multiplication of numbers. For example, we have the following results:

Proposition 2.2.8 [Distributive Laws]. *Let X_1, X_2, and Y be sets. Then*

(1) $(X_1 \cup X_2) \times Y = (X_1 \times Y) \cup (X_2 \times Y)$,

(2) $Y \times (X_1 \cup X_2) = (Y \times X_1) \cup (Y \times X_2)$,

(3) $(X_1 \cap X_2) \times Y = (X_1 \times Y) \cap (X_2 \times Y)$,

(4) $Y \times (X_1 \cap X_2) = (Y \times X_1) \cap (Y \times X_2)$,

(5) $(X_1 - X_2) \times Y = (X_1 \times Y) - (X_2 \times Y)$,

(6) $Y \times (X_1 - X_2) = (Y \times X_1) - (Y \times X_2)$.

Proof: We prove only the first equation. For an ordered pair (x,y), $(x,y) \in (X_1 \cup X_2) \times Y$ iff $x \in X_1 \cup X_2$ and $y \in Y$, iff either $x \in X_1$ or $x \in X_2$ and $y \in Y$, iff either $x \in X_1$ and $y \in Y$ or $x \in X_2$ and $y \in Y$, iff either $(x,y) \in X_1 \times Y$ or $(x,y) \in X_2 \times Y$, iff $(x,y) \in (X_1 \times Y) \cup (X_2 \times Y)$. ■

A subset $f \subset X \times Y$ is called a function or mapping from X into Y if for any $x \in X$ and any y_1 and $y_2 \in Y$, $(x,y_1),(x,y_2) \in f$ implies that $y_1 = y_2 = y$. In this case, the element y is denoted $f(x)$. The function f is written $f : X \to Y$. When each $x \in X$ belongs to a pair in f, the function f is defined on X. Otherwise, f is called a partial function from X to Y.

Theorem 2.2.4. *If $f : X \to Y$ is a function, then $f^{-1} = \{(x,y) : (y,x) \in f\}$ is a partial function from Y into X iff f is one-to-one; that is, if (x_1,y) and $(x_2,y) \in f$ then $x_1 = x_2$.*

Proof: By definition, the subset f^{-1} is a partial function from Y into X iff for any y,x_1 and x_2, (y,x_1) and $(y,x_2) \in f^{-1}$ implies that $x_1 = x_2$. That is, if $f(x_1) = y = f(x_2)$, then $x_1 = x_2$. Thus, f is one-to-one. ■

2.3. Arithmetic of Cardinal Numbers

Let X and Y be two sets. The notation $X \sim Y$ indicates that the sets are equipollent. The following result says that the equipollence relation is reflexive, symmetric, and transitive.

Theorem 2.3.1. *For arbitrary sets A, B, and C the following holds:*

(1) Reflexive property: $A \sim A$.

(2) Symmetric property: $A \sim B$ implies $B \sim A$.

(3) Transitive property: $(A \sim B)$ and $(B \sim C)$ imply $A \sim C$.

The proof is straightforward and is omitted.

Example 2.3.1. Since the concept of equipollence is of fundamental importance, we illustrate it by some examples.

(a) The sets of the points of the intervals $[0,1]$, $[0,1)$, $(0,1]$, and $(0,1)$ are equipollent to each other.

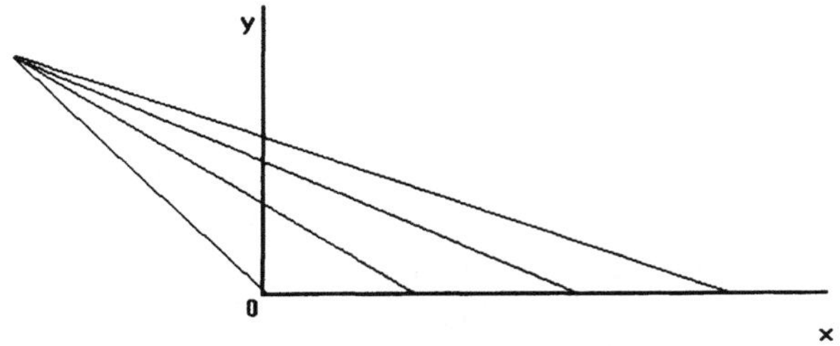

Figure 2.1. A half-line obtained by central projection.

First, let us prove that $(0,1] \sim (0,1)$. We denote the points of the first interval by x and those of the second by y, and set up the following correspondence:

$$\begin{aligned}
y &= \frac{3}{2} - x; \text{ for } \frac{1}{2} < x \le 1, \text{ then } \frac{1}{2} \le y < 1 \\
y &= \frac{3}{4} - x; \text{ for } \frac{1}{4} < x \le \frac{1}{2}, \text{ then } \frac{1}{4} \le y < \frac{1}{2} \\
y &= \frac{3}{8} - x; \text{ for } \frac{1}{8} < x \le \frac{1}{4}, \text{ then } \frac{1}{8} \le y < \frac{1}{4} \\
&\vdots
\end{aligned}$$

It is evident that we have already set up a 1–1 correspondence between the two intervals. This proves our assertion.

We can show analogously that $[0,1) \sim (0,1)$. From this it follows finally that $[0,1) \sim [0,1]$.

(b) A half-line and an entire line are equipollent to an interval.

A half-line can be obtained by central projection from an interval erected at right angles to it and open above (Fig. 2.1). A full line can be obtained similarly from a bent, open interval (Fig. 2.2).

(c) Two finite sets are equipollent if, and only if, they contain equally many elements.

(d) An infinite set can be equipollent to one of its proper subsets. This is shown by the equipollence of the sets $\{1,2,3,\dots\}$ and $\{2,4,6,\dots\}$. Infinite sets

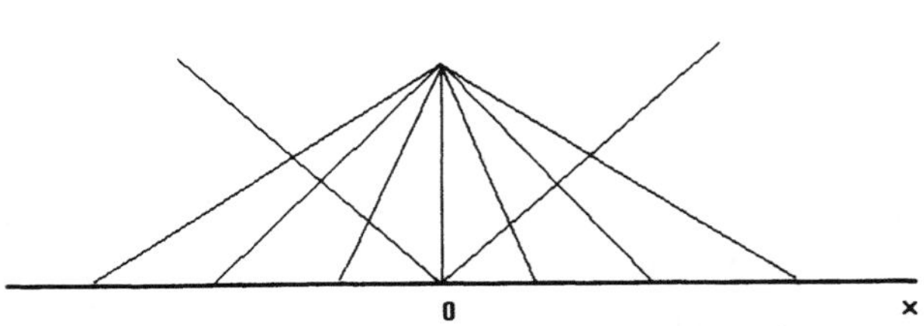

Figure 2.2. A full line obtained from a bent open interval.

thus exhibit in this respect an entirely different behavior from that of finite sets [see example (c)]. This property can therefore be used to distinguish finite and infinite sets independently of any enumeration.

(e) The set of all real functions defined on the interval $[0, 1]$ is neither denumerable nor equipollent to the continuum.

The proof of this fact is obtained by a procedure which is again essentially the second diagonal method. It is clear not only that there are infinitely many distinct functions on $[0, 1]$ but also that there exist uncountably many. For, at the point $x = 0$ alone, the functions can already assume the uncountably many values between 0 and 1. Thus, there remains to be proved only that the set of functions is not equipollent to the continuum.

Suppose that the set of functions were equipollent to the continuum. Then it would be possible to make the function $f(x)$ and the points z of $[0, 1]$ correspond in a 1–1 manner. Denote by $f_z(x)$ the function thus assigned to the point z. Now, construct a function $g(z)$ defined on the interval $0 \leq z \leq 1$ with the property that, at every point z, $g(z) \neq f_z(z)$. Since $g(z)$ is also a function defined on $[0, 1]$, it must coincide with an $f_u(x)$; hence, in particular, $g(u) = f_u(u)$. This, however, is excluded by the definition of $g(z)$. The assumption that the set of our functions is equipollent to the continuum has thus led to a contradiction.

When two sets X and Y are given, precisely one of the following cases can occur:

(a) X is equipollent to a subset of Y.

(b) X is equipollent to no subsets of Y.

Likewise, exactly one of the following possibilities can take place:

(1) Y is equipollent to a subset of X,

(2) Y is equipollent to no subsets of X.

There are four combinations of these two pairs of cases, vis., (a)(1), (a)(2), (b)(1), (b)(2). In the two middle cases $|X| < |Y|$ and $|Y| < |X|$. Case (b)(2) means that X and Y are not comparable. We will show that this case never occurs. Case (a)(1) is discussed next.

Theorem 2.3.2 [Bernstein's Equipollence Theorem]. *If each of two sets X and Y is equipollent to a subset of the other, then $X \sim Y$.*

Proof: Assume that $f : X \to Y$ and $g : Y \to X$ are both 1–1 functions. Let $A_0 = X - g(Y)$, $B_0 = Y - f(X)$, and define inductively

$$A_{n+1} = g(B_n), \; B_{n+1} = f(A_n), \quad \text{for each } n \in \mathbb{N}$$

where $\mathbb{N} = \{0, 1, 2, \dots\}$. Then by induction on n we have

$$A_n \cap A_m = \emptyset, \; B_{n+1} \cap B_{m+1} = \emptyset, \quad \text{if } n \neq m \tag{2.24}$$

If $A_n \cap A_m = \emptyset$, then $B_{n+1} \cap B_{m+1} = \emptyset$ since f is 1–1 and if $B_n \cap B_m = \emptyset$, then $A_{n+1} \cap A_{m+1} = \emptyset$ since g is 1–1. So, since $A_0 \cap A_n = \emptyset$ and $B_0 \cap B_n = \emptyset$, by induction, for each $n \neq 0$, Eq. (2.24) follows. Now, define $A = \bigcup_{n \in \mathbb{N}} A_n$ and $B = \bigcup_{n \in \mathbb{N}} B_n$ and define a 1–1 function h from A onto B by

$$h(z) = \begin{cases} f(z), & \text{if } z \in A_m \text{ and } m \text{ is even} \\ g^{-1}(z), & \text{if } z \in A_m \text{ and } m \text{ is odd} \end{cases} \tag{2.25}$$

Then if $t \in B_m$ and m is even, $h^{-1}(t) = g(t)$, and if m is odd then $h^{-1}(t) = f^{-1}(t)$. So, h is from A onto B.

In addition, f is 1–1 onto between $X - A$ and $Y - B$ since if $z \in X - A$, then $f(z) \notin B$ (since $f(z) \in B_m$ implies that $m \neq 0$ and $z \in A_{m-1}$); similarly, if $t \in Y - B$, then $t = f(z)$ for some $z \in X$ (otherwise $t \in B_0$), and $z \in A$ implies $t \in B$. Now, we extend h as follows:

$$h(z) = f(z), \quad \text{if } z \in X - A \tag{2.26}$$

Then h is 1–1 from X onto Y, which establishes the equipollence between X and Y. ■

We know from Section 2.1 that to each set X a cardinality (or cardinal number) is assigned. Corollary 2.1.1 says that besides the finite cardinal numbers, the number $\aleph_0$, and the continuum c, there exist infinitely many other cardinal numbers. In the sequel, we use $p(X)$ to indicate the power set of the set X.

Theorem 2.3.3. *Let A be an arbitrary set. Then no two of the sets*

$$A, \quad p(A), \quad p(p(A)), \quad p(p(p(A))) \tag{2.27}$$

are equipollent.

Proof: Let p_κ be the κth set in the sequence (2.27) and suppose that there exist κ and ι such that $\kappa > \iota$ and p_κ is equipollent to a subset of p_ι. The set $p_{\kappa-1}$ is clearly equipollent to a subset of p_κ, namely to the subset of singletons $\{x\}$, where $x \in p_{\kappa-1}$. Thus, the set $p_{\kappa-1}$ is equipollent to a subset of p_ι. Repeating this argument, we conclude that each of the sets $p_\kappa, p_{\kappa-1}, \dots, p_{\iota+1}$ is equipollent to some subset of p_ι, but this contradicts Corollary 2.1.1 and Theorem 2.3.2 because $p_{\iota+1} = p(p_\iota)$. ■

Theorem 2.3.4. *Let the family A have the property that for every $X \in A$ there exists a set $Y \in A$ which is not equipollent to any subset of X. Then the union $\bigcup_{X \in A} X$ is not equipollent to any $X \in A$ nor to any subset of $X \in A$.*

Proof: Assume that there exists an $X \in A$ such that $\bigcup_{X \in A} X \sim X_1 \subset X$. It follows that there exists a 1–1 function f such that $f(\bigcup_{X \in A} X) = X_1$. By assumption of the theorem there is a set $Y \in A$ which is not equipollent to any subset of X. Since $Y \subset \bigcup_{X \in A} X$, we have $f(Y) \subset f(\bigcup_{X \in A} X)$; that is, $f(Y) \subset X_1$ and consequently $Y \sim f(Y) \subset X$. The contradiction thus shows that $\bigcup_{X \in A} X \not\sim X_1$. ■

Theorems 2.3.3 and 2.3.4 give us some ideas about how many distinct infinite cardinal numbers exist. Starting with the set $\mathbb{N}$ of nonnegative integers which has power $\aleph_0$, we can construct the sets

$$\mathbb{N}, \quad p(\mathbb{N}), \quad p(p(\mathbb{N})), \quad p(p(p(\mathbb{N}))), \quad p(p(p(p(\mathbb{N})))), \quad \dots \tag{2.28}$$

no two of which are equipollent by Theorem 2.3.3. In this way we obtain infinitely many distinct cardinal numbers. Theorem 2.3.4 says the union $S = \bigcup_{n \in \mathbb{N}} p^n(\mathbb{N})$, where $p^0(\mathbb{N}) = \mathbb{N}$, $p^{n+1}(\mathbb{N}) = p(p^n(\mathbb{N}))$, for each $n \in \mathbb{N}$, has a cardinal number different from each of the sets in sequence (2.28) and from each of their subsets. Again applying Theorem 2.3.3 we obtain the sequence of sets

$$S, \quad p(S), \quad p(p(S)), \quad p(p(p(S))), \quad \dots \tag{2.29}$$

no two of which are equipollent and none of which is equipollent to any of the sets in sequence (2.28). We obtain in this way a new infinite number of distinct cardinalities.

We may still obtain other cardinal numbers by constructing the family $B = \bigcup_{n \in \mathbb{N}} p^n(\mathbb{N}) \cup \bigcup_{n \in \mathbb{N}} p^n(S)$ and a new sequence

$$B, \quad p(B), \quad p(p(B)), \quad p(p(p(B))), \quad \dots \tag{2.30}$$

This procedure continues infinitely. We see that the hierarchy of distinct infinite cardinal numbers is comparably richer than the hierarchy of finite cardinalities, which coincides with $\mathbb{N}$.

We now begin to define the operations of addition, multiplication, and exponentiation for cardinal numbers. Let m and n be cardinal numbers and X and Y two disjoint sets such that $|X| = m$ and $|Y| = n$. Then define

$$m + n = |X \cup Y| \tag{2.31}$$

For the definition to have a meaning, it is necessary to show that the sum $m + n$ does not depend on the choice of the sets X and Y. Therefore, let X^* and Y^* be two disjoint sets with $|X^*| = |X|$ and $|Y^*| = |Y|$. Then it is not hard to see that $X^* \cup Y^* \sim X \cup Y$.

Theorem 2.3.5. *The addition of cardinal numbers is commutative and associative; that is, for arbitrary cardinal numbers n_1, n_2, and n_3,*

$$n_1 + n_2 = n_2 + n_1 \tag{2.32}$$

$$n_1 + (n_2 + n_3) = (n_1 + n_2) + n_3 \tag{2.33}$$

The proof follows from the definition of addition of cardinal numbers.

Theorem 2.3.6. *Let m_i and n_i be cardinal numbers, $i = 1,2$, such that $m_i \leq n_i$. Then*

$$m_1 + m_2 \leq n_1 + n_2 \tag{2.34}$$

Proof: Let X_i and Y_i, $i = 1,2$, be four pairwise disjoint sets such that $|X_i| = m_i$ and $|Y_i| = n_i$. Then by hypothesis there exist 1–1 functions $f_i : X_i \to Y_i$, $i = 1,2$. Define a new function $g : X_1 \cup X_2 \to Y_1 \cup Y_2$ by

$$g(x) = f_i(x), \quad \text{if } x \in X_i \tag{2.35}$$

We then check that g is a one-to-one function from $X_1 \cup X_2$ into $Y_1 \cup Y_2$. This gives $m_1 + m_2 \leq n_1 + n_2$. ■

We cannot, however, always infer from $m_1 < n_1$ and $m_2 \leq n_2$ that $m_1 + m_2 < n_1 + n_2$. For example, for every $n \in \mathbb{N}$ we have $n < \aleph_0$ and $\aleph_0 \leq \aleph_0$, but $n + \aleph_0 = \aleph_0 = \aleph_0 + \aleph_0$.

The cardinal number m is the product of n_1 and n_2, denoted $m = n_1 \cdot n_2$, if every set of power m is equipollent to the Cartesian product $A_1 \times A_2$, where $|A_i| = n_i$, $i = 1,2$. Thus, $|A_1| \cdot |A_2| = |A_1 \times A_2|$.

Theorem 2.3.7. *The multiplication of cardinal numbers is commutative, associative, and distributive over addition:*

$$n_1 \cdot n_2 = n_2 \cdot n_1 \tag{2.36}$$

$$n_1 \cdot (n_2 \cdot n_3) = (n_1 \cdot n_2) \cdot n_3 \tag{2.37}$$

$$n_1 \cdot (n_2 + n_3) = n_1 \cdot n_2 + n_1 \cdot n_3 \tag{2.38}$$

Proof: Equation (2.36) is an immediate consequence of the relation

$$A \times B \sim B \times A$$

Equation (2.37) follows from

$$[A \times (B \times C)] \sim [(A \times B) \times C]$$

Equation (2.38) is shown by the following:

$$A_1 \times (A_2 \cup A_3) = (A_1 \times A_2) \cup (A_1 \times A_3)$$
$$[A_2 \cap A_3 = \emptyset] \text{ implies } [(A_1 \times A_2) \cap (A_1 \times A_3) = \emptyset]$$ ■

Denote the n-fold product $m \cdot m \cdots m$ by m^n. Then m^n is the power of the set of all sequences of n-elements $(a_1, a_2, \ldots, a_n)$, where $a_1, \ldots, a_n$ are elements of a set A of power m. In other words, $|A|^n = |A^n|$. Generalizing this idea, let m, n, and p be cardinal numbers. Then

$$m = n^p \tag{2.39}$$

(read as m equals n raised to the pth power) if every set of power m is equipollent to the set A^B, where $|A| = n$, $|B| = p$, and $A^B = \{f : f \text{ is a function from } B \text{ into } A\}$.

Theorem 2.3.8. *For arbitrary cardinal numbers n, p, and q,*

$$n^{p+q} = n^p \cdot n^q \tag{2.40}$$
$$(n \cdot p)^q = n^q \cdot p^q \tag{2.41}$$
$$(n^p)^q = n^{p \cdot q} \tag{2.42}$$
$$n^1 = n \tag{2.43}$$
$$1^n = 1 \tag{2.44}$$

Proof: Suppose that X, Y, and Z are pairwise disjoint sets with $|X| = n$, $|Y| = p$, and $|Z| = q$. Then, showing Eq. (2.40)–(2.44) is equivalent to showing

$$X^{Y \cup Z} \sim X^Y \times X^Z \tag{2.45}$$
$$(X \times Y)^Z \sim X^Z \times Y^Z \tag{2.46}$$
$$(X^Y)^Z \sim X^{Y \times Z} \tag{2.47}$$
$$X^{\{0\}} \sim X \tag{2.48}$$
$$\{0\}^X \sim \{0\} \tag{2.49}$$

The proofs of Eqs. (2.48) and (2.49) are straightforward. We give only the arguments of the important formulas Eqs. (2.45)–(2.47).

To prove Eq. (2.45), we associate with every function $f \in X^{Y \cup Z}$ the ordered pair of restricted functions $(f|Y, f|Z)$. Then it is not difficult to show that this correspondence is a bijection from the set $X^{Y \cup Z}$ onto the set $X^Y \times X^Z$.

For the proof of Eq. (2.46), notice that if $f \in (X \times Y)^Z$, then $f(z)$ is, for every $z \in Z$, an ordered pair $(g(z), h(z))$, where $g(z) \in X$ and $h(z) \in Y$. Then $g \in X^Z$

and $h \in Y^Z$. It is easy to show that this correspondence of the function f to the pair (g, h) determines a 1–1 mapping from the set $(X \times Y)^Z$ onto the set $X^Z \times Y^Z$.

Finally, let us prove Eq. (2.47). Let $f \in X^{Y \cup Z}$; hence f is a function of two variables y and z, where y ranges through the set Y and z through the set Z and where f takes values in X. For each fixed z, the function g_z defined by $g_z(y) = f(y,z)$ is a function from Y into Y; i.e., $g_z \in X^Y$. The function F defined by $F(z) = g_z$ associates with every $z \in Z$ an element of the set X^Y, so $F \in (X^Y)^Z$.

If f_1 and f_2 are distinct functions from $X^{Y \cup Z}$, then the corresponding functions F_1 and F_2 are also distinct. In fact, if $f_1(y_0, z_0) \neq f_2(y_0, z_0)$, then the elements $F_1(z_0)$ and $F_2(z_0)$ of X^Y are distinct. Each function $F \in (X^Y)^Z$ corresponds in a manner described before to some function $f \in X^{Y \times Z}$, namely, $f(y,z) = g_z(y)$, where $g_z = F(z)$. It follows that the correspondence of the function $f \in X^{Y \times Z}$ to the function $F \in (X^Y)^Z$ establishes the equipollence set $X^{Y \times Z}$ with $(X^Y)^Z$, which proves Eq. (2.47). ■

Theorem 2.3.9. *If A has the power m, then the power set $p(A)$ has power 2^m; i.e., $2^m = |p(A)| = |2^A|$.*

Proof: By definition, 2^m is the power of the set $\{0, 1\}^A$, consisting of all functions f whose values are the numbers 0 and 1 and whose domain is the set A. Each such function is uniquely determined by the set X_f of those a $\in A$ for which $f(a) = 1$. To distinct functions f_1 and f_2 correspond distinct sets X_{f_1} and X_{f_2}. Thus, associating with the function $f \in \{0, 1\}^A$ the set $X_f \subset A$, we obtain a bijection from the set $\{0, 1\}^A$ onto the set $p(A)$. ■

Theorem 2.3.10. *The relation "$\leq$" between cardinal numbers possesses the following properties:*

$$m \leq n \text{ and } n \leq p \quad \text{imply} \quad m \leq p \tag{2.50}$$

$$m \leq n \quad \text{implies} \quad m + p \leq n + p \tag{2.51}$$

$$m \leq n \quad \text{implies} \quad mp \leq np \tag{2.52}$$

$$m \leq n \quad \text{implies} \quad m^p \leq n^p \tag{2.53}$$

$$m \leq n \quad \text{implies} \quad p^m \leq p^n \tag{2.54}$$

Proof: Equation (2.50) expresses the transitivity of the relation $\leq$. Equations (2.51)–(2.54) express the monotonicity of addition, multiplication and exponentiation with respect to $\leq$. The proof is straightforward and is omitted here. ■

In the arithmetic of natural numbers, the laws converse to Eqs. (2.51)–(2.54) are called the cancellation laws for the relation $\leq$ with respect to the operations of addition, multiplication, and exponentiation. These theorems hold in arithmetic provided that $p > 1$. In the arithmetic of arbitrary cardinal numbers all of the cancellation laws fail to fold: it suffices to let $m = 2$, $n = 3$, and $p = \aleph_0$ to obtain a counterexample. On the other hand, the cancellation laws with respect to addition, multiplication, and exponentiation hold for the relation $<$. They follow without difficulty from the law of trichotomy, which now we state without proof.

Theorem 2.3.11. *For arbitrary cardinal numbers m and n, either $m \leq n$ or $n \leq m$.*

We now study a few properties of the cardinal numbers $\aleph_0$ and c.

(1) $\mathbf{c = c + c}$:
In fact, let X and Y be intervals $(0,1)$ and $[1,2)$. Then $|X| = |Y| = c$ and $|X \cup Y| = c + c = |(0,2)| = c$.

(2) **For each natural number $\boldsymbol{n}$, $\boldsymbol{n < \aleph_0 < c}$:**
The inequality follows from Theorem 2.1.3.

(3) **For each natural number $\boldsymbol{n}$, $\boldsymbol{n + c = \aleph_0 + c = c}$:**
In fact, $c \leq n + c \leq \aleph_0 + c \leq c + c = c$. So, by property 1 and Bernstein's theorem, the result follows.

(4) $\mathbf{c = c \cdot c}$:

Proof: Let X be the set of all points in $(0,1]$. Then $X \times X = \{(x,y) : x,y \in (0,1]\}$ has power $c \cdot c$. The assertion is proved as soon as we show that the sets X and $X \times X$ are equipollent.

Let $z \in (0,1)$ and $(x,y) \in X \times X$. We write the numbers x, y and z by nonterminating decimal fractions. Since the representation of our numbers by means of such decimal fractions is unique, it is necessary to show that the set of decimal fraction pairs x, y can be mapped on the set of decimal fractions z. Let the pair of decimal fractions x and y be given. Now we write the first x-complex, then the first y-complex, then the second x-complex, followed by the second y-complex, etc. Since neither x nor y exhibits only zeros from a certain point on, the process can be continued without ending and gives rise to a nonterminating decimal fraction. This decimal fraction shall be made to correspond to the number pair x,y. Every number pair x,y thus determines precisely one z; conversely, every z determines exactly one number pair x,y, and, in fact, precisely that pair which gave rise to z. A one-to-one correspondence between the number pairs x,y and the numbers z is thereby established, and this completes the proof. ■

(5) **For each natural number n, $n \cdot c = \aleph_0 \cdot c = c$:**
In fact, from Theorem 2.3.10, we have

$$c \leq n \cdot c \leq \aleph_0 \cdot c \leq c \cdot c \tag{2.55}$$

Thus, it follows from Bernstein's theorem and property 4 that $n \cdot c = \aleph_0 \cdot c = c$.

(6) **For each natural number n, $c^n = c$:**
By induction and property 4 we obtain the proof.

(7) Under the hypothesis that $c = 2^{\aleph_0}$, this is the continuum hypothesis. We have the following: for each natural number $n > 1$,

$$n^{\aleph_0} = \aleph_0^{\aleph_0} = c^{\aleph_0} = c \tag{2.56}$$

In fact, from Theorem 2.3.10, it follows that

$$c = 2^{\aleph_0} \leq n^{\aleph_0} \leq \aleph_0^{\aleph_0} = c^{\aleph_0} = (2^{\aleph_0})^{\aleph_0} = 2^{\aleph_0 \aleph_0} = 2^{\aleph_0} = c$$

whence property 7 follows by applying Bernstein's theorem.

2.4. Ordered Sets

Since the elements of sets need by no means be numbers or points on the real number line, and even if the elements are numbers, it is not at all necessary for them to be ordered according to magnitude. The concept of ordered sets will be formulated abstractly as follows:

A set X is called an ordered set if a binary relation, called an order relation and denoted by $\leq$, exists with its field X such that

(1) For all $x \in X$, $x = x$.

(2) For $x, y \in X$, if $x \leq y$ and $y \leq x$, then $x = y$; i.e., the relation is antisymmetric.

(3) For $x, y, z \in X$, if $x \leq y$ and $y \leq z$, then $x \leq z$; i.e., the relation is transitive.
The ordered set X will be denoted by the ordered pair $(X, \leq)$.

The notation $x > y$ means the same as $y < x$ and is read either "y precedes x" or "x succeeds y." The field of the order relation $\leq$ is often said to be ordered without explicitly mentioning $\leq$. It is necessary to remember that an ordering is by no means an intrinsic property of the set. The same set may be ordered by many different order relations.

Let X be a set ordered by a relation $\leq$. If x and y are elements of X and either $x \leq y$ or $y \leq x$, then we say that the elements x and y are comparable; otherwise they are incomparable. If $Y \subset X$ and any two elements of Y are comparable, then we call Y a chain in X; if any two elements of Y are incomparable, then Y is called an antichain in X. An order relation $\leq$ of a set X is called a linear order relation if X ordered by $\leq$ becomes a chain.

Example 2.4.1.

(1) Every family of sets is ordered by the inclusion relation.

(2) The set of natural numbers is ordered by the relation of divisibility.

(3) The set of natural numbers in the order indicated by $\{0,2,4,\ldots,1,3,5,\ldots\}$ is a linearly ordered set; i.e., first come the even nonnegative integers ordered according to increasing magnitude, and then the odd integers.

(4) The set $\{0,2,4,6,\ldots,7,5,3,1\}$, as written, is linearly ordered; that is, first come the even nonnegative integers ordered according to increasing magnitude, then the odd integers ordered according to decreasing magnitude.

The next result is obvious.

Proposition 2.4.1. *Let X be an ordered set with ordering $\leq$. Then every subset of X is also ordered by $\leq$. Hence, the subsets of ordered sets can always be regarded as ordered.*

Two ordered sets $(X, \leq_X)$ and $(Y, \leq_Y)$ are equal,

$$(X, \leq_X) = (Y, \leq_Y) \tag{2.57}$$

if $X = Y$ and for any $x, y \in X$, $x \leq_X y$ is equivalent to $x \leq_Y y$. Without confusion, we write $X = Y$ instead. Thus, the sets in Examples 2.4.1(3) and 2.4.1(4) are to be considered distinct in the sense of ordered sets. Without regard to an order, however, the two sets are identical.

Let $(X, \leq)$ be an ordered set and $a \in X$. The element a is called a maximal (respectively, minimal) element in the ordered set X provided that no element $x \in X$ satisfies $x > a$ (respectively, $x < a$). The element a is called the maximum (respectively, minimum) element of X provided that for every $x \in X$, $x \leq a$ (respectively, $a \leq x$).

Proposition 2.4.2. *An ordered set contains at most one maximum element and at most one minimum element.*

Proof: Let a and a^* be two maximum elements of a set ordered by an ordering $\leq$. Then $a^* \leq a$ and $a \leq a^*$, which implies that $a = a^*$. The same argument shows that an ordered set contains at most one minimum element. ■

An ordered set, however, need have neither a maximum nor a minimum element.

An ordered set A is cofinal with its subset B if for every $x \in A$ there exists $y \in B$ such that $x \leq y$. The subset B is a cofinal subset in A. Analogously, we can define coinitial sets and subsets.

Example 2.4.2. The set of all real numbers is cofinal and coinitial with the set of all integers.

An ordered set $(X, \leq_X)$ is similar to an ordered set $(Y, \leq_Y)$,

$$(X, \leq_X) \simeq (Y, \leq_Y) \tag{2.58}$$

if there exists a bijection $h : X \to Y$ such that for any $x_1, x_2 \in X$, $x_1 \leq_X x_2$ implies $h(x_1) \leq_Y h(x_2)$. Without causing confusion, we write $X \simeq Y$ instead of Eq. (2.58). The bijection h is called a similarity mapping from the ordered set $(X, \leq_X)$ onto the ordered set $(Y, \leq_Y)$.

The following four fundamental properties result immediately from the definition:

(1) $(X, \leq_X) \simeq (X, \leq_X)$; i.e., every ordered set is similar to itself.

(2) $(X, \leq_X) \simeq (Y, \leq_Y)$ implies $(Y, \leq_Y) \simeq (X, \leq_X)$.

(3) If $(X, \leq_X) \simeq (Y, \leq_Y)$ and $(Y, \leq_Y) \simeq (Z, \leq_Z)$, then $(X, \leq_X) \simeq (Z, \leq_Z)$.

(4) $(X, \leq_X) \simeq (Y, \leq_Y)$ implies $X \sim Y$.

Theorem 2.4.1. *If a set X is equipollent to an ordered set $(Y, \leq_Y)$, then X can be ordered in such a way that $X \simeq (Y, \leq_Y)$.*

Proof: Let h be a bijection from X onto Y. We define an order relation $\leq_X$ on X as follows: For any $x_1, x_2 \in X$, $x_1 \leq_X x_2$ if and only if $h(x_1) \leq_Y h(x_2)$. ■

Theorem 2.4.2. *If two ordered sets are similar, then both possess maximum (minimum) elements or neither set has such an element.*

Proof: Let $(X, \leq_X)$ and $(Y, \leq_Y)$ be two similar ordered sets with a similarity mapping $h : X \to Y$. If x_0 is a minimum element from X, then for every $x \in X$,

$x\ {}_X\!\ge x_0$. So $h(x)\ {}_Y\!\ge h(x_0)$. This implies that for every $y \in Y$, $h(x_0) \le_Y y$. So Y also has a minimum element. ■

The family of all ordered sets is divided, by similarity relation, into pairwise disjoint groups, each of which contains all sets similar to each other. Two ordered sets $(X, \le_X)$ and $(Y, \le_Y)$ are of the same order type if they are similar. Each order type will be denoted by a small Greek letter, and the order type of $(X, \le_X)$ will also be denoted by $\overline{(X, \le_X)}$, or $\overline{X}$ without confusion.

The order type of the set of natural numbers, ordered according to increasing magnitude, is denoted by ω. If, on the other hand, these numbers are ordered according to decreasing magnitude, then the order type of the set is denoted by $^*\omega$. Generally, $^*\mu$ denotes that order type which results from μ when the order of succession of the elements is reversed; i.e., when a new order relation is derived from the relation $<$ defined for the original set, by setting $^* <$ equal to $>$.

Let $(X, \le_X)$ and $(Y, \le_Y)$ be two disjoint ordered sets. An order relation $\le$ on the union $X \cup Y$ is defined as follows: For any s_1 and $s_2 \in X \cup Y$, if s_1 and $s_2 \in X$, we let $s_1 \le s_2$ if and only if $s_1 \le_X s_2$; if s_1 and $s_2 \in Y$, we let $s_1 \le s_2$ if and only if $s_1 \le_Y s_2$; if $s_1 \in X$ and $s_2 \in Y$, then we let $s_1 \le s_2$. The order relation $\le$ thus defined on the union is obviously transitive, and it therefore orders $X \cup Y$. The ordered sum $X + Y$ of the ordered sets $(X, \le_X)$ and $(Y, \le_Y)$ is now understood to be the ordered set $(X \cup Y, \le)$.

Let μ and γ be two order types and $(X, \le_X)$ and $(Y, \le_Y)$ be two disjoint ordered sets such that $\overline{X} = \mu$ and $\overline{Y} = \gamma$. The sum of the order types μ and γ is defined by

$$\mu + \gamma = \overline{(X \cup Y, \le)} \tag{2.59}$$

It is again easy to see that the sum $\mu + \gamma$ is independent of the choice of the ordered sets $(X, \le_X)$ and $(Y, \le_Y)$.

The addition of order types need not be commutative. For example, if

$$X = \{1, 2, 3, \dots\} \quad \text{and} \quad Y = \{0\} \tag{2.60}$$

then

$$X + Y = \{1, 2, 3, \dots, 0\} \quad \text{and} \quad Y + X = \{0, 1, 2, 3, \dots\} \tag{2.61}$$

so that $1 + \omega = \omega$, but $\omega + 1 \ne \omega$, where 1 is the order type of Y.

The associative law of order types μ, γ, and π is valid:

$$(\mu + \gamma) + \pi = \mu + (\gamma + \pi) \tag{2.62}$$

The concept of addition of order types can be defined for arbitrary many order types. In fact, let an ordered set $(I, \le_I)$ be given, and to each $k \in I$ there corresponds an ordered set $(X_k, \leq_k)$, where the sets X_k are pairwise disjoint. Then the ordered sum

$$\sum_{k \in I}(X_k, \leq_k) = (\bigcup_{k \in I} X, \leq) \tag{2.63}$$

is ordered in the following manner: If two elements s_1 and s_2 of $\bigcup_{k \in I} X_k$ belong to the same X_k, then we let $s_1 \leq s_2$ if and only if $s_1 \leq_k s_2$; if, however, s_1 and s_2 belong to different X_k's, say $s_1 \in X_{k_1}$ and $s_2 \in X_{k_2}$, then we let $s_1 < s_2$ if and only if $k_1 < k_2$.

Let μ_k be an order type for each $k \in I$ and $(X_k, \leq_k)$ an ordered set such that $\mu_k = \overline{X}_k$, where $\{X_k : k \in I\}$ consists of pairwise disjoint sets. The sum of the order types μ_k is defined by

$$\sum_{k \in I} \mu_k = \overline{\sum_{k \in I}(X_k, \leq_k)} \tag{2.64}$$

It is easy to see once more that this sum is independent of the choice of the ordered sets $(X_k, \leq_k)$.

In the previous definition, if $\mu_k = \mu$ for each $k \in I$ and $\overline{(I, \leq_I)} = \gamma$, then the product of the order types $\mu \cdot \gamma$ will equal the sum thus determined; i.e.,

$$\mu \cdot \gamma = \sum_{k \in I} \mu \tag{2.65}$$

For $\gamma = 0$ we let $\mu \cdot \gamma = 0$. The concept of product of order types can also be defined as follows: Let κ and μ be two ordered types of nonempty sets $(X, \leq_X)$ and $(Y, \leq_Y)$ such that $\overline{X} = \kappa$ and $\overline{Y} = \mu$. Define an ordered set $(X \times Y, \leq)$ by letting $(x_1, y_1) < (x_2, y_2)$ if and only if $x_1 < x_2$, or when $x_1 = x_2$ then $y_1 < y_2$. Then the product $\mu \cdot \kappa = \overline{(X \times Y, \leq)}$. The order relation defined here is called a lexicographical order.

This second definition follows immediately from the first. It is also evident from the second definition that the product of order types is independent of their particular set representatives.

For multiplication there is no commutative law either. For if we choose $\kappa = \omega$ and $\mu = 2$, where 2 is the order type of $\{0, 1\}$ ordered by $0 < 1$, then $2 \cdot \omega = \omega$ and $\omega \cdot 2$ is the order type of the ordered set

$$\{1, 3, 5, \ldots; 2, 4, 6, \ldots\} \tag{2.66}$$

This set is not certainly similar to any ordered set of type ω. Hence, $2 \cdot \omega \neq \omega \cdot 2$.

We do have, however, for multiplication, as for addition, the associative law, which is so important for calculation:

$$(\mu \cdot \gamma) \cdot \pi = \mu \cdot (\gamma \cdot \pi) \tag{2.67}$$

In fact, let $(X,\leq_X)$, $(Y,\leq_Y)$, and $(Z,\leq_Z)$ be ordered sets such that $\overline{X}=\mu$, $\overline{Y}=\gamma$, and $\overline{Z}=\pi$. Then, according to the second definition, the order type $(\mu\cdot\gamma)\cdot\pi$ is the order type of the ordered set $(Z\times(Y\times X),\leq_1)$, where for any $(z_i,(y_i,x_i))\in Z\times(Y\times X)$, $i=1,2$,

$$(z_1,(y_1,x_1)) <_1 (z_2,(y_2,x_2)) \tag{2.68}$$

if $z_1<z_2$, or $z_1=z_2$, $y_1<y_2$, or $z_1=z_2$ and $y_1=y_2$ and $x_1<x_2$. Likewise, $\mu\cdot(\gamma\cdot\pi)$ is the order type of the ordered set $((Z\times Y)\times X,\leq_2)$, where for any $((z_i,y_i),x_i)\in(Z\times Y)\times X$, $i=1,2$,

$$((z_1,y_1),x_1) <_2 ((z_2,y_2),x_2) \tag{2.69}$$

if $z_1<z_2$, or $z_1=z_2, y_1<y_2$, or $z_1=z_2, y_1=y_2, x_1<x_2$. If we now associate, for the same elements x, y, and z, the element $(z,(y,x))$ with the element $((z,y),x)$, we define a similarity correspondence between $(Z\times(Y\times X),\leq_1)$ and $((Z\times Y)\times X,\leq_2)$. This proves the assertion.

The right distributive law does not hold. Namely,

$$(\mu+\gamma)\cdot\pi\neq\mu\cdot\pi+\gamma\cdot\pi \tag{2.70}$$

For example,

$$\begin{aligned}(\omega+1)\cdot 2&=(\omega+1)+(\omega+1)\\&=\omega+(1+\omega)+1\\&=\omega+\omega+1\\&=\omega\cdot 2+1\end{aligned} \tag{2.71}$$

and this differs from

$$(\omega+1)\cdot 2=\omega\cdot 2+2 \tag{2.72}$$

The left distributive law, however, is valid; that is,

$$\pi\cdot(\mu+\gamma)=\pi\cdot\mu+\pi\cdot\gamma \tag{2.73}$$

In fact, let $(X,\leq_X)$, $(Y,\leq_Y)$, and $(Z,\leq_Z)$ be disjoint ordered sets such that $\overline{X}=\mu$, $\overline{Y}=\gamma$, and $\overline{Z}=\pi$. Then $\pi\cdot(\mu+\gamma)$ is the order type of the ordered set $((X\cup Y)\times Z,\leq_1)$, where for any $(q_i,z_i)\in(X\cup Y)\times Z$, $i=1,2$,

$$(q_1,z_1) <_1 (q_2,z_2) \tag{2.74}$$

if $q_1\in X$ and $q_2\in Y$, or $q_1,q_2\in X$, $q_1<_X q_2$, or $q_1,q_2\in Y$, $q_1<_Y q_2$, or $q_1=q_2$, $z_1<_Z z_2$.

Now it is evident that the same ordered set is obtained with order type $\pi \cdot \mu + \pi \cdot \gamma$. This completes the proof of the statement.

It does not yet follow from the results derived thus far that to every cardinal number $\aleph$ there corresponds at least one order type μ with $|\mu| = \aleph$, where $|\mu|$ means the cardinality of any of those ordered sets with type μ. In other words, it does not follow from the results derived that every set can be ordered. Let $\mathfrak{T}_{\aleph}$ be the set of all order types which belong to a given cardinal number $\aleph$, and call this set the type class belonging to $\aleph$.

Theorem 2.4.3. *For any cardinal number $\aleph$, the type class $\mathfrak{T}_{\aleph}$ satisfies*

$$|\mathfrak{T}_{\aleph}| \leq 2^{\aleph \cdot \aleph} \tag{2.75}$$

Proof: Let X be a set with cardinality $\aleph$. Let μ be any order type in $\mathfrak{T}_{\aleph}$. Then it is always possible to order X so that $\overline{X} = \mu$. The set of all order types of class $\mathfrak{T}_{\aleph}$ is equipollent to the set of all possible orderings of our set X. For each order relation $<$ on the set X, let $S_{<} = \{(x_1, x_2) \in X^2 : x_1 < x_2\}$. Then it can be seen that different order relations $<_1$ and $<_2$ on X correspond to different subsets $S_{<_1}$ and $S_{<_2}$ of X^2. Therefore,

$$|\mathfrak{T}_{\aleph}| \leq \left|2^{X^2}\right| = 2^{|X^2|} = 2^{\aleph \cdot \aleph} \tag{2.76}$$

■

Theorem 2.4.4. *For the class $\mathfrak{T}_{\aleph_0}$ of all order types of denumerable sets, $\left|\mathfrak{T}_{\aleph_0}\right| = 2^{\aleph_0}$.*

Proof: The previous theorem implies that

$$\left|\mathfrak{T}_{\aleph_0}\right| \leq 2^{\aleph_0} \tag{2.77}$$

It remains to show that $\left|\mathfrak{T}_{\aleph_0}\right| \geq 2^{\aleph_0}$. Let $\gamma = {}^*\omega + \omega$. With any sequence of natural numbers

$$(a_1, a_2, a_3, \dots) \tag{2.78}$$

we form the order type

$$\alpha = a_1 + \gamma + a_2 + \gamma + a_3 + \gamma + \cdots \tag{2.79}$$

Then distinct sequences (2.78) always give rise to distinct order types (2.79). In fact, let

$$(b_1, b_2, b_3, \dots) \tag{2.80}$$

also be a sequence of natural numbers, and with it we form the order type

$$\beta = b_1 + \gamma + b_2 + \gamma + b_3 + \gamma + \dots$$

Suppose that $\alpha = \beta$. Then we are to show that the sequences (2.78) and (2.80) coincide. This follows by induction from the following auxiliary consideration:

Let X_i and Y_i be distinct ordered sets, $i = 1,2$. Then the relations

$$X_1 \simeq X_2, \quad Y_1 \simeq Y_2 \tag{2.81}$$

follow from

(a) $X_1 + Y_1 \simeq X_2 + Y_2$, Y_i, $i = 1,2$, have no first element, and X_i, $i = 1,2$, are finite sets,

(b) $Y_1 + X_1 \simeq Y_2 + X_2$ and $\overline{Y}_1 = \overline{Y}_2 = \gamma$.

In fact, for the case (a), under a similarity mapping of $X_1 + Y_1$ onto $X_2 + Y_2$, no element of Y_2 can correspond to an element of X_1 because every element of X_1 is preceded by only a finite number of elements in $X_1 + Y_1$, whereas every element of Y_2 is preceded by infinitely many elements in $X_2 + Y_2$. It follows, likewise, that no element of Y_1 can correspond to an element of X_2. Consequently, X_1 is mapped on X_2 and Y_1 is mapped on Y_2. In (b) under a similarity mapping of $Y_1 + X_1$ onto $Y_2 + X_2$, an element x_1 of X_1 cannot correspond to an element y_2 of Y_2; otherwise the subset Y_1 preceding x_1 in $Y_1 + X_1$ would have to be mapped on a part Y_2^* of the elements of Y_2, which is impossible, because Y_1 does not have a last element, whereas Y_2^* does. In like manner we see that no element of Y_1 can correspond to an element of X_2. Hence, also in this case, X_1 is mapped on X_2 and Y_1 is mapped on Y_2.

If we now apply (a) to the equal sums $\alpha = \beta$, we get

$$a_1 = b_1$$

and

$$\gamma + a_2 + \gamma + a_3 + \gamma + \cdots = \gamma + b_2 + \gamma + b_3 + \gamma + \cdots$$

so that, by (b),

$$a_2 + \gamma + a_3 + \gamma + \cdots = b_2 + \gamma + b_3 + \gamma + \cdots$$

and hence, by (a),

$$a_2 = b_2 \text{ and } \gamma + a_3 + \gamma + \cdots = \gamma + b_3 + \gamma + \cdots$$

etc. Thus, by induction it follows that the sequences (2.78) and (2.80) coincide.

Since the totality of distinct sequences (2.78) gives rise to distinct order types (2.79), and since (2.79) implies that any set with order type (2.79) has cardinality $\leq \aleph_0 \cdot \aleph_0 = \aleph_0$, $\mathfrak{T}_{\aleph_0}$ has at least the power of the set of all sequences (2.78). The power of these sequences of numbers is $\aleph_0^{\aleph_0} = 2^{\aleph_0}$. ■

A set A is densely ordered by an order relation $<$ if $<$ is linear and for any two elements x and $y \in A$ there exists an element $z \in A$ between x and y. We say also that A is dense. In a densely ordered set no element has either a direct successor or a direct predecessor. An ordered set is called unbordered if it is not empty and has no maximal or minimal elements.

All one-element sets, as well as the empty set, are densely ordered. All other densely ordered sets contain infinitely many elements.

Theorem 2.4.5 [G. Cantor]. *All unbordered, dense, denumerable sets are similar to one another.*

Proof: Let $(X, \leq_X)$ and $(Y, \leq_Y)$ be two sets of the kind mentioned in the theorem. Since they are denumerable, they can be written as sequences

$$X = \{x_0, x_1, x_2, \dots\} \tag{2.82}$$

and

$$Y = \{y_0, y_1, y_2, \dots\} \tag{2.83}$$

We now show that X and Y with their orders $\leq_X$ and $\leq_Y$ can be mapped onto each other. To this end, let x_0 correspond to any element n_0 of Y, say $n_0 = y_0$. We then look for the element in $Y - \{n_0\}$ with the smallest index, denoted by n_1, such that $n_0 <_Y n_1$ when $x_0 <_X x_1$, or $n_0 \;_Y\!> n_1$ when $x_0 \;_X\!> x_1$. This is possible because Y is unbordered. We associate x_1 with n_1.

Now x_2 may appear before, after, or between x_0 and x_1 under $\leq_X$. In each case, since $(Y, \leq_Y)$ is unbordered and dense, there is an element n_2 with the smallest index in sequence (2.83) which has the same position relative to n_0 and n_1 as x_2 has relative to x_0 and x_1.

In this way, a mapping h from X into Y is defined. In fact, h is onto. Suppose that h is not onto; then $Y - h(X) \neq \emptyset$. Let y_n be the first element in (2.83) such that $y_n \in Y - h(X)$. Then there are three possibilities:

(1) $y_n <_Y y_i$, for all $i < n$.

(2) $y_n \;_Y\!> y_i$, for all $i < n$.

(3) there are $i, j < n$ such that $y_i <_Y y_n <_Y y_j$ and no other y_k, for $k < n$, is between y_i and y_j.

If (1) is true, and since $(X, \leq_X)$ is unbordered, it follows that there exists $x \in X$ such that $x <_X h^{-1}(y_i)$, $i < n$. Let x_m be the first element in sequence (2.82) such that $x_m <_X h^{-1}(y_i)$, $i < n$. Then the definition of h implies that $h(x_m) = y_n$, contradiction. A similar argument shows that (2) and (3) do not hold. Therefore, h is a similarity mapping from $(X, \leq_X)$ onto $(Y, \leq_Y)$. ■

The proof of Theorem 2.4.5 yields the following result, the proof of which is left for the reader.

Theorem 2.4.6. *Given any dense set with at least two elements and any linearly ordered denumerable set, there is always a subset of the former which is similar to the latter.*

Let $(A, \leq_A)$ be a linearly ordered set. A pair (X, Y) is a cut in the ordered set A, if $Y = \{a \in A : x \leq_A a, \text{ for all } x \in X\}$ and $X = \{a \in A : y\ {}_A\!\geq a, \text{ for all } y \in Y\}$. The set X is called the lower section, and Y the upper section, of the cut.

Proposition 2.4.3. *Let (X, Y) be a cut in a linearly ordered set A. Then $X \cap Y$ contains at most one element.*

Proof: If there are x and $y \in X \cap Y$, then it must be both $x \leq y$ and $y \leq x$. Thus, $x = y$. ■

A set A is continuously ordered if A is linearly ordered and each proper cut (X, Y) in A (i.e., $X \neq \emptyset \neq Y$) has a nonempty intersection $X \cap Y$. We also say that A is continuous.

Theorem 2.4.7. *Every continuous set $(A, \leq_A)$ contains a subset which is similar to the set of all real numbers in their natural order.*

Proof: According to Theorem 2.4.6, $(A, \leq_A)$ contains a subset Q which is similar to the set of all rational numbers. The set Q ordered by $\leq_A$ has gaps. In other words, there is at least one proper cut (X, Y) in $(Q, \leq_Q)$, where $\leq_Q$ is the restriction of $\leq_A$ on the set Q, defined by $\leq_Q = \leq_A \cap Q^2$, such that $X \cap Y = \emptyset$. Let

$$Q = X + Y \tag{2.84}$$

be a decomposition of Q where the cut (X, Y) determines a gap in Q. Then $(A, \leq_A)$ contains an element a which succeeds all elements of X while preceding all elements in Y. In fact, let X^* be the totality of elements of A which belong to X or precede an element of X, and let Y^* be the totality of elements of A which belong to Y or succeed an element of Y. If (X^*, Y^*) is a cut with a gap, the set $(A, \leq_A)$ would have a gap, contrary to the assumption that $(A, \leq_A)$ is continuous. Hence, $X^* + Y^*$ lacks at least one element a of A, and, according to the construction of

$X^* + Y^*$, this element succeeds X^* and precedes Y^*, so it also succeeds X and precedes Y.

Every cut (2.84) thus determines at least one element a of A. For every gap (2.84), we choose one of these elements $a_{(X,Y)}$ and keep it fixed. Let

$$R^* = Q \cup \{a_{(X,Y)} : (X,Y) \text{ is a proper cut in } Q \text{with a gap}\}$$

This set is also ordered by $\leq_A$ and is similar to the set of all rational numbers together with their gap elements in the domain of real numbers. In other words, according to Dedekind's introduction of the real numbers, R^* is similar precisely to the set of all real numbers, which proves the theorem. ■

Theorem 2.4.8. *Let $(A, \leq_A)$ be an ordered set with the following properties:*

(1) $(A, \leq_A)$ is unbordered.

(2) $(A, \leq_A)$ is continuous.

(3) $(A, \leq_A)$ contains a denumerable set U such that between every pair of elements of A there exists at least one element of U.

Then $(A, \leq_A)$ is similar to the naturally ordered set of all real numbers.

Proof: The set U is unbordered, because before and after every element of U there are still elements of A, and hence, due to (3), elements of U. Likewise, it follows from (2) and (3) that U is dense also. Consequently, U is similar to the set of all rational numbers ordered naturally. Every gap in U determines, according to the proof of Theorem 2.4.7, at least one element a of A and, because of (3), not more than one element of A. This, however, implies the assertion. For the set of all rational numbers can be mapped on U, whereby with every gap in the domain of rational numbers there is associated precisely one gap of U and, consequently, precisely one element of A. We have thus obtained a similarity mapping of the set of all real numbers on the set A. ■

2.5. Well-Ordered Sets

An ordered set $(X, \leq_X)$ is well-ordered if it is linearly ordered and every nonempty subset of X contains a first element with respect to the relation $\leq_X$.

Example 2.5.1. Every finite linearly ordered set is well-ordered and so is the set $\mathbb{N} = \{0,1,2,3,\dots\}$. The set $\{\dots,3,2,1,0\}$ is not well-ordered because it does not have a first element. The set of all real numbers in the interval $[0,1]$, in their natural order, is not well-ordered because the subset $(0,1]$ does not have a first element.

The following two theorems are simple consequences of the definition.

Theorem 2.5.1. *In every well-ordered set there exists a first element. Every element except the last element (if such exists) has a direct successor.*

Theorem 2.5.2. *No subset of a well-ordered set is of type* $^*\omega$.

Theorem 2.5.3. *If a linearly ordered set A is not well-ordered, it contains a subset of type* $^*\omega$.

Proof: Let P be a nonempty subset of A which contains no first element. Let

$$Q(x) = \{y \in P : y < x\} \tag{2.85}$$

We have $Q(x) \neq \emptyset$ for every $x \in P$. Let p_0 be any element in P. We define by induction a sequence $p_0, p_1, \dots, p_n, \dots$ by letting $p_n \in Q(p_{n-1})$ for $n > 0$. Since $p_n < p_{n-1}$, for $n > 1$, this sequence is of type $^*\omega$. ■

Theorems 2.5.2 and 2.5.3 imply the next results.

Theorem 2.5.4. *In order that a linearly ordered set be well-ordered, it is necessary and sufficient that it contain no subset of type* $^*\omega$.

Theorem 2.5.5 [Principle of transfinite induction]. *If a set A is well-ordered, $B \subset A$, and if for every $x \in A$ the set B satisfies the condition*

$$[\{y \in A : y < x\} \subset B] \to (x \in B) \tag{2.86}$$

then $B = A$.

Proof: Suppose that $A - B \neq \emptyset$. Then there exists a first element in $A - B$. This means that if $y < x$, then $y \in B$. This shows that $\{y \in A : y < x\} \subset B$. Now it follows from (2.86) that $x \in B$, which contradicts the hypothesis that $x \notin B$. ■

Let $(A, \leq)$ be a linearly ordered set. A function $f : A \to A$ is an increasing function if f satisfies the condition

$$x < y \to f(x) < f(y) \tag{2.87}$$

Theorem 2.5.6. *If a function $f : A \to A$ is increasing, where $(A, \leq_A)$ is a well-ordered set, then for every $x \in A$ we have $x \leq_A f(x)$.*

Proof: Let $B = \{x \in A : x \leq_A f(x)\}$ and $O(x) = \{y \in A : y <_A x\} \subset B$. We now show that $x \in B$. In fact, let $y \in O(x)$; that is, $y <_A x$. By (2.87) it follows that $f(y) <_A f(x)$. Since $y \in B$, we have $y \leq_A f(y)$ and thus $y <_A f(x)$. This means that the element $f(x)$ occurs after every element y of $O(x)$, that is, $f(x) \in A - O(x)$. Since x is the first element of $A - O(x)$, we have $x \leq_A f(x)$ and finally $x \in B$. Hence, from Theorem 2.5.4 the set B equals A. ■

Corollary 2.5.1. *If the well-ordered sets A and B are similar, then there exists only one similarity mapping from A onto B.*

Proof: Suppose that the sets A and B are well ordered by $\leq_A$ and $\leq_B$ and that there exist two functions f and g establishing similarity from A onto B.

The function $g^{-1} \circ f$ is clearly increasing in A. By Theorem 2.5.5 we thus have $x \leq_A g^{-1} \circ f(x)$ for every $x \in A$. Hence, $g(x) \leq_B f(x)$. Consequently, $f^{-1} \circ g$ instead of $g^{-1} \circ f$. By the same argument $f(x) \leq_B g(x)$. This implies $f(x) = g(x)$. ■

Corollary 2.5.2. *No well-ordered set is similar to any of its initial segments.*

Proof: Let $x \in A$, where $(A, \leq_A)$ is a well-ordered set. The initial segment $Q_{\leq_A}(x)$ is the set defined by

$$Q_{\leq_A}(x) = \{y \in A : y <_A x\} \tag{2.88}$$

Suppose that A and $Q_{\leq_A}(x)$ are similar. Then the function f establishing the similarity from A onto $Q_{\leq_A}(x)$ would be increasing and would satisfy $f(x) \in Q_{\leq_A}(x)$; that is, $f(x) <_A x$. But this contradicts Theorem 2.5.5. ■

Theorem 2.5.7. *Let A and B be two well-ordered sets. Then,*

(i) A and B are similar, or

(ii) A is similar to a segment of B, or

(iii) B is similar to a segment of A.

Proof: Let $\leq_A$ and $\leq_B$ be relations which well-order A and B, respectively, and let

$$Z = \{x \in A : \text{ there is } y \in B \text{ such that } Q_{\leq_A}(x) \simeq Q_{\leq_B}(y)\} \tag{2.89}$$

By Corollary 2.5.2, for given $x \in Z$ there exists only one segment $Q_{\leq_B}(y)$ such that $Q_{\leq_B}(y) \simeq Q_{\leq_A}(x)$. Let a function $f : A \to B$ be defined by $f(x) = y$ for each $x \in A$, such that $Q_{\leq_A}(x) \simeq Q_{\leq B}(f(x))$.

First we show that either $Z = A$ or Z is a segment of A that is, there exists an $a \in A$ such that $Z = Q_{\leq_A}(x)$. In fact, let $x <_A x' \in Z$. Since $Q_{\leq_A}(x)$ is a segment of $Q_{\leq_A}(x')$, the function f also maps $Q_{\leq_A}(x)$ onto a segment of B. Hence, $x \in Z$.

Similarly, either $f(Z) = B$ or else $f(Z)$ is a segment of $B : f(Z) = Q_{\leq_B}(b)$. To show this it suffices to observe that

$$f(Z) = \{y \in B : Q_{\leq_B}(y) \simeq Q_{\leq_A}(x), \text{ for some } x \in Z\} \tag{2.90}$$

Finally, we observe that f establishes the similarity between Z and $f(Z)$. Indeed, we have just shown that $x < x' \in Z$ implies that $Q_{\leq_B}(f(x))$ is a segment of $Q_{\leq_B}(f(x'))$; therefore, $f(x) < f(x')$.

Therefore, we have one of the following four possibilities:

(1) $Z = A$ and $f(Z) = B$.

(2) $Z = A$ and $f(Z) = Q_{\leq_B}(b)$.

(3) $Z = Q_{\leq_A}(a)$ and $f(Z) = B$.

(4) $Z = Q_{\leq_A}(a)$ and $f(Z) = Q_{\leq_B}(b)$.

The first three possibilities correspond to those stated in the theorem. Case (4) is impossible, because then $Q_{\leq_A}(a) \simeq Q_{\leq_B}(b)$ and, thus, by definition of Z, $a \in Z$; that is, $a \in Q_{\leq_A}(a)$, which contradicts the definition of a segment. ■

Corollary 2.5.3. *If A and B are well-ordered, then $|A| \leq |B|$ or $|B| \leq |A|$.*

By ordinal numbers (or ordinals or ordinalities) we understand the order types of well-ordered sets. Theorem 2.5.7 implies that we can define a "less than" relation for ordinals. We say that an ordinal α is less than an ordinal β if any set of type α is similar to a segment of a set of type β. This relation will be denoted $\alpha < \beta$ or $\beta > \alpha$. We write $\alpha \leq \beta$ instead of $\alpha < \beta$ or $\alpha = \beta$.

Theorem 2.5.8. *For any ordinals α and β one and only one of the following formulas holds:*

$$\alpha < \beta, \quad \alpha = \beta, \quad \alpha > \beta \tag{2.91}$$

This theorem is a direct consequence of Theorem 2.5.7.

Theorem 2.5.9. *If α, β, and γ are ordinals such that $\alpha < \beta$ and $\beta < \gamma$, then $\alpha < \gamma$.*

Proof: Let A, B, and C be well-ordered sets of types α, β, γ, respectively. By assumption, A is similar to a segment of B and B is similar to a segment of C. Thus, A is similar to a segment of C. ■

The following formulas can be proved without difficulty:

$$(\alpha \le \beta) \text{ and } (\beta \le \alpha) \rightarrow (\alpha = \beta), \qquad (\alpha \le \beta) \text{ and } (\beta \le \gamma) \rightarrow (\alpha \le \gamma)$$

Theorem 2.5.10. *If the well-ordered sets A and B are of types α and β and if A is similar to a subset of B, then $\alpha \le \beta$.*

Proof: Let $h : A \rightarrow B$ be a mapping such that h establishes the similarity between A and $h(A)$. If the theorem were not true, then we would have $\beta < \alpha$ and then B would be similar to a segment of $h(A)$. This contradicts Theorem 2.5.5. ■

Theorem 2.5.11. *The set $W(\alpha)$ consisting of all ordinals less than α is well-ordered by the relation $\le$. Moreover, the type of $W(\alpha)$ is α.*

Proof: Let A be a well-ordered set of type α. Associating the type of the segment $O(a)$ with the element $a \in A$, we obtain a one-to-one mapping of A onto $W(\alpha)$. It is easily seen that the following conditions are equivalent:

(1) a_1 precedes a_2 or $a_1 = a_2$.

(2) $O(a_1)$ is a segment of $O(a_2)$ or $O(a_1) = O(a_2)$.

(3) The type of $O(a_1)$ is not greater than the type of $O(a_2)$.

This shows that the relation $\le$ indeed orders $W(\alpha)$ in type α. ■

Theorem 2.5.12. *Every nonempty set of ordinals has a first element under the relation $\le$.*

Proof: Let A be a nonempty set of ordinals and $\alpha \in A$. If α is not the smallest ordinal of A, then $A \cap W(\alpha) \neq \emptyset$. Then in the set $A \cap W(\alpha)$ there exists a smallest number β because $W(\alpha)$ is well-ordered. At the same time, β is the smallest ordinal in A. ■

An ordinal number is a limit ordinal if it has no direct predecessor. All order types of finite well-ordered sets are denoted by the corresponding natural numbers. For example, $\overline{\{0\}} = 1$, $\overline{\{0,1\}} = 2$, $\overline{\{0,1,2,\}} = 3, \ldots$. Then the order type 0 of the empty set is a limit ordinal.

Theorem 2.5.13. *Each ordinal can be represented in the form $\lambda + n$, where λ is a limit ordinal and n is a finite ordinal (natural number).*

Proof: Let α be an ordinal and A a set of type α. Every set of the form $A - O(a)$ is a remainder of A. Clearly,

$$A - O(a_1) \subset A - O(a_2) \text{ if and only if } (a_1 \geq a_2) \tag{2.92}$$

This implies that there exists no infinite increasing sequence of distinct remainders. So there exists only a finite number of $m \in \mathbb{N}$ such that there exists a remainder of power m. If n is the greatest such number and if $A - O(a)$ is a remainder of power n, then the segment $O(a)$ has no last element. Thus $\overline{O(a)}$ is a limit ordinal λ. This implies that $\alpha = \lambda + n$. ■

Let α be an ordinal number. A transfinite sequence of type α (or an α-sequence) is a function φ whose domain is $W(\alpha)$. If the values of φ are ordinals, the sequence is also called a sequence of ordinal numbers. In particular, an α-sequence φ of ordinals will be called an increasing sequence if for any μ, $\gamma \in W(\alpha)$, $\mu < \gamma$ always implies $\varphi(\mu) < \varphi(\gamma)$; it is a decreasing sequence, if $\mu < \gamma$ invariably implies $\varphi(\mu) > \varphi(\gamma)$.

Theorem 2.5.14. *Every set of ordinal numbers can be written as an increasing sequence.*

This is a consequence of Theorem 2.5.12.

Theorem 2.5.15. *Every decreasing sequence of ordinal numbers contains only a finite number of ordinal numbers.*

This result follows from Theorem 2.5.12.

Theorem 2.5.16. *To every set M of ordinal numbers μ, there is an ordinal number which is greater than every ordinal number μ in M.*

Proof: By Theorem 2.5.14, we can assume that M is an increasing sequence. Let $M^* = \{\mu + 1 : \mu \in M\}$ and $\sigma = \overline{\bigcup\{W(\mu+2) : \mu \in M\}}$. Then every $\mu + 1 \leq \sigma$; hence, $\mu < \sigma$, for every $\mu \in M$. ■

Let φ be a λ-sequence of ordinals where λ is a limit ordinal. By Theorem 2.5.16 there exist ordinals greater than all ordinals $\varphi(\gamma)$ where $\gamma < \lambda$. The smallest such ordinal (see Theorem 2.5.12) is called the limit of the λ-sequence $\varphi(\gamma)$ for $\gamma < \lambda$ and is denoted by $\lim_{\gamma<\lambda} \varphi(\gamma)$.

We say that an ordinal λ is cofinal with a limit ordinal α if λ is the limit of an increasing α-sequence

$$\lambda = \lim_{\xi<\alpha} \varphi(\xi) \tag{2.93}$$

Theorem 2.5.17. *An ordinal λ is cofinal with the limit ordinal α if and only if $W(\lambda)$ contains a subset of type α cofinal with $W(\lambda)$.*

Proof: Let A be a subset of $W(\lambda)$ cofinal with $W(\lambda)$ and such that $\overline{A} = \alpha$. For every ordinal $\xi < \alpha$ there exists an ordinal $\varphi(\xi)$ in A such that the set $\{\eta \in A : \eta < \varphi(\xi)\}$ is of type ξ. The sequence $\varphi(\xi)$ is clearly increasing and $\varphi(\xi) < \lambda$ for $\xi < \alpha$ because $\varphi(\xi) \in A \subset W(\lambda)$. If $\mu < \lambda$, there exists an ordinal $\xi < \lambda$ such that $\mu < \xi$ because the sets A and $W(\lambda)$ are cofinal. Thus, $\mu < \xi \leq \varphi(\eta)$ for some $\eta < \alpha$. This shows that λ is the least ordinal greater than all ordinals $\varphi(\xi)$, $\xi < \alpha$, which proves Eq. (2.93).

Suppose, in turn, that Eq. (2.93) holds. Let A be the set of all values of φ. We have $\eta < \lambda$ for all $\eta \in A$ and, consequently, since λ is a limit ordinal, there exists $\xi \in W(\lambda)$ such that $\eta < \xi$. Conversely, if $\xi \in W(\lambda)$ then $\xi < \lambda$, and by the definition of limit there exists $\xi' < \alpha$ such that $\xi < \varphi(\xi')$. This means that some ordinal in A is greater than ξ. Hence, the sets A and $W(\lambda)$ are cofinal. ∎

We now consider the arithmetic of ordinal numbers.

Theorem 2.5.18. *The sum and product of two ordinal numbers are ordinal numbers.*

Theorem 2.5.19. *The ordered sum of ordinal numbers, where the index set is also well-ordered, is an ordinal number.*

These theorems are consequences of the definitions in the previous section.

Theorem 2.5.20 [The First Monotonic Law for Addition].

$$(\alpha < \beta) \rightarrow (\gamma + \alpha < \gamma + \beta) \tag{2.94}$$

Proof: Let B and C be disjoint sets such that $\overline{B} = \beta$ and $\overline{C} = \gamma$. Since $\alpha < \beta$, B contains a segment A of type α. The ordered sum $C + A$, which is of type $\gamma + \alpha$, is a segment of $C + B$, which is of type $\gamma + \beta$. Thus, $\gamma + \alpha < \gamma + \beta$. ∎

Theorem 2.5.21 [The Second Monotonic Law for Addition].

$$(\alpha \leq \beta) \rightarrow (\alpha + \gamma \leq \beta + \gamma) \tag{2.95}$$

Proof: Let A, B, and C be the sets defined in the proof of Theorem 2.5.20. Then by applying Theorem 2.5.10 to the ordered sums $A+C$ and $B+C$ we obtain (2.95). ■

Theorem 2.5.22. *Let α and β be ordinals. If $\alpha \geq \beta$, then there exists exactly one ordinal γ such that $\alpha = \beta + \gamma$.*

Proof: Let $\overline{A} = \alpha$, B be a segment of A of type β, and $\gamma = \overline{A - B}$. Clearly, $\alpha = \beta + \gamma$. To prove the uniqueness of γ suppose that $\beta + \gamma_1 = \beta + \gamma_2$. By (2.94) this implies that $\gamma_1 \geq \gamma_2$ and $\gamma_2 \geq \gamma_1$. Thus, $\gamma_1 = \gamma_2$ by Theorem 2.5.8. ■

Theorem 2.5.23 [The First Monotonic Law for Multiplication].

$$(\alpha < \beta) \rightarrow (\gamma\alpha < \gamma\beta), \quad \textit{for } \gamma > 0 \tag{2.96}$$

Proof: In fact, $\gamma\beta$ is the type of the Cartesian product $B \times C$ ordered lexicographically, where $\overline{B} = \beta$ and $\overline{C} = \gamma$. Let $\overline{A} = \alpha$, and let A be a segment of B. The Cartesian product $A \times C$ ordered lexicographically is a segment of $B \times C$. ■

Theorem 2.5.24 [The Second Monotonic Law for Multiplication].

$$(\alpha \leq \beta) \rightarrow (\alpha\gamma \leq \beta\gamma) \tag{2.97}$$

Proof: Let A, B, and C be well-ordered sets of type α, β, and γ, respectively, and we assume that $A \subset B$. Thus, $C \times A \subset C \times B$, which proves Eq. (2.97). ■

It follows from the identity $1 \cdot \omega = 2 \cdot \omega$ that $\leq$ in Eq. (2.97) cannot be replaced by $<$.

The operation of ordinal exponentiation is defined by transfinite induction as follows:

$$\gamma^0 = 1 \tag{2.98}$$

$$\gamma^{\xi+1} = \gamma^{\xi} \cdot \gamma \tag{2.99}$$

$$\gamma^{\lambda} = \lim_{\xi < \lambda} \gamma^{\xi} \tag{2.100}$$

where λ is a limit ordinal. We say that γ^{α} is the power of γ, γ is the base, and α the exponent.

Theorem 2.5.25. *For any ordinals α, β, and γ with $\gamma > 1$,*

$$\alpha < \beta \to \gamma^{\alpha} < \gamma^{\beta} \tag{2.101}$$

This follows from the definition.

Theorem 2.5.26. *For any ordinals ξ, η, and γ, we have*

$$\gamma^{\xi+\eta} = \gamma^{\xi} \cdot \gamma^{\eta} \tag{2.102}$$

Proof: Given an ordinal ξ, let B denote the set of those $\varsigma \in W(\eta + 1)$ for which $\gamma^{\xi+\varsigma} = \gamma^{\xi} \cdot \gamma^{\varsigma}$. We shall show that if $\varsigma \leq \eta$, then

$$W(\varsigma) \subset B \to \varsigma \in B \tag{2.103}$$

In fact, the following three cases are possible: (i) $\varsigma = 0$; (ii) ς is not a limit ordinal; (iii) ς is a limit ordinal > 0. In case (i), $\varsigma \in B$, because $\gamma^{\xi+0} = \gamma^{\xi} = \gamma^{\xi} \cdot 1 = \gamma^{\xi} \cdot \gamma^{0}$. In case (ii), $\varsigma = \varsigma_1 + 1$ where $\varsigma_1 \in W(\varsigma)$; thus, by assumption, $\varsigma_1 \in B$. Hence. $\gamma^{\xi+\varsigma_1} = \gamma^{\xi} \cdot \gamma^{\varsigma_1}$ and therefore

$$\begin{aligned}\gamma^{\xi+\eta} &= \gamma^{\xi+(\varsigma_1+1)} = \gamma^{(\xi+\varsigma_1)+1} = \gamma^{(\xi+\varsigma_1)} \cdot \gamma = (\gamma^{\xi} \cdot \gamma^{\varsigma_1}) \cdot \gamma \\ &= \gamma^{\xi} \cdot (\gamma^{\varsigma_1} \cdot \gamma) = \gamma^{\xi} \cdot \gamma^{\varsigma_1+1} = \gamma^{\xi} \cdot \gamma^{\varsigma}\end{aligned}$$

which shows that $\varsigma \in B$. Finally, in case (iii), $\xi + \varsigma$ is a limit ordinal; thus,

$$\gamma^{\xi+\varsigma} = \lim_{\alpha<\xi+\varsigma} \gamma^{\alpha} \tag{2.104}$$

Because the sequence γ^{α}, $\alpha < \xi + \varsigma$, is increasing when $\gamma > 1$,

$$\gamma^{\xi+\varsigma} = \lim_{\alpha<\varsigma} \gamma^{\xi+\alpha} \tag{2.105}$$

Since for $\alpha < \varsigma$ we have $\alpha \in W(\varsigma)$, it follows that $\alpha \in B$; i.e., $\gamma^{\xi+\alpha} = \gamma^{\xi} \cdot \gamma^{\alpha}$. Thus,

$$\gamma^{\xi+\varsigma} = \lim_{\alpha<\varsigma}(\gamma^{\xi} \cdot \gamma^{\alpha}) = \gamma^{\xi}(\lim_{\alpha<\varsigma} \gamma^{\alpha}) = \gamma^{\xi} \cdot \gamma^{\alpha}$$

which implies that $\varsigma \in B$. Hence, implication (2.103) is proved. By induction it follows that $B = W(\eta + 1)$. Thus, $\eta \in B$. This proves Eq. (2.102). ■

Theorem 2.5.27. *For any ordinals ξ, η, and γ,*

$$(\gamma^{\xi})^{\eta} = \gamma^{\xi\eta} \tag{2.106}$$

Proof: This argument is analogous to that of the previous theorem. Let B denote the set of those $\varsigma \in W(\eta+1)$ for which $(\gamma^{\xi})^{\varsigma} = \gamma^{\xi\eta}$. It suffices to show that implication (2.103) holds. As before we consider cases (i)–(iii). In case (i), $\varsigma \in B$, because $(\gamma^{\xi})^0 = 1 = \gamma^0 = \gamma^{\xi \cdot 0}$. In case (ii) we have $\varsigma = \varsigma_1 + 1$, where ς_1 satisfies the condition that $(\gamma^{\xi})^{\varsigma_1} = \gamma^{\xi\varsigma_1}$. This formula in turn implies

$$\begin{aligned}(\gamma^{\xi})^{\varsigma} &= (\gamma^{\xi})^{\varsigma_1+1} = (\gamma^{\xi})^{\varsigma_1} \cdot (\gamma^{\xi}) \\ &= \gamma^{\xi\varsigma_1} \cdot \gamma^{\xi} = \gamma^{\xi\varsigma_1+\xi} \\ &= \gamma^{\xi(\varsigma_1+1)} = \gamma^{\xi\varsigma}\end{aligned}$$

which shows that $\varsigma \in B$. Finally, if ς is a limit ordinal > 0, then

$$\begin{aligned}(\gamma^{\xi})^{\varsigma} &= \lim_{\alpha<\varsigma} (\gamma^{\xi})^{\alpha} = \lim_{\alpha<\varsigma} (\gamma^{\xi\alpha}) \\ &= \lim_{\alpha<\xi\varsigma} \gamma^{\alpha} = \gamma^{\xi\varsigma}\end{aligned}$$

we have $\varsigma \in B$. Hence Eq. (2.106) is proved. ∎

Theorem 2.5.28. *For any given ordinal $\gamma > 1$, every ordinal number α can be uniquely represented in the form*

$$\alpha = \gamma^{\eta_1}\beta_1 + \gamma^{\eta_2}\beta_2 + \cdots + \gamma^{\eta_n}\beta_n$$

where n is a natural number, and η_i and β_i, $i = 1, 2, \ldots, n$, are ordinals such that $\eta_1 > \eta_2 > \cdots > \eta_n$, and $0 \leq \beta_i < \gamma$, for $i = 1, 2, \ldots, n$.

Proof: By transfinite induction it can be shown that for any ordinal α,

$$\alpha \leq \gamma^{\alpha}$$

We now let ς be the smallest ordinal such that $\alpha < \gamma^{\varsigma}$. If ς were a limit ordinal, then $\gamma^{\varsigma} = \lim_{\lambda<\varsigma} \gamma^{\lambda}$ and $\gamma^{\lambda} \leq \alpha$ for $\lambda < \varsigma$, which implies $\gamma^{\varsigma} \leq \alpha$. This contradicts the definition of ς. Thus, $\varsigma = \eta_1 + 1$, where

$$\gamma^{\eta_1} \leq \alpha < \gamma^{\eta_1+1}$$

Notice that $\alpha = \overline{W(\gamma) \times W(\gamma^{\eta_1})}$; there exists an ordinal pair $(\beta_1, \sigma_1) \in W(\gamma) \times W(\gamma^{\eta_1})$ such that $\overline{O((\beta_1, \sigma_1))} = \alpha$, where $O((\beta_1, \sigma_1)) = O(\beta_1) \times W(\gamma^{\eta_1}) + \{\beta_1\} \times O(\sigma_1)$. Therefore,

$$\alpha = \gamma^{\eta_1}\beta_1 + \sigma_1$$

where $\beta_1 < \gamma$ and $\sigma_1 < \gamma^{\eta_1}$. For the same reason, there exist ordinals $\eta_2 < \eta_1$, $\beta_2 < \gamma$, and $\sigma_2 < \gamma^{\eta_2}$ such that

$$\sigma_1 = \gamma^{\eta_2}\beta_2 + \sigma_2$$

This procedure must stop within a finite steps because, otherwise, $\{\eta_1, \eta_2, \dots\}$ would be a sequence of type $^*\omega$ of ordinals numbers, and this contradicts Theorem 2.5.12. Therefore, there exists a natural number n such that

$$\alpha = \gamma^{\eta_1}\beta_1 + \gamma^{\eta_2}\beta_2 + \cdots + \gamma^{\eta_n}\beta_n$$

with the property that $\eta_1 > \eta_2 > \cdots > \eta_n$ and $0 \leq \beta_i < \gamma$ for $i = 1, 2, \dots, n$.

Now we show the uniqueness of the expression. If

$$\gamma^{\eta_1}\beta_1 + \gamma^{\eta_2}\beta_2 + \cdots + \gamma^{\eta_n}\beta_n = \gamma^{\xi_1}\theta_1 + \gamma^{\xi_2}\theta_2 + \cdots + \gamma^{\xi_p}\theta_p$$

where $\eta_1 > \eta_2 > \cdots > \eta_n$, $\xi_1 > \xi_2 > \cdots > \xi_p$, $0 < \beta_i, \theta_j < \gamma$, $i = 1, 2, \dots, n$, and $j = 1, 2, \dots, p$. By the definition of η_1 and ξ_1, we must have $\eta_1 = \xi_1$, $\beta_1 = \theta_1$, and

$$\gamma^{\eta_2}\beta_2 + \cdots + \gamma^{\eta_n}\beta_n = \gamma^{\xi_2}\theta_2 + \cdots + \gamma^{\xi_p}\theta_p$$

Repeating this procedure, we have $n = p$, $\eta_k = \xi_k$, and $\beta_k = \theta_k$ for $k \leq n$. ■

Well-ordered sets owe their importance mainly to the fact that for each set there exists a relation which well-orders the set. This theorem is also of the greatest importance for the theory of cardinal numbers because it shows that all cardinal numbers are comparable with each other.

Theorem 2.5.29 [Well-Ordering Theorem]. *Every set can be well-ordered.*

Proof: Let f be a function defined on $p(A) - \{\emptyset\}$, where A is a nonempty set, such that for each nonempty subset $B \subset A, f(B) \in B$. We can extend this function to the whole power set $p(A)$ by letting $f(\emptyset) = p$, where p is any fixed element which does not belong to A.

Now let C denote the set of all relations $R \subset A \times A$ well-ordering their field. Then there exists a set of all ordinals β, where $\beta =$ the order type of some $R \in C$. Let α be the smallest ordinal greater than every ordinal β in this set.

We now define a transfinite sequence of type α by transfinite induction such that

$$\varphi_0 = f(A) \tag{2.107}$$

$$\varphi_\xi = f(A - \{\varphi_\eta : \eta < \xi\}) \tag{2.108}$$

for each $\xi < \alpha$. If $\varphi_\xi \neq p$, then $\varphi_\xi \in A - \{\varphi_\eta : \eta < \xi\}$ and $\varphi_\xi \neq \varphi_\eta$ for $\eta < \xi$. If for all $\xi < \alpha$ we had $\varphi_\xi \neq p$, then there would exist a transfinite sequence of type α with distinct values belonging to A. This implies that there exists a relation on A well-ordering a subset of A into type α. But this contradicts the definition of α. Therefore, there exists a smallest ordinal β such that $\varphi_\beta = p$. This implies that $A = \{\varphi_\eta : \eta < \beta\}$; thus, A is the set of all terms of a transfinite sequence of type β whose terms are all distinct. Consequently, there exists a relation well-ordering A into type β. ■

Theorem 2.5.30. *If A and B are two arbitrary sets, precisely one of the following holds:*

$$|A| < |B|, \quad |A| = |B|, \quad |A| > |B|$$

i.e., any two cardinal numbers are comparable with each other.

The proof follows from the well-ordering theorem and Theorem 2.5.8.

By Theorem 2.5.30, every set of cardinal numbers can be ordered according to the increasing magnitude of its elements. This ordered set will prove to be actually well-ordered. For this and similar considerations concerning cardinal numbers, it is practical to represent the cardinal numbers by means of ordinal numbers. According to the well-ordering theorem, every cardinal number can be represented by a well-ordered set; briefly, every cardinal number can be represented by an ordinal number. The totality of ordinal numbers which represents the same cardinal number $\aleph$ is designated the number class $Q(\aleph)$, where $Q(\aleph)$ contains a smallest ordinal, which is called the initial number of $Q(\aleph)$ or the initial number belonging to $\aleph$.

Theorem 2.5.31. *Every transfinite initial number is a limit number.*

Proof: If some transfinite initial number were not a limit number, it would be immediately preceded by an ordinal number γ, and the initial number would be $\gamma + 1$. But then γ and $\gamma + 1$ would have the same cardinal number; i.e., $\gamma + 1$ would not be the smallest ordinal number in its number class, contradiction. ■

Theorem 2.5.32. *Every set of cardinal numbers, ordered according to increasing magnitude, is well-ordered. There exists a cardinal number which is greater than every cardinal number in the set.*

Proof: For the first part of the assertion we have to show that every nonempty subset of cardinal numbers contains a smallest cardinal number. This is, however, clear if every cardinal number is represented by the initial number belonging to it. Among these initial numbers then there is, as in every nonempty set of ordinal numbers, a smallest one, and the corresponding cardinal number is the smallest cardinal number in the set.

The second part of the assertion follows from Theorems 2.5.11 and 2.1.4. ■

If the family of all transfinite cardinal numbers is ordered according to increasing magnitude, then we denote the smallest transfinite cardinal number by $\aleph_0$, the next by $\aleph_1$, etc. In general, every transfinite cardinal number receives as index the ordinal number of the set of transfinite cardinal numbers which precede the cardinal number in question. Therefore,

$$\aleph_\mu < \aleph_\gamma, \quad \text{for } \mu < \gamma \tag{2.109}$$

The initial number which belongs to $\aleph_\mu$ is denoted by ω_μ. Then ω_0 is the smallest transfinite limit number and is therefore the same as the ω introduced before.

Theorem 2.5.33. *For any ordinal* $\mu > 0$,

$$\aleph_\mu \cdot \aleph_\mu = \aleph_\mu \tag{2.110}$$

Proof: The set $W(\omega_\mu)$ has order type ω_μ and is therefore a set of power $\aleph_\mu$. Consequently, the cardinal number $\aleph_\mu \cdot \aleph_\mu$ is the power of the set $W(\omega_\mu) \times W(\omega_\mu)$. We now show that this set has at most the cardinal number $\aleph_\mu$; i.e., $\aleph_\mu \cdot \aleph_\mu \leq \aleph\mu$.

Let $(\xi, \eta) \in W(\omega_\mu) \times W(\omega_\mu)$. Then by Theorem 2.5.28 and adding whatever necessary terms with coefficient 0, we can represent ξ and η uniquely in the form

$$\begin{aligned} \xi &= \omega^{\gamma_1} m_1 + \omega^{\gamma_2} m_2 + \cdots + \omega^{\gamma_k} m_k \\ \eta &= \omega^{\gamma_1} n_1 + \omega^{\gamma_2} n_2 + \cdots + \omega^{\gamma_k} n_k \end{aligned} \tag{2.111}$$

where $\gamma_1 > \gamma_2 > \cdots > \gamma_k$ and for all $i \leq k$ either $m_i > 0$ or $n_i > 0$. Define a polynomial

$$\varsigma = \sigma(\xi, \eta) = \omega^{\gamma_1}(m_1 + n_1) + \omega^{\gamma_2}(m_2 + n_2) + \cdots + \omega^{\gamma_k}(m_k + n_k)$$

Since only a finite number of nonzero terms appear, it follows that

$$\sigma(\xi, \eta) < \omega^{\gamma_1}(m_1 + n_1 + 1)$$

and the proof of Theorem 2.5.28 implies that

$$\omega^{\gamma_1}(m_1 + n_1 + 1) < \omega_\mu$$

Consequently,

$$\varsigma = \sigma(\xi, \eta) < \omega_\mu$$

We therefore certainly obtain all number pairs ξ, η by solving the equation $\sigma(\xi, \eta) = \varsigma$ for ξ, η for every $\varsigma < \omega_\mu$. But for every ς this equation has only a finite number of solutions. Therefore, the cardinal number of the set of all solutions for all $\varsigma < \omega_\mu$ is at most $\aleph_0 \cdot \aleph_\mu$. We now need only show that $\aleph_0 \cdot \aleph_\mu \leq \aleph_\mu$.

Theorem 2.5.28 implies that for any ordinal α,

$$\alpha = \omega\beta + \sigma \tag{2.112}$$

where $\sigma < \omega$. Therefore, when α is a limit number, we have $\alpha = \omega\beta$. This implies that $\omega_\mu = \omega\beta$ for some fixed ordinal β. Therefore,

$$\begin{aligned}\aleph_\mu &= |W(\omega_\mu)| = |W(\omega) \times W(\beta)| \\ &= |W(\omega)| \cdot |W(\beta)| \\ &= \aleph_0 \cdot |W(\beta)| = \aleph_0 \cdot \aleph_0 \cdot |W(\beta)| \\ &= \aleph_0 \cdot |W(\omega) \times W(\beta)| = \aleph_0 \cdot \aleph_\mu\end{aligned}$$

Thus, $\aleph_\mu \cdot \aleph_\mu \le \aleph_0 \cdot \aleph_\mu = \aleph_\mu$. ■

Theorem 2.5.34. *For* $\aleph_\mu \le \aleph_\gamma$, $\aleph_\mu + \aleph_\gamma = \aleph_\mu \cdot \aleph_\gamma = \aleph_\gamma$*; in particular,* $\aleph_\gamma + \aleph_\gamma = \aleph_\gamma$.

Proof: $\aleph_\gamma \le \aleph_\mu + \aleph_\gamma \le 2\aleph_\gamma \le \aleph_\gamma \cdot \aleph_\gamma = \aleph_\gamma$, and $\aleph_\gamma \le \aleph_\mu \cdot \aleph_\gamma \le \aleph_\gamma \cdot \aleph_\gamma = \aleph_\gamma$. ■

Theorem 2.5.35. *For* $\aleph_\rho < \aleph_\sigma$ *and* $\aleph_\mu < \aleph_\gamma$,

$$\aleph_\rho + \aleph_\mu < \aleph_\sigma + \aleph_\gamma \text{ and } \aleph_\rho \cdot \aleph_\mu < \aleph_\sigma \cdot \aleph_\gamma$$

Proof: Without loss of generality, let $\aleph_\rho \le \aleph_\mu$. Then

$$\aleph_\rho + \aleph_\mu = \aleph_\rho \cdot \aleph_\mu = \aleph_\mu < \aleph_\gamma$$ ■

Theorem 2.5.36. *(1)* $\sum_{\rho \le \mu} \aleph_\rho = \aleph_\mu$*, and (2) if* μ *is a limit number,* $\sum_{\rho<\mu} \aleph_\rho = \aleph_\mu$.

Proof: We have invariably

$$\aleph_\mu \le \sum_{\rho \le \mu} \aleph_\rho \le \aleph_\mu \cdot \aleph_\mu = \aleph_\mu$$

If μ is a limit number, then, on the one hand,

$$\sum_{\rho<\mu} \aleph_\rho \le \aleph_\mu$$

and, on the other hand, for every $\gamma < \mu$ we also have $\gamma + 1 < \mu$, so $\sum_{\rho<\mu} \aleph_\rho \ge \aleph_{\gamma+1} > \aleph_\gamma$; i.e., the sum is greater than every cardinal number which is less than $\aleph_\mu$. These two results then yield the assertion in this case, too. ■

Theorem 2.5.37. *For ordinals γ and μ with $\mu \leq \gamma + 1$,*

$$2^{\aleph_\gamma} = \aleph_\mu^{\aleph_\gamma}$$

Proof: According to Theorem 2.3.9 and the hypothesis

$$2^{\aleph_\gamma} \geq \aleph_{\gamma+1} \geq \aleph_\mu$$

and hence

$$2^{\aleph_\gamma} = 2^{\aleph_\gamma \cdot \aleph_\gamma} = (2^{\aleph_\gamma})^{\aleph_\gamma} \geq (\aleph_\mu)^{\aleph_\gamma} \geq 2^{\aleph_\gamma}$$ ■

Theorem 2.5.38. *The number class $Q(\aleph_\mu)$, ordered according to increasing magnitude of its elements, has order type $\omega_{\mu+1}$ and therefore has cardinality $\aleph_{\mu+1}$.*

Proof: The elements of $Q(\aleph_\mu)$ are the ordinal numbers γ with

$$\omega_\gamma \leq \gamma < \omega_{\mu+1}$$

Hence, in the sense of addition of well-ordered sets, we have

$$W(\omega_\mu) + Q(\aleph_\mu) = W(\omega_{\mu+1}) \tag{2.113}$$

This implies that $|W(\omega_\mu)| + |Q(\aleph_\mu)| = |W(\omega_{\mu+1})|$; i.e.,

$$\aleph_\mu + |Q(\aleph_\mu)| = \aleph_{\mu+1}$$

From this we have $|Q(\aleph_\mu)| \leq \aleph_{\mu+1}$. But Theorem 2.5.35 implies that we cannot have $|Q(\aleph_\mu)| < \aleph_{\mu+1}$. So, $|Q(\aleph_\mu)| = \aleph_{\mu+1}$.

Let ς be the order type of $Q(\aleph_\mu)$. Then Eq. (2.113) implies that $\varsigma \leq \omega_{\mu+1}$, and $|Q(\aleph_\mu)| = \aleph_{\mu+1}$ implies that $\varsigma \geq \omega_{\mu+1}$. Hence, $\varsigma = \omega_{\mu+1}$. ■

2.6. Crises of Naive Set Theory

Careful readers may have already found some inconsistencies in the development of naive set theory. These contradictions are connected with the following concepts.

(a) The set of all cardinal numbers. For example, let A be the set of all cardinal numbers. Then Theorem 2.5.16 implies that there exists a cardinal number $\aleph$ greater than all numbers in A. That is, $\aleph > \aleph$. This contradicts Theorem 2.5.33.

(b) The set of all ordinal numbers. For example, Theorem 2.5.16 implies that to every set of ordinal numbers there exists a still greater ordinal number. Accordingly, the concept of "set of all ordinal numbers" is a meaningless concept. This is the so-called Burali-Forti paradox.

(c) The set of all sets which do not contain themselves as elements (Zermelo–Russell paradox or simply Russell's paradox). For this set X, as for every set, only the following two cases are conceivable: either X contains itself as an element, or it does not. The first case cannot occur because otherwise X would have an element, namely, X, which contained itself as an element. The second assumption, however, also leads to a contradiction. For in this case, X would be a set which did not contain itself as an element, and for this very reason X would be contained in the totality of all such sets, i.e., in X.

(d) The set of all sets. For if X is such a set, it would have to contain the set of all its subsets. The set of all subsets of X, however, has a greater power than X, according to Theorem 2.1.4.

From the list of paradoxes above, we can see that two concepts, or words, appear in each of them. One of the two words is "set" and the other is "all." The concept "set" was introduced by Cantor as "a collection into a whole of definite, well-distinguished objects." The meaning of this definition might be too wide for us to avoid these paradoxes. The concept "all" actually was never introduced in the context, and the meaning of "all" we have been using contradicts some results we get in set theory. Our analysis on the appearance of the paradoxes shows that if we can narrow the meaning of the concept of "set" and if we can avoid using the word "all," we may be in good shape to establish a set theory without contradictory propositions. In the next chapter, we will try along this line to study properties of abstract sets and to set up a new theory of sets, called the axiomatic set theory on ZFC.

For further study of naive set theory, the reader is advised to consult (Kuratowski and Mostowski, 1976).

CHAPTER 3

Axiomatic Set Theory

This chapter investigates how naive set theory can be rebuilt so that the resulting set theory will not contain the paradoxes in Section 2.6. Axiomatic set theory, based on the ZFC axiom system (i.e., Zermelo–Fränkel axioms and the Axiom of Choice) is introduced.

In Section 3.1 we introduce the language of axiomatic set theory, list the axioms of ZFC, and discuss some related philosophical problems. Section 3.2 sketches the development of the ordinal and cardinal number systems, based on the axioms of ZFC, excluding regularity. Section 3.3 covers some special topics in combinatorial set theory. Martin's axiom (MA) is introduced to study some related problems. Section 3.4 develops the class **WF** of all well-founded sets and shows that all mathematics takes place within **WF**.

3.1. Some Philosophical Issues

Most scientific workers have little need for a precise codification of the concepts of set theory. It is generally understood that some principles are correct and some are questionable. In this chapter, we establish a set theory which makes more precise the basic principles of naive set theory. To accomplish this, we build the theory on a few axioms.

The idea of establishing a theory upon a set of axioms harks back to the time of Euclid. Usually, in science, axioms, principles, and laws are stated in an informal language, i.e., a language with many grammatical inconsistencies, such as English. A major disadvantage of using an informal language is the possible misunderstanding of a sentence or word. For example, in the past few decades, the word "system" has been understood in many different ways (see Chapter 4 for a detailed discussion). In order to establish a rigorous set theory, we state our axioms in a formal, or artificial, language, called first-order predicate calculus. The characteristic of a formal language is that there are precise rules of formation for linguistic objects.

The basic symbols of our formal language include $\wedge$, $\neg$, $\exists$, (,), $\in$, $=$, and v_j for each natural number j. Their grammatical meanings are as follows:

- $\wedge$ means "and."
- $\neg$ means "not."
- $\exists$ means "there exists."
- $\in$ means membership.
- $=$ means equality.
- v_j, for each natural number j, are variables.

The parentheses are used for phrasing. An expression is any finite sequence of basic symbols. The intuitive interpretation of the symbols indicates which expressions are meaningful; these are called formulas. Precisely, a formula is defined to be any expression constructed by the rules:

(a) $v_i \in v_j$, $v_i = v_j$ are formulas for any natural numbers i,j.

(b) If ϕ and ψ are formulas, so are $\phi \wedge \psi$, $\neg\phi$, and $\exists v_i(\phi)$ for any i.
To save ourselves the work of always writing long expressions, we will use the following abbreviations.

(c) $\forall v_i(\psi)$ for $\neg(\exists v_i(\neg(\psi)))$.

(d) $\phi \vee \psi$ for $\neg((\neg\psi) \wedge (\neg\psi))$.

(e) $\phi \rightarrow \psi$ for $(\neg\phi) \vee \psi$.

(f) $\phi \leftrightarrow \psi$ for $(\phi \rightarrow \psi) \wedge (\psi \rightarrow \phi)$.

(g) $v_i \neq v_j$ for $\neg(v_i = v_j)$.

(h) $v_i \notin v_j$ for $\neg(v_i \in v_j)$.

Other letters from English, Greek, and Hebrew are used for variables. In this chapter, we follow standard mathematical usage of writing expressions mostly in English, augmented by logical symbols when this seems necessary. For example, we may say "there are sets x,y,z such that $x \in y \wedge y \in z$," rather than

$$\exists v_0(\exists v_1(\exists v_2((v_0 \in v_1) \wedge (v_1 \in v_2)))) \tag{3.1}$$

A subformula of a formula ϕ is a constructive sequence of symbols of ϕ which forms a formula. The scope of an occurrence of a quantifier $\exists v_i$ is the unique subformula beginning with that $\exists v_i$. For example, the scope of the quantifier $\exists v_2$ in formula (3.1) is $\exists v_2((v_0 \in v_1) \wedge (v_1 \in v_2))$. An occurrence of a variable in a formula is called bound if and only if it lies in the scope of a quantifier acting on that variable, otherwise it is called free. For example, in formula (3.2),

$$(\exists v_0(v_0 \in v_1)) \wedge (\exists v_1(v_2 \in v_1)) \tag{3.2}$$

the first occurrence of v_1 is free, but the second is bound, whereas v_0 is bound at its occurrences and v_2 is free at its occurrence.

Note that since $\forall v_i$ is an abbreviation for $\neg\exists v_i\neg$, it also binds its variable v_i, whereas the abbreviations $\vee, \rightarrow, \leftrightarrow$ are defined in terms of other propositional connections and do not bind variables.

We often write a formula as $\phi(x_1, \ldots, x_n)$ to emphasize its dependence on the variables $x_1, \ldots, x_n$. Then, if $y_1, \ldots, y_n$ are different variables, we use $\phi(y_1, \ldots, y_n)$ to indicate the formula resulting from substituting a y_i for each occurrence of the free variable x_i. Such a substitution is called free, or legitimate, if no free occurrence of an x_i is in the scope of a quantifier $\exists y_i$. The idea is that the formula $\phi(y_1, \ldots, y_n)$ says about variables $y_1, \ldots, y_n$ what the formula $\phi(x_1, \ldots, x_n)$ said about $x_1, \ldots, x_n$, but this will not be the case if the substitution is not free and some y_i gets bound by a quantifier of ϕ. In general, we will always assume that our substitutions are legitimate. The use of the notation $\phi(x_1, \ldots, x_n)$ does not imply that each x_i actually occurs free in $\phi(x_1, \ldots, x_n)$; also, $\phi(x_1, \ldots, x_n)$ may have other free variables, which in this discussion we are not emphasizing.

A sentence is a formula with no free variables. Intuitively, a sentence states an assertion which is either true or false. If S is a set of sentences and ϕ is a sentence, then $S \vdash \phi$ means that ϕ is provable from S by a purely logical argument that may quote sentences in S as axioms but may not refer to any intended "meaning" of $\in$.

If $S \vdash \phi$, where S is the empty set of sentences, we write $\vdash \phi$ and say that ϕ is logically valid. If $\vdash (\phi \leftrightarrow \psi)$, we say that ϕ and ψ are logically equivalent.

Let ϕ be a formula; a universal closure of ϕ is a sentence obtained by universally quantifying all free variables of ϕ. For example, let ϕ be the formula

$$x = y \rightarrow \forall z(z \in x \leftrightarrow z \in y) \tag{3.3}$$

Then $\forall x \forall y \phi$ and $\forall y \forall x \phi$ are universal closures of ϕ. All universal closures of a formula are logically equivalent. In general, when we assert ϕ, we mean to assert its universal closure. Formally, if S is a set of sentences and ϕ is a formula, we define $S \vdash \phi$ to mean that the universal closure of ϕ is derivable from S.

We call a formula ϕ logically valid if its universal closure is logically valid; we say two formulas ϕ and ψ are logically equivalent if $\phi \leftrightarrow \psi$ is logically valid.

A set S of sentences is consistent, denoted by $\mathrm{Con}(S)$, if for no formula ϕ does $S \vdash \phi$ and $S \vdash \neg\phi$. We can see easily that for any sentence ϕ, $S \vdash \phi$ if and only if $S \cup \{\neg\phi\}$ is inconsistent.

In order to use formal deduction, we need the following important facts: (a) If $S \vdash \phi$, then there exists a finite $S_0 \subset S$ such that $S_0 \vdash \phi$. (b) If S is inconsistent, there is a finite $S_0 \subset S$ such that S_0 is inconsistent.

As an exercise, we now list all axioms in the ZFC axiom system.

Axiom 1. (Set Existence)

$$\exists x(x = x)$$

Axiom 2. (Extensionality)

$$\forall x \forall y(\forall z(z \in x \leftrightarrow z \in y) \to x = y)$$

Axiom 3. (Regularity)

$$\forall x\,[\exists y(y \in x) \to \exists y(y \in x \land \neg \exists z(z \in x \land z \in y))]$$

Axiom 4. (Comprehension Scheme) For each formula ϕ with free variables among $x, z, w_1, \ldots, w_n$,

$$\forall z \forall w_1, \ldots, w_n \exists y \forall x(x \in y \leftrightarrow x \in z \land \phi)$$

Axiom 5. (Pairing)

$$\forall x \forall y \exists z(x \in z \land y \in z)$$

Axiom 6. (Union)

$$\forall \mathcal{F} \exists A \forall Y \forall x(x \in Y \land Y \in \mathcal{F} \to x \in A)$$

Axiom 7. (Replacement Scheme) For each formula ϕ with free variables among $x, y, A, w_1, \ldots, w_n$,

$$\forall A \forall w_1, \ldots, w_n\,[\forall x \in A \exists! y \phi \to \exists Y \forall x \in A \exists y \in Y \phi]$$

where the symbol $\exists!$ means "there exists exactly one."

Axiom 8. (Infinity)

$$\exists x(\emptyset \in x \land \forall y \in x(S(y) = y \cup \{y\} \in x))$$

Axiom 9. (Power Set)

$$\forall x \exists y \forall z(z \subset x \to z \in y)$$

Axiom 10. (Choice (Axiom of choice or AC))

$$\forall A \exists R(R \text{ well orders } A)$$

In the following, we write the meanings of the axioms in English.

Axiom 1. (Set Existence) There exists a set.

Axiom 2. (Extensionality) Two sets x and y are identical if they contain the same elements.

Axiom 3. (Regularity) Each nonempty set x has an element y such that $x \wedge y = \emptyset$.

Axiom 4. (Comprehension Scheme) For each given formula ϕ and each set z, there exists a set y such that $y = \{x \in z : \phi(x)\}$.

Axiom 5. (Pairing) For any sets x and y, there exists a set z such that $x \in z$ and $y \in z$.

Axiom 6. (Union) For any given set $\mathcal{F}$ of sets there exists a set A which consists of all elements belonging to at least one element set in $\mathcal{F}$.

Axiom 7. (Replacement Scheme) For each formula ϕ and any set A, if to each element $x \in A$ there corresponds exactly one y satisfying $\phi(x,y)$, then there exists a set Y consisting of all y satisfying $\phi(x,y)$, for some $x \in A$.

Axiom 8. (Infinity) There exists a set x satisfying (1) $\emptyset \in x$, and (2) for each $y \in x$ there exists a set $S(y) \in x$ which contains all elements in y and y itself.

Axiom 9. (Power Set) For any set x, there exists a set y, called the power set of the set x and denoted by $p(x)$, such that all subsets of x are contained in y.

Axiom 10. (Choice) Any set can be well-ordered.

An interpretation of the language of set theory is defined by specifying a nonempty domain of disclosure, over which the variables are intended to vary, together with a binary relation defined on that domain, which is the interpretation of the membership relation $\in$. If ϕ is a sentence in the language of set theory, ϕ is either true or false under a specified interpretation.

In the intended interpretation, under which the axioms of ZFC are presumed true, the symbol $x \in y$ is interpreted to mean that x is a member of y. Since we wish to talk only about sets and their elements in our domain of discourse, all elements of such a set should be sets also. Repeating this, we can see that the domain of our discourse consists of those x such that

$$x \text{ is a set, and}$$
$$\forall y(y \in x \rightarrow y \text{ is a set}), \text{ and}$$
$$\forall z \forall y(y \in x \wedge z \in y \rightarrow z \text{ is a set}), \text{ etc.}$$

We call such an x a hereditary set.

If S is a set of sentences, we may show that S is consistent by producing an interpretation under which all sentences of S are true. Usually, $\in$ will still be interpreted as membership, but the domain of discourse will be some subdomain of the hereditary sets.

The justification for this method of producing consistency proofs is the easy direction of the Gödel completeness theorem, which says that if S holds in some interpretation, then S is consistent. The reason this theorem is true is that the rules of formal deduction are set up so that if $S \vdash \phi$, then ϕ must be true under any interpretation which makes all sentences in S true. If we fix an interpretation in which S holds, then any sentence that is false in that interpretation is not provable from S. Since $\neg\phi$ and ϕ cannot both hold in a given interpretation, S cannot prove both ϕ and $\neg\phi$; thus S is consistent.

We now conclude this section with a few extremes of mathematical thought. For a more serious discussion, see (Fraenkel et al., 1973; Kleene, 1952; Kreisel and Krivine, 1967).

Platonists believe that the set-theoretic universe has an existence outside the spirit of human beings. From this point of view, the axioms in ZFC are merely certain obvious facts in the universe, and we might have failed to list some other obvious but basic facts in the ZFC axiom system. Platonists are still interested in locating principles of the universe which are not provable from ZFC.

A finitist believes only in finite objects. It is meaningless to consider any sort of concept like that of "whole." If, say, "human knowledge" does not exist, the set of all rational numbers is not a practical concept; hence, the theory of infinite sets is a piece of meaningless garbage. There is some merit in the finitist's position, since all objects in known physical reality are finite, so infinite sets may be discarded as figments of the mathematician's imagination. Unfortunately, this point of view also discards a big portion of modern mathematics.

The formalist can hedge his bets. The formal development of ZFC makes sense from a strictly finitistic point of view: the axioms of ZFC do not say anything, but are merely certain finite sequences of symbols. The assertion ZFC $\vdash \phi$ means that there is a certain kind of finite sequence of finite sequences of symbols, namely, a formal proof of ϕ. Even though ZFC contains infinitely many axioms, notions like ZFC $\vdash \phi$ will make sense only when a particular sentence is recognized as an axiom of ZFC. A formalist can thus do mathematics just as a platonist, but if challenged about the validity of handling infinite objects, he can reply that all he is really doing is juggling finite sequences of symbols.

In this chapter, we will develop ZFC from a platonistic point of view. Thus, to establish that ZFC $\vdash \phi$, we simply produce an argument that ϕ is true based on the assumption that ZFC contains true principles.

3.2. Constructions of Ordinal and Cardinal Number Systems

The axiom on set existence says that our domain of discourse is not empty. Let z be any set. Axiom 3 (comprehensive scheme) implies that

$$\{x \in z : x \neq x\} \tag{3.4}$$

defines the existence of the empty set. By extensionality, such a set is unique.

The Comprehension Axiom is intended to formalize the construction of sets of the form $\{x : P(x)\}$, where $P(x)$ denotes a property of x. Since the notion of property is made rigorous via formulas, it is tempting to set forth as axioms statements of the form

$$\exists y \forall x (x \in y \leftrightarrow \phi) \tag{3.5}$$

where ϕ is a formula. Unfortunately, such a scheme is inconsistent by the famous Zermelo–Russell paradox (Section 2.6): Let ϕ be $x \neq x$ then this axiom gives a y such that

$$\forall x (x \in y \leftrightarrow x \notin x) \tag{3.6}$$

whence $y \in y \leftrightarrow y \notin y$. To avoid this contradiction the Comprehension Axiom is given as follows: For each formula ϕ without y free, the universal closure of the following is an axiom:

$$\exists y \forall x (x \in y \leftrightarrow x \in z \wedge \phi)$$

We can also prove, not simply to avoid this paradox, that there is no universal set.

Theorem 3.2.1. *$\neg \exists z \forall x (x \in z)$.*

Proof: If $\forall x (x \in z)$, by the Comprehension Axiom we can form $\{x \in z : x \notin x\} = \{x : x \notin x\}$, which would yield a contradiction by the Zermelo–Russell paradox. ∎

We let $A \subset B$ abbreviate $\forall x(x \in A \to x \in B)$. So, $A \subset A$ and $\emptyset \subset A$.

The example $\{x : x = x\}$ shows that for a given formula $\phi(x)$, there need not exist a set $\{x : \phi(x)\}$. Generally, this collection (or class) may be too big to form a set. The Comprehension Axiom says that if the collection is a subcollection of a given set, then it is a set. Axioms 4–8 of ZFC say that certain collections do form sets.

By the Pairing Axiom, for given x and y we may let z be any set such that $x \in z \wedge y \in z$; then $\{v \in z : v = x \vee v = y\}$ is the unique (by extensionality) set whose elements are precisely x and y; we call this set $\{x,y\}$. Note that $(x,y) = \{\{x\},\{x,y\}\}$ is the ordered pair of x and y.

In the Union Axiom, we think of $\mathcal{F}$ as a family of sets and postulate the existence of a set A such that each member Y of $\mathcal{F}$ is a subset of A. This justifies our defining the union of the family, denoted by $\bigcup \mathcal{F}$, by

$$\bigcup \mathcal{F} = \{x : \exists\, Y \in \mathcal{F}\ (x \in Y)\} \tag{3.7}$$

When $\mathcal{F} \neq \emptyset$, we let

$$\bigcap \mathcal{F} = \{x : \forall\, Y \in \mathcal{F} (x \in Y)\} \tag{3.8}$$

If $\mathcal{F} = \emptyset$, then $\bigcup \mathcal{F} = \emptyset$ and $\bigcap \mathcal{F}$ "should be" the set of all sets, which does not exist. Finally, we let $A \cup B = \bigcup \{A,B\}$, $A \cap B = \bigcap \{A,B\}$, and $A - B = \{x \in A : x \notin B\}$.

For any sets A and B, we define the Cartesian product

$$A \times B = \{(x,y) : x \in A \wedge y \in B\} \tag{3.9}$$

To justify this definition, we must apply replacement twice. First for any $y \in B$, we have

$$\forall x \in A \exists! z (z = (x,y))$$

so by replacement (and comprehension) we can define

$$\text{prod}(A,y) = \{z : \exists\, x \in A\ (z = (x,y))\}$$

Now

$$\forall y \in B \exists! z (z = \text{prod}(A,y))$$

so by replacement we may define

$$\text{prod}(A,B) = \{\text{prod}(A,y) : y \in B\}$$

Finally, we define $A \times B = \bigcup \text{prod}'(A,B)$.

A relation H is a set all of whose elements are ordered pairs. Let

$$D(H) = \{x : \exists\, y\ (x,y) \in H\} \tag{3.10}$$

and

$$R(H) = \{y : \exists x((x,y) \in H)\} \tag{3.11}$$

We define $H^{-1} = \{(x,y) : (y,x) \in H\}$, so $(H^{-1})^{-1} = H$.

A function (or mapping) is a relation f such that

$$\forall x \in D(f) \exists ! y \in R(f)((x,y) \in f) \tag{3.12}$$

A total ordering (or strict total ordering or linear ordering) is a pair (A,H) such that H totally orders A; that is, A is a set, H is a relation such that

$$\forall x,y,z \in A(xHy \wedge yHz \rightarrow xHz) \tag{3.13}$$

$$\forall x,y \in A(x = y \vee xHy \vee yHx) \tag{3.14}$$

and

$$\forall x \in A(\neg(xHx)) \tag{3.15}$$

where we write xHy for $(x,y) \in H$. Note that our notation does not assume $H \subset A \times A$, so if (A,H) is a total ordering so is (B,H) whenever $B \subset A$.

Whenever H and K are relations and A and B are sets, we say $(A,H) \simeq (B,K)$ if there exists a bijection $f : A \rightarrow B$ such that $\forall x,y \in A(xHy \leftrightarrow f(x)Kf(y))$. This mapping f is called an isomorphism from (A,H) to (B,K).

We say H well-orders A, or (A,H) a well-ordering, if (A,H) is a total ordering and every nonempty subset of A has an H-least element; i.e., $\forall B \subset A(B \neq \emptyset \rightarrow \exists x \in B \forall y \in B(xHy))$. If $x \in A$, let $\text{pred}(A,x,H) = \{y \in A : yHx\}$.

Lemma 3.2.1. *If (A,H) is a well-ordering, then*

$$(A,H) \not\simeq (\text{pred}(A,x,H),H) \quad \textit{for each } x \in A$$

Proof: If $f : A \rightarrow \text{pred}(A,x,H)$ were an isomorphism, let z be the H-least element of $\{y \in A : f(y) \neq y\}$. Then for each $y \in A$ with $yHz, f(y) = y$. Since f is an isomorphism it follows that $f(z) = z$. This contradicts the definition of z. ■

Lemma 3.2.2. *If (A,H) and (B,K) are isomorphic well-orderings, then the isomorphism between them is unique.*

Proof: If f and g were different isomorphisms, let z be the H-least $y \in A$ such that $f(y) \neq g(y)$. Then a similar contradiction as in the proof of Lemma 3.2.1 will occur. ■

Theorem 3.2.2. *Let (A,H) and (B,K) be two well-orderings. Then exactly one of the following holds:*

(a) $(A,H) \simeq (B,K)$.

(b) $\exists y \in B((A,H) \simeq (\text{pred}(B,y,K),K))$.

(c) $\exists x \in A((\text{pred}(A,x,R) \simeq (B,K))$.

Proof: Let

$$f = \{(v,w) : v \in A \wedge w \in B \wedge (\text{pred}(A,v,H),H) \simeq (\text{pred}(B,w,K),K)\}$$

Note that f is an isomorphism from some initial segment of A onto some initial segment of B, and that these segments cannot both be proper. ■

A set x is transitive if every element of x is a subset of x. Examples of transitive sets are $\emptyset, \{\emptyset\}, \{\emptyset, \{\emptyset\}\}, \ldots$. The set $\{\{\emptyset\}\}$ is not transitive. A set x is an ordinal if x is transitive and well-ordered by the relation $\in$. More formally, the assertion that x is well-ordered by $\in$ means that $(x, \in_x)$ is a well-ordering, where $\in_x = \{(y,z) \in x \times x : y \in z\}$. Examples of ordinals are $\emptyset, \{\emptyset\}, \{\emptyset, \{\emptyset\}\}$. If $x = \{x\}$, then x is not ordinal since we have defined orderings to be strict. Without causing confusion, we write $x \simeq (A,H)$ for $(x, \in_x) \simeq (A,H)$ and $y \in x$, $\text{pred}(x,y)$ for $\text{pred}(x,y,\in_x)$.

Theorem 3.2.3.

(1) If x is an ordinal and $y \in x$, then y is an ordinal and $y = \text{pred}(x,y)$.

(2) If x and y are ordinals and $x \simeq y$, then $x = y$.

(3) If x and y are ordinals, then exactly one of the following is true: $x = y$, $x \in y$, $y \in x$.

(4) If x, y, and z are ordinals, $x \in y$ and $y \in z$ implies that $x \in z$.

(5) If C is an nonempty set of ordinals, then $\exists x \in C \forall y \in C(x \in y \vee x = y)$.

Proof: For (3), we use (1), (2), and Theorem 3.2.2 to show that at least one of the three conditions holds. That no more than one holds follows from the fact that no ordinal can be a member of itself, since $x \in x$ would imply that $(x, \in_x)$ is not a strict total ordering (since $x \in_x x$). For (5), we note that the conclusion is, by (3), equivalent to $\exists x \in C(x \cap C = \emptyset)$. Let $x \in C$ be arbitrary. If $x \cap C \neq \emptyset$, then since x is well-ordered by $\in$, there is an $\in$-least element x^1 of $x \cap C$, and then $x^1 \cap C = \emptyset$. ■

Theorem 3.2.3 implies that the set of all ordinals, if it existed, would be an ordinal, and thus cannot exist, because if such an ordinal existed, denoted by η, then Theorem 3.2.3(5) implies that $\eta \cap \{\eta\} = \emptyset$, but the definition of η implies $\eta \in \eta \cap \{\eta\}$, contradiction. This fact takes care of the Burali-Forti paradox (see Section 2.6 for details).

Lemma 3.2.3. *Let A be a set of ordinals and $\forall x \in A \forall y \in x(y \in A)$ then A is an ordinal.*

Theorem 3.2.4. *If (A,H) is a well-ordering, then there exists a unique ordinal C such that $(A,H) \simeq C$.*

Proof: The uniqueness of the ordinal C follows from Theorem 3.2.3(2). to prove existence, let

$$B = \{a \in A : \exists x(x \text{ is an ordinal} \wedge (\text{pred}(A,a,H),H) \simeq x)\} \tag{3.16}$$

Let f be the function with domain B such that for each $a \in B, f(a) =$ the unique ordinal x such that $(\text{pred}(A,a,H),H) \simeq x$, and let $C = R(f)$. Now, we check that C is an ordinal (from Lemma 3.2.3), that f is an isomorphism from (B,H) onto C, and that either $B = A$ (in which case we are done), or $B = \text{pred}(A,b,H)$ for some $b \in A$ (in which case $b \in B$, and hence a contradiction). ■

Theorem 3.2.4 implies that ordinals can be used as representatives of the order types of well-ordered sets. Thus, we have the following: If (A,H) is a well-ordering, the type (A,H) is defined to be the unique ordinal C such that $(A,H) \simeq C$. From now on we use Greek letters $\alpha, \beta, \gamma, \ldots$ to indicate ordinals. We may thus say, e.g., $\forall \alpha \ldots$ instead of $\forall x(x \text{ is an ordinal} \rightarrow \ldots)$. Since $\in$ orders the ordinals, we write $\alpha < \beta$ for $\alpha \in \beta$ and use $\alpha \geq \beta$ to indicate $\beta \in \alpha \vee \beta = \alpha$.

Lemma 3.2.4.

(1) $\forall \alpha, \beta(\alpha \leq \beta \rightarrow \alpha \subset \beta)$.

(2) If X is a set of ordinals, then $\bigcup X$ is the least ordinal $\geq$ all elements in X, and, if $X \neq \emptyset$, $\bigcap X$ is the least ordinal in X.

We use natural numbers to count finite sets. The Axiom of Choice tells that every set can be counted by an ordinal. Therefore, we use natural numbers to indicate finite ordinals.

Let α be an arbitrary ordinal. The successor $S(\alpha)$ of the ordinal is defined by $S(\alpha) = \alpha \cup \{\alpha\}$. Then $S(\alpha)$ is also an ordinal, and we have

$$0 = \emptyset, \quad 1 = S(\emptyset), \quad 2 = S(1), \quad 3 = S(2), \quad \ldots \tag{3.17}$$

Intuitively, natural numbers are those ordinals obtained by applying S to $\emptyset$ a finite number of times.

An ordinal α is a successor ordinal (or isolated ordinal) if $\exists \beta(\alpha = S(\beta))$. It is a limit ordinal if $\alpha \neq 0$ and α is not a successor ordinal. For example, let ω be the set of all natural numbers and 0. Then ω is an ordinal and all smaller ordinals are successor ordinals or 0. So ω is a limit ordinal, since if not it would be a natural number, and hence ω is the least limit ordinal. Actually, the Axiom of Infinity is equivalent to postulating the existence of a limit ordinal.

The following theorem shows that the elements of ω are the real natural numbers.

Theorem 3.2.5. *The Peano postulates are*

(1) $0 \in \omega$.

(2) $\forall n \in \omega(S(n) \in \omega)$.

(3) $\forall n, m \in w(n \neq m \rightarrow S(n) \neq S(m))$.

(4) (Induction) $\forall X \subset w[(0 \in X \wedge \forall n \in X(S(n) \in X)) \rightarrow X = \omega]$.

Proof: For (4), if $X \neq \omega$, let γ be the least element of $\omega \backslash X$, and show that γ is a limit ordinal $< \omega$, contradiction. ■

Let α and β be two ordinals. Define $\alpha + \beta = \text{type}(\alpha \times \{0\} \cup \beta \times \{1\}, H)$, where

$$\begin{aligned} H = \{((\xi,0),(\eta,0)) : \xi < \eta < \alpha\} \bigcup \{((\xi,1),(\eta,1)) : \xi < \eta < \beta\} \\ \bigcup [(\alpha \times \{0\}) \times (\beta \times \{1\})] \end{aligned}$$

Theorem 3.2.6. *For any* α, β, γ*:*

(1) $\alpha + (\beta + \gamma) = (\alpha + \beta) + \gamma$.

(2) $\alpha + 0 = \alpha$.

(3) $\alpha + 1 = S(\alpha)$.

(4) $\alpha + S(\beta) = S(\alpha + \beta)$.

(5) If β *is a limit ordinal,* $\alpha + \beta = \bigcup\{\alpha + \xi : \xi < \beta\}$.

Proof: The results are straightforward from the definition of addition. ∎

Recall that the addition of ordinals is not commutative. Let α and β be two ordinals. We define the multiplication $\alpha \cdot \beta = \text{type}(\beta \times \alpha, H)$, where H is the lexicographic order on $\beta \times \alpha$:

$$(\xi, \eta) H (\xi^1, \eta^1) \leftrightarrow (\xi < \xi^1 \vee (\xi = \xi^1 \wedge \eta < \eta^1)) \tag{3.18}$$

Again, we get directly from the definition the basic properties of the multiplication of ordinals.

Theorem 3.2.7. *For any* α, β, γ*:*

(1) $\alpha \cdot (\beta \cdot \gamma) = (\alpha \cdot \beta) \cdot \gamma$.

(2) $\alpha \cdot 0 = 0$.

(3) $\alpha \cdot 1 = \alpha$.

(4) $\alpha \cdot S(\beta) = \alpha \cdot \beta + \alpha$.

(5) If β *is a limit ordinal,* $\alpha \cdot \beta = \bigcup\{\alpha \cdot \xi : \xi < \beta\}$.

(6) $\alpha \cdot (\beta + \gamma) = \alpha \cdot \beta + \alpha \cdot \gamma$.

We, again, need to remember that the multiplication of ordinals is not commutative and the distributive law (6) fails for multiplication on the right.

Let A be a set. We define A^n to be the set of all functions from n to A, $A^{<w} = \bigcup\{A^n : n \in \omega\}$, and $(x_0, \ldots, x_{n-1})$ is the function s with domain n such that $s(0) = x_0, s(1) = x_1, \ldots, s(n-1) = x_{n-1}$. Under these definitions, A^2 and $A \times A$ are not the same, but there is an obvious 1–1 correspondence between them. In general, if s is a function with domain(s) $= I$, we may think of I as an index set and s as a sequence indexed by I. In this case, we write s_i for $s(i)$. When $D(s)$ is an ordinal α, we may think of s as a sequence of length α. If $D(t) = \beta$, we may concatenate the sequences s and t to form a sequence $\widehat{st}$ of length $\alpha + \beta$ as follows: the function $\widehat{st}$ with domain $\alpha + \beta$ is defined by $(\widehat{st})(\eta) = s(\eta)$, for each $\eta < \alpha$ and $(\widehat{st})(\alpha + \xi) = t(\xi)$ for all $\xi < \beta$.

We have seen that there need not exist a set of the form $\{x : \phi(x)\}$. But there is nothing wrong with thinking about such collections, and they sometimes provide useful motivation. Informally, we call any collection of the form $\{x : \phi(x)\}$ a class. A proper class is a class which is not a set (because it is too "big"). We use V and **ON** to indicate the following two classes

$$V = \{x : x = x\} \tag{3.19}$$

$$\mathbf{ON} = \{x : x \text{ is an ordinal}\} \tag{3.20}$$

Theorem 3.2.8 [Transfinite Induction on ON]. *If $C \subset \mathbf{ON}$ and $C \neq \emptyset$, then C has a least element.*

Proof: Fix $\alpha \in C$. If α is not the least element of C, let β be the least element of $\alpha \cap C$. Then β is the least element of C. ∎

Theorem 3.2.9 [Transfinite Recursion on ON]. *If $F : V \to V$, then there exists a unique $G : \mathbf{ON} \to V$ such that*

$$\forall\alpha[G(\alpha) = F(G|\alpha)] \tag{3.21}$$

Proof: For uniqueness, if G_1 and G_2 both satisfy Eq. (3.21), we can show that $\forall\alpha(G_1(\alpha) = G_2(\alpha))$ by transfinite induction on α.

To establish existence, we call g a δ-approximation if g is a function with domain g and

$$\forall\alpha < \delta[g(\alpha) = F(g|\alpha)] \tag{3.22}$$

As in the proof of uniqueness, if g is a δ-approximation and g' is a δ'-approximation, then $g|(\delta \cap \delta') = g'|(\delta \cap \delta')$. Next, by transfinite induction on δ, we can show that for each δ there is a δ-approximation (which then is unique). Now, let $G(\alpha)$ be the value $g(\alpha)$, where g is the δ-approximation for some $\delta > \alpha$. ∎

A useful application of recursion is in defining ordinal exponentiation: Let α and β be two ordinals. Then α^β is defined by recursion on β by

(1) $\alpha^0 = 1$.

(2) $\alpha^{\beta+1} = \alpha^\beta \cdot \alpha$.

(3) If β is a limit, $\alpha^\beta = \bigcup\{\alpha^\xi : \xi < \beta\}$.

As in Chapter 2, 1–1 functions are used to compare the size of sets. Generally, if A can be well-ordered, then $A \simeq \alpha$ for some α (Theorem 3.2.4), and there is then a least such α, which we will call the cardinality of the set A, denoted by $|A|$. In any statement involving $|A|$, we imply that A can be well-ordered. Under the Axiom of Choice, $|A|$ is defined for every set A. Regardless of the Axiom of Choice, $|\alpha|$ is defined and is $\leq \alpha$ for every ordinal α.

Theorem 3.2.10. *If* $|\alpha| \leq \beta \leq \alpha$*, then* $|\beta| = |\alpha|$.

Proof: $\beta \subset \alpha$ implies $|\beta| \leq |\alpha|$, and $\alpha \sim |\alpha| \subset \beta$, so $|\alpha| \leq |\beta|$. Thus, by Bernstein's equipollent theorem in the previous chapter, we have $|\alpha| = |\beta|$. ■

Theorem 3.2.11. *If* $n \in \omega$*, then*

(1) $n \not\sim n+1$.

(2) $\forall \alpha(\alpha \sim n \to \alpha = n)$.

Proof: Statement (1) can be proved by induction on n and (2) follows by using Theorem 3.2.6. ■

Corollary 3.2.1. ω *is a cardinal and each* $n \in \omega$ *is a cardinal.*

A set A is finite if $|A| < \omega$ and to be countable if $|A| \leq \omega$. Infinite means not finite. Uncountable means not countable.

Cardinal multiplication and addition are defined in the following. They must be distinguished from ordinal multiplication and addition. Let κ and λ be two cardinals we define $\kappa \oplus \lambda = |\kappa \times \{0\} \cup \lambda \times \{1\}|$ and $\kappa \oplus \lambda = |\kappa \times \lambda|$.

Theorem 3.2.12. *For* $n, m \in \omega$, $n \oplus m = n + m \prec \omega$ *and* $n \otimes m = n \cdot m \prec \omega$.

Proof: First show $n + m < \omega$ by induction on m. Then show $n \cdot m < \omega$ by induction on m. The rest follows from Theorem 3.2.11 (2). ■

Theorem 3.2.13. *Every infinite cardinal is a limit ordinal.*

Proof: If κ is an infinite cardinal and $\kappa = \alpha + 1$, then since $1 + \alpha = \alpha$, $\kappa = |\kappa| = |1 + \alpha| = |\alpha|$. This contradicts the definition of κ. ■

Theorem 3.2.14. *If* κ *is an infinite cardinal,* $\kappa \times \kappa = \kappa$.

Proof: We prove the theorem by transfinite induction on κ. Assume this holds for smaller cardinals. Then for $\alpha < \kappa$, $|\alpha \times \alpha| = |\alpha| \otimes |\alpha| < \kappa$. Define a well-ordering $\lhd$ on $\kappa \times \kappa$ by $(\alpha, \beta) \lhd (\gamma, \delta)$ if and only if

$$\begin{aligned} &\max\{\alpha, \beta\} < \max\{\gamma, \delta\} \vee [\max\{\alpha, \beta\} \\ &\qquad = \max\{\gamma, \delta\} \wedge (\alpha, \beta) \text{ precedes } (\gamma, \delta) \text{ lexicographically}] \end{aligned} \tag{3.23}$$

Each $(\alpha, \beta) \in \kappa \times \kappa$ has no more than $|(\max\{\alpha, \beta\} + 1) \times (\max\{\alpha, \beta\} + 1)| < \kappa$ predecessors under the ordering $\lhd$, so type$(\kappa \times \kappa, \lhd) \leq \kappa$, whence $|\kappa \times \kappa| = \kappa$. ■

Corollary 3.2.2. *Let κ and λ be infinite cardinals then*

(1) $\kappa \oplus \lambda = \kappa \otimes \lambda = \max\{\kappa, \lambda\}$.

(2) $|\kappa^{<\omega}| = \kappa$.

Proof: For (2), use the proof of Theorem 3.2.13 to define a 1–1 mapping $f_n : \kappa^n \to \kappa$. This gives a 1–1 mapping $f : \bigcup_{n\in\omega} \kappa^n \to \omega \times \kappa$. Therefore, $|\kappa^{<\omega}| \leq \omega \otimes \kappa = \kappa$. ■

Theorem 3.2.15. $\forall\alpha\exists\kappa(\kappa \succ \alpha$ *and* κ *is a cardinal).*

Proof: Assume $\alpha \geq \omega$. Let $W = \{H : H \in P(\alpha \times \alpha)$ and H well-orders $\alpha\}$. Let $S = \{\text{type}(H) : H \in W\}$. Then $\bigcup S$ is a cardinal $> \alpha$. ■

Let α be an ordinal and α^+ the least cardinal $> \alpha$. A cardinal number κ is a successor cardinal if there exists some ordinal α such that $\kappa = \alpha^+$. The cardinal κ is a limit cardinal if $\kappa > \omega$ and is not a successor cardinal. Now, the sequence of transfinite cardinals $\aleph_\alpha = \omega_\alpha$ is defined by transfinite recursion on α by

(1) $\omega_0 = \omega$.

(2) $\omega_{\alpha+1} = (\omega_\alpha)^+$.

(3) For any limit ordinal γ, $\omega_\gamma = \bigcup\{\omega_\alpha : \alpha < \gamma\}$.

Theorem 3.2.16.

(1) Every infinite cardinal is equal to ω_α for some α.

(2) $\alpha < \beta \to \omega_\alpha < \omega_\beta$.

(3) ω_α is a limit cardinal if and only if α is a limit ordinal.

(4) ω_α is a successor cardinal if and only if α is a successor ordinal.

We now turn to cardinal exponentiation. Let $A^B = {}^BA = \{f : f \text{ is a function} \wedge D(f) = B \wedge R(f) \subset A\}$. We define $\kappa^\lambda = |{}^\lambda\kappa|$.

Theorem 3.2.17. *If $\lambda \geq \omega$ and $2 \leq \kappa \leq \lambda$, then ${}^\lambda\kappa \sim {}^\lambda 2 \sim P(\lambda)$.*

Proof: ${}^\lambda 2 \sim P(\lambda)$ follows by identifying sets with their characteristic functions. Then

$$ {}^\lambda 2 \preceq {}^\lambda k \preceq {}^\lambda\lambda \preceq P(\lambda \times \lambda) \sim P(\lambda) \sim 2^\lambda \tag{3.24} $$

where $X \preceq Y$ implies that there exists a 1–1 function from X into Y. ■

Cardinal exponentiation is not the same as ordinal exponentiation. For example, the ordinal 2^ω is ω, but the cardinal $2^\omega = |P(\omega)| > \omega$.

Theorem 3.2.18. *If* κ, λ, *and* σ *are any cardinals, then* $\kappa^{\lambda\oplus\sigma} = \kappa^\lambda \otimes \kappa^\sigma$ *and* $(\kappa^\lambda)^\sigma = \kappa^{\lambda\otimes\sigma}$.

Proof: We can check that

$$^{(B\cup C)}A \sim {}^{B}A \times {}^{C}A \quad (\text{if } B\cap C = \emptyset)$$

and

$$^{C}({}^{B}A) \sim {}^{C\times B}A.$$

■

The Continuum Hypothesis (CH) is the conjecture that $2^\omega = \omega_1$. The Generalized Continuum Hypothesis is the conjecture that $\forall\alpha(2^{\omega_\alpha} = \omega_{\alpha+1})$.

If α and β are ordinals and $f : \alpha \to \beta$ is a mapping. The mapping f maps α cofinally if $D(f)$ is unbounded in β; i.e., $\forall b \in \beta \exists a \in D(f)(b \leq a)$. The cofinality of β, denoted by $\mathrm{cf}(\beta)$, is the least α such that there is a mapping from α cofinally into β. Therefore, $\mathrm{cf}(\beta) \leq \beta$; if β is a successor, $\mathrm{cf}(\beta) = 1$.

Theorem 3.2.19. *There is a cofinal mapping* $f : \mathrm{cf}(\beta) \to \beta$ *which is strictly increasing* $(\xi < \eta \to f(\xi) < f(\eta))$.

Proof: Let $g : \mathrm{cf}(\beta) \to \beta$ be any cofinal mapping, and define a mapping f recursively by

$$f(\eta) = \max\{g(\eta), \bigcup\{f(\xi)+1 : \xi < \eta\}\}.$$

■

Theorem 3.2.20. *If* α *is a limit ordinal and* $f : \alpha \to \beta$ *is a strictly increasing cofinal mapping, then* $\mathrm{cf}(\alpha) = \mathrm{cf}(\beta)$.

Proof: $\mathrm{cf}(\beta) \leq \mathrm{cf}(\alpha)$ follows by composing a cofinal mapping from $\mathrm{cf}(\alpha)$ into α with the mapping f. To show $\mathrm{cf}(\alpha) \leq \mathrm{cf}(\beta)$, let $g : \mathrm{cf}(\beta) \to \beta$ be a cofinal mapping, and let $h(\xi)$ be the least η such that $f(\eta) > g(\xi)$; then $h : \mathrm{cf}(\beta) \to \alpha$ is a cofinal mapping. ■

Corollary 3.2.3. $\mathrm{cf}(\mathrm{cf}(\beta)) = \mathrm{cf}(\beta)$.

An ordinal β is regular if β is a limit ordinal and $\mathrm{cf}(\beta) = \beta$. Thus, Corollary 3.2.3 implies that $\mathrm{cf}(\beta)$ is regular for all limit ordinals β.

Theorem 3.2.21. *If an ordinal β is regular, then β is a cardinal.*

Theorem 3.2.22. *κ^+ is regular.*

Proof: If f mapped α cofinally into κ^+ for some $\alpha < \kappa^+$, then

$$\kappa^+ = \bigcup \{f(\xi) : \xi < \alpha\} \tag{3.25}$$

but the union of $\leq \kappa$ sets each of cardinality $\leq \kappa$ must have cardinality $\leq \kappa \otimes \kappa = \kappa$ by Theorem 3.2.14. ■

Theorem 3.2.23 [König's Theorem]. *If κ is infinite and* $\mathrm{cf}(\kappa) \leq \lambda$, *then* $\kappa^\lambda > \kappa$.

Proof: Fix a cofinal mapping $f : \lambda \to \kappa$. Let $G : \kappa \to {}^\lambda\kappa$. We show that G cannot be an onto mapping. Define $h : \lambda \to \kappa$ so that $h(\alpha)$ is the least element of

$$\kappa - \{(G(\mu))(\alpha) : \mu < f(\alpha)\} \tag{3.26}$$

Then $h \notin R(G)$. ■

Theorem 3.2.24. *Assume* AC *and* GCH. *Let κ and $\lambda \geq 2$ and at least one of them be infinite. Then*

(1) $\kappa \leq \lambda \to \kappa^\lambda = \lambda^+$.

(2) $\kappa > \lambda \geq \mathrm{cf}(\kappa) \to \kappa^\lambda = \kappa^+$.

(3) $\lambda < \mathrm{cf}(\kappa) \to \kappa^\lambda = \kappa$.

Proof: (1) follows from Theorem 3.2.17. For (2), $\kappa^\lambda > \kappa$ by Theorem 3.2.23, but $\kappa^\lambda \leq \kappa^\kappa = 2^\kappa = \kappa^+$. For (3), $\lambda < \mathrm{cf}(\kappa)$ implies that ${}^\lambda\kappa = \bigcup\{{}^\lambda\alpha : \alpha < \kappa\}$ and each $|{}^\lambda\alpha| \leq (\max\{\alpha,\lambda\})^+ \leq \kappa$. ■

We conclude this section by showing that all of classical mathematics can be built upon ZFC.

Let $\mathbb{Z}$ be the ring of integers, $\mathbb{Q}$ the field of rational numbers, $\mathbb{R}$ the field of real numbers, and $\mathbb{C}$ the field of complex numbers. It suffices to show that there is a way, based on ZFC, to construct the sets of $\mathbb{Z}$, $\mathbb{Q}$, $\mathbb{R}$, and $\mathbb{C}$. Any reasonable way of defining these sets from the natural numbers will do, but for definiteness we take $\mathbb{Z} = \omega \times \omega / \sim$, where (n,m) is intended to represent $n - m$ the equivalence relation $\sim$ is defined by the following: $(a,b) \sim (c,d)$ if and only if $a + d = c + b$.

$\mathbb{Z}$ is the set of equivalence classes of the relation $\sim$, and operations $+$ and $\cdot$ are defined as follows:

$$[(a,b)]+[(c,d)]=[(a+c,b+d)] \tag{3.27}$$

$$[(a,b)]\cdot[(c,d)]=[(ac+bd,ad+bc)] \tag{3.28}$$

where $[(a,b)]$ is the equivalence class of the relation $\sim$ containing (a,b).

Let $\mathbb{Q}=(\mathbb{Z}\times(\mathbb{Z}-\{0\}))/\simeq$, where (x,y) is intended to represent x/y, and $\mathbb{R}=\{X\in P(\mathbb{Q}):X\neq\emptyset\wedge X\neq\mathbb{Q}\wedge\forall x\in X\forall y\in\mathbb{Q}(y<x\rightarrow y\in X)\}$. So $\mathbb{R}$ is the set of left sides of all Dedekind cuts. $\mathbb{C}=\mathbb{R}\times\mathbb{R}$, with field operations defined in the usual way.

We will need the following definitions.

(a) ${}^{<\beta}A=A^{<\beta}=\bigcup\{{}^{\alpha}A:\alpha<\beta\}$.

(b) $\kappa^{<\lambda}=\left|{}^{<\lambda}\kappa\right|$.

3.3. Families of Sets

For an infinite cardinal κ we say that two sets X and Y are κ-disjoint if $|X\cap Y|<\kappa$.

Theorem 3.3.1. *If $\kappa\geq\omega_0$ and $|A_0|=\kappa$, then there exists $S_0\subset p(A_0)$ of power κ^+ consisting of m-disjoint sets and such that $A_0=\bigcup S_0$.*

Proof: It will be sufficient to find just one set A which can be decomposed as in the theorem and has power κ. A 1–1 mapping of A onto A_0 will then enable us to obtain the desired decomposition of A_0.

Let $\kappa=\aleph_\alpha$ and $F=\omega_\alpha^{\omega_\alpha}$. Two functions f,g in F (conceived as sets of ordered pairs) are $\aleph_\alpha$-disjoint if $|\{\xi<\omega_\alpha:f(\xi)=g(\xi)\}|<\aleph_\alpha$.

We will construct a sequence of type $\omega_{\alpha+1}$ of $\aleph_\alpha$ disjoint functions. To carry out this construction we need the following lemma.

Lemma 3.3.1. *If φ is a transfinite sequence of type $\gamma<\omega_{\alpha+1}$ with range contained in F, then there exists a function $f\in F$ which is $\aleph_\alpha$-disjoint from each term φ_ξ of the sequence.*

Proof of lemma: To see this we notice that the range of φ has power $\leq\aleph_\alpha$ and thus can be represented as the range of a function $\Phi:\omega_\alpha\rightarrow F$. If $\xi<\omega_\alpha$, $|\{\Phi_\eta(\xi):\eta<\xi\}|<\omega_\alpha$. Let $f(\xi)$ be the first ordinal in $\omega_\alpha-\{\Phi_\eta(\xi):\eta<\xi\}$. Then $f\in F$ which is $\aleph_\alpha$-disjoint from every Φ_η, because $f(\xi)=\Phi_\eta(\xi)$ can hold only if $\eta\geq\xi$. ■

Let $\Gamma : p(F) \to F \cup \{\rho\}$, where $\rho \notin F$, such that $\forall A \in p(F)[A \neq \emptyset \to \Gamma(A) \in A) \wedge (A = \emptyset \to \Gamma(A) = \rho)$. For each transfinite sequence φ denote by $B(\varphi)$ the set of all functions in F which are $\aleph_\alpha$-disjoint from all terms of φ. By transfinite induction, we obtain a function $g : \omega_{\alpha+1} \to F \cup \{\rho\}$ satisfying the equation

$$g_\xi = \Gamma(B(g|\xi)) \tag{3.29}$$

for each $\xi < \omega_{\alpha+1}$.

Thus, $\forall \xi < \omega_{\alpha+1}, g_\xi \in F$ and is $\aleph_\alpha$-disjoint from all g_η, $\eta < \xi$, which belong to F, provided that such functions exist; otherwise $g_\xi = \rho$.

We claim that $g_\xi \neq \rho$ for each $\rho < \omega_{\alpha+1}$. Otherwise there would be a smallest $\gamma < \omega_{\alpha+1}(g_\gamma = \rho)$ and $g|\gamma$ would be a sequence satisfying the assumptions of Lemma 3.3.1. This contradicts the assumption that $\rho \notin F$. Thus, $|R(g)| = \omega_{\alpha+1}$. Let $A = \bigcup R(g)$. Since each element of A is an ordered pair of ordinals $< \omega_\alpha$, we infer that $|A| \leq \omega_\alpha$; on the other hand, $|A| \geq |g_0| = \omega_\alpha$. Thus, $|A| = \omega_\alpha$. ■

For an infinite cardinal κ two sets $x, y \subset \kappa$ are almost disjoint if $|x \cap y| < \kappa$. An almost disjoint family $\mathcal{F} \subset p(\kappa)$ such that $\forall x \in \mathcal{F}(|x| = \kappa)$ and any two distinct elements of $\mathcal{F}$ are almost disjoint. A maximal almost disjoint family is an almost disjoint family with no almost disjoint family $\mathcal{B}$ properly containing it.

Theorem 3.3.2. *Let $\kappa \geq \omega_0$ then*

(a) If $\mathcal{A} \subset p(\kappa)$ is an almost disjoint family and $|\mathcal{A}| = \kappa$, then $\mathcal{A}$ is not maximal.

(b) There is a maximal almost disjoint family $\mathcal{B} \subset p(\kappa)$ of cardinality $\geq \kappa^+$.

The proof follows from that of Theorem 3.3.1.

Theorem 3.3.3. *If $\kappa \geq \omega_0$ and $2^{<\kappa} = \kappa$, then there exists an almost disjoint family $\mathcal{A} \subset p(\kappa)$ with $|\mathcal{A}| = 2^\kappa$.*

Proof: Let $I = \{x \subset \kappa : \bigcup x < \kappa\}$. Since $2^{<\kappa} = \kappa$, $|I| = \kappa$. If $X \subset \kappa$, let $A_x = \{X \cap \alpha : \alpha < \kappa\}$. If $|X| = \kappa$, then $|A_x| = \kappa$ also. If $X \neq Y$, then $|A_x \cap A_y| < \kappa$, since if we fix β such that $\neg(\beta \in X \leftrightarrow \beta \in Y)$, then

$$A_x \cap A_y \subset \{X \cap \alpha : \alpha \leq \beta\} \tag{3.30}$$

Let $\mathcal{A} = \{A_x : X \subset \kappa \wedge |X| = \kappa\}$; then $|\mathcal{A}| = 2^\kappa$ and is an almost disjoint family of subsets of I. If we let f be a 1–1 function from I onto κ, $\{f(A) : A \in \mathcal{A}\}$ is an almost disjoint family of 2^κ subsets of κ. ■

A family $\mathcal{A}$ of sets is called a Δ-system, or a quasi-disjoint family, if there is a fixed set r, called the root of the Δ-system, such that a $\bigcap b = r$ whenever a and b are disjoint members of $\mathcal{A}$.

Theorem 3.3.4. *(Δ-System Lemma). Let κ be an infinite cardinal. Let $\theta > \kappa$ be regular and satisfy $\forall\alpha < \theta(|\alpha^{<\kappa}| < \theta)$. Assume $|\mathcal{A}| \geq \theta$ and $\forall x \in \mathcal{A}(|x| < \kappa)$ then there is a $\mathcal{B} \subset \mathcal{A}$ such that $|\mathcal{B}| = \theta$ and $\mathcal{B}$ forms a Δ-system.*

Proof: By shrinking $\mathcal{A}$ if necessary, we may assume $|\mathcal{A}| = \theta$. Then $|\bigcup\mathcal{A}| \leq \theta$. Since what the elements of $\mathcal{A}$ are as individuals is irrelevant, we may assume $\bigcup\mathcal{A} \subset \theta$. Then each $x \in \mathcal{A}$ has some order type $< \kappa$ as a subset of θ. Since θ is regular and $\theta > \kappa$, there is some $\rho < \kappa$ such that $\mathcal{A}_1 = \{x \in \mathcal{A} : x \text{ has type } \rho\}$ has cardinality θ. We now fix such a ρ and deal with $\mathcal{A}_1$.

For each $\alpha < \theta$, $|\alpha^{<\kappa}| < \theta$ implies that fewer than θ elements of $\mathcal{A}_1$ are subsets of α. Thus, $\bigcup\mathcal{A}_1$ is unbounded in θ. If $x \in \mathcal{A}_1$ and $\xi < \rho$, let $x(\xi)$ be the ξth element of x. Since θ is regular, there is some ξ such that $\{x(\xi) : x \in \mathcal{A}_1\}$ is unbounded in θ. Now fix ξ_0 to be the least such ξ (ξ_0 may be 0). Let

$$\alpha_0 = \bigcup\{x(\eta) + 1 : x \in \mathcal{A}_1 \wedge \eta < \xi_0\} \tag{3.31}$$

Then $\alpha_0 < \theta$ and $x(\eta) < \alpha_0$ for all $x \in \mathcal{A}_1$ and all $\eta < \xi_0$.

By transfinite recursion on $\mu < \theta$, pick $x_\mu \in \mathcal{A}_1$ so that $x_\mu(\xi_0) > \alpha_0$ and $x_\mu(\xi_0)$ is above all elements of earlier x_γ; i.e.,

$$x_\mu(\xi_0) > \max(\alpha_0, \bigcup\{x_\gamma(\eta) : \eta < \rho \wedge \gamma < \mu\}) \tag{3.32}$$

Let $\mathcal{A}_2 = \{x_\mu : \mu < \theta\}$. Then $|\mathcal{A}_2| = \theta$ and $x \cap y \subset \alpha_0$ whenever x and y are distinct elements in $\mathcal{A}_2$. Since $|\alpha_0^{<\kappa}| < \theta$, there is an $r \subset \alpha_0$ and $\mathcal{B} \subset \mathcal{A}_2$ with $|\mathcal{B}| = \theta$ and $\forall x \in \mathcal{B}(x \cap \alpha_0 = r)$, whence $\mathcal{B}$ forms a Δ-system with root r. ■

Under the assumption of CH or GCH, we calculated many exponentiations of cardinals in Section 3.2. If we now assume that CH fails, questions arise about the various infinite cardinals $\kappa < 2^\omega$. For example, we have

Question 3.3.1. If $\kappa < 2^\omega$, does $2^\kappa = 2^\omega$?

Question 3.3.2. If $\kappa < 2^\omega$, does every almost disjoint family $A \subset p(\omega)$ of size κ fail to be maximal?

Since the answer to these questions is clearly "yes" when $\kappa = \omega$, they are only of interest for $\omega < \kappa < 2^\omega$. For such κ, it is known by the method of forcing that neither question can be settled under the axioms ZFC + $\neg$CH. In the following we show how these questions can be settled by using a new axiom, called Martin's axiom. Martin's axiom is known to have numerous important consequences in combinatorics, set-theoretic topology, algebra, and analysis. For examples other than those given here, please consult (Martin and Solovay, 1970; Rudin, 1977; Shoenfield, 1975).

A partial order is a pair $(P,\leq)$ such that $P\neq\emptyset$ and $\leq$ is a relation on P which is transitive and reflexive $(\forall p\in P(p\leq p))$. $p\leq q$ is read "p extends q." Elements of p are called conditions. A partial order $(P,\leq)$ is a partial order in the strict sense, if it, in addition, satisfies

$$\forall p,q(p\leq q\wedge q\leq p\rightarrow p=q) \tag{3.33}$$

We often abuse notation by referring to "the partial order P" or "the partial order $\leq$" if $\leq$ or P is clear from context.

If $(P,\leq)$ is a partial order, p and $q\in P$ are compatible if

$$\exists r\in P(r\leq p\wedge r\leq q) \tag{3.34}$$

They are compatible, denoted by $p\perp q$ if $\neg\exists r\in P(r\leq p\wedge r\leq q)$. An antichain in P is a subset $A\subset P$ such that $\forall p,q\in A(p\neq q\rightarrow p\perp q)$. A partial order $(P,\leq)$ has the countable chain condition (ccc) if every antichain in P is countable.

A subset $G\subset P$ is called a filter in P if

(a) $\forall p,q\in G\exists r\in G(r\leq p\wedge r\leq q)$.

(b) $\forall p\in G\forall q\in P(q\geq p\rightarrow q\in G)$.

Axiom 3.3.1 [Martin's Axiom]. MA(κ) is the statement: Whenever $(P,\leq)$ is a nonempty ccc partial order and $\mathcal{D}$ is a family of at most κ many coinitial subsets of P, then there is a filter G in P such that $\forall D\in\mathcal{D}(G\cap D\neq\emptyset)$. MA is the statement $\forall\kappa<2^{\omega}(\mathrm{MA}(\kappa))$.

Example 3.3.1. Let P be the set of all finite partial functions from ω to 2; i.e.

$$P=\{p:p<\omega\times 2\wedge|p|<\omega\wedge p\text{ is a function}\} \tag{3.35}$$

Define $p\leq q$ by $q\subset p$; i.e., the function p extends the function q. Now p and q are compatible if and only if they have the same values on the set $D(p)\cap D(q)$, in which case $p\cup q$ is a common extension of both p and q.

The partial order $(P,\leq)$ has ccc. In fact, let $D\subset P$ be an antichain in P then $|D|\leq|P|=\omega$. If G is a filter in P, the elements of G are pairwise compatible. Hence, if we let $f=f_G=\bigcup G$, then f_G is a function with $D(f_G)\subset\omega$. In this example, we are more interested in f than G. If $p\in P$, we think of p as a finite approximation to f, and we say intuitively that p forces "$p\subset f$" in the sense that if $p\in G$, then $p\subset f_G$. So p says what f is restricted to $D(p)$. If $q\leq p$, then q says more about f than p does.

$D(f_G)$ could be a very small subset of ω; for example, $\emptyset$ is a filter and $\bigcup\emptyset=\emptyset$, the empty function. However, by requiring that G intersect many coinitial sets, we can make f_G representative of typical functions. For example, let

$$D_n=\{p\in P:n\in D(p)\} \tag{3.36}$$

for each $n \in \omega$. Since each $p \in P$ can be extended to a condition with n in its domain, D_n is coinitial in P. If $\forall n \in \omega(G \cap D_n \neq \emptyset)$, then f_G has a domain equal to ω. Let $E = \{p \in P : \exists n \in D(p)(p(n) = 1)\}$. E is coinitial in P; and if $G \cap E \neq \emptyset$, then f_G takes the value 1 somewhere.

Let $h : \omega \to 2$ be a fixed mapping. Define

$$E_h = \{p \in P : \exists n \in D(p)(p(n) \neq h(n))\} \tag{3.37}$$

Then E_h is coinitial in P; and if $G \cap E_h \neq \emptyset$, then $f_G \neq h$. Let $\mathcal{D} = \{D_n : n \in \omega\} \cup \{E_h : h \in {}^{\omega}2\}$; then $|\mathcal{D}| = 2^{\omega}$. If G is a filter in P and $G \cap D \neq \emptyset \forall D \in \mathcal{D}$, then f_G is a function from ω to 2, which is impossible.

This fact shows the next result.

Theorem 3.3.5. MA(2^{ω}) *is false.*

Theorem 3.3.6.

(a) If $\kappa < \kappa'$, *then* MA$(\kappa') \to$ MA(κ).

(b) MA(ω) *is true.*

Proof: (a) is clear. For (b), let $\mathcal{D} = \{D_n : n \in \omega\}$ and define, by induction on n, $P_n \in P$ so that P_0 is an arbitrary element of P (since $P \neq \emptyset$) and P_{n+1} is any extension of P_n such that $P_{n+1} \in D_n$. This is possible since D_n is dense. So $P_0 \geq P_1 \geq P_2 \geq \cdots$. Let G be the filter generated by $\{P_n : n \in \omega\}$; i.e., $G = \{q \in P : \exists n(q \geq P_n)\}$. Then G is a filter in P and $G \cap D_n \neq \emptyset$ for each $n \in \omega$. ■

Theorem 3.3.6(b) implies that MA follows from CH, because we can view MA as saying that all infinite cardinals $< 2^{\omega}$ have properties similar to ω.

Since the proof of Theorem 3.3.6(b) did not need the fact that P was ccc, we might attempt to strengthen MA(k) by dropping this requirement, but for $k > \omega$ this strengthening becomes inconsistent by the next example.

Example 3.3.2. Let

$$P = \{p \subset \omega \times \omega_1 : |p| < \omega \wedge p \text{ is a function}\} \tag{3.38}$$

If G is a filter in P, then, as in Example 3.3.1, $\bigcup G$ is a function with $D(\bigcup G) \subset \omega$ and $R(\bigcup G) \subset \omega_1$. If $\alpha < \omega_1$, let

$$D_{\alpha} = \{p \in P : \alpha \in R(p)\} \tag{3.39}$$

Then D_{α} is coinitial in P. We claim that no filter G in P could intersect each D_{α}, $\alpha < \omega_1$, since that would mean that $R(\bigcup G) = \omega_1$, contradiction. Of course, P here is not ccc since the conditions $\{(0, \alpha) : \alpha \in \omega_1\}$ are pairwise incompatible.

We now proceed to show how MA is applied to answer Questions 3.3.1 and 3.3.2.

Theorem 3.3.7. *Assume* MA(κ). *Let* $\mathcal{A},\mathcal{C} \subset p(\omega)$, *where* $|\mathcal{A}|,|\mathcal{C}| \leq \kappa$, *and assume that for all* $y \in \mathcal{C}$ *and all finite* $F \subset \mathcal{A}$, $|y - \cup F| = \omega$. *Then there is* $d \subset \omega$ *such that* $\forall x \in \mathcal{A}(|d \cap x| < \omega)$ *and* $\forall y \in \mathcal{C}(|d \cap y| = \omega)$.

Proof: Define a partial order $P_{\mathcal{A}}$ by

$$P_{\mathcal{A}} = \{(s,F) : s \subset \omega \wedge |s| < \omega \wedge F \subset \mathcal{A} \wedge |F| < \omega\} \tag{3.40}$$

where $(s^1,F^1) \leq (s,F)$ if and only if

$$s \subset s^1 \wedge F \subset F^1 \wedge \forall x \in F(x \cap s^1 \subset s) \tag{3.41}$$

Then the partial order $(P_{\mathcal{A}},\leq)$ has the ccc. In fact, for (s_1,F_1) and $(s_2,F_2) \in P_{\mathcal{A}}$, they are compatible if and only if

$$\forall x \in F_1(x \cap s_2 \subset s_1) \wedge \forall x \in F_2(x \cap s_1 \subset s_2) \tag{3.42}$$

Suppose that (s_ξ,F_ξ), for $\xi < \omega_1$, were pairwise incompatible. By Eq. (3.42), the sets s_ξ would all be distinct, which is impossible.

For each $x \in \mathcal{A}$, define

$$D_x = \{(s,F) \in P_{\mathcal{A}} : x \in F\} \tag{3.43}$$

Then D_x is coinitial in $P_{\mathcal{A}}$. In fact, for each $(s,F) \in P_{\mathcal{A}}$, $(s,F \cup \{x\}) \subset (s,F)$.

For any $y \in \mathcal{C}$ and $n \in \omega$, let

$$E_{n,y} = \{(s,F) \in P_{\mathcal{A}} : s \cap y \not\subset n\} \tag{3.44}$$

$E_{n,y}$ is coinitial in $P_{\mathcal{A}}$, since for each $(s,F) \in P_{\mathcal{A}}$, $|y - \cup F| = \omega$. If we pick $m \in y - \cup F$ with $m > n$, then $(s \cup \{m\},F)$ is an extension of (s,F) in $E_{n,y}$.

By MA(k), there is a filter G in $P_{\mathcal{A}}$ intersecting all sets in

$$\{D_x : x \in \mathcal{A}\} \cup \{E_{n,y} : y \in \mathcal{C} \text{ and } n \in \omega\} \tag{3.45}$$

Let $d_G = \bigcup\{s : \exists F((s,F) \in G)\}$. Then $\forall x \in \mathcal{A}(|d_G \cap x| < \omega)$ and $\forall y \in \mathcal{C}(|d_G \cap y| = \omega)$. In fact, $\forall x \in \mathcal{A}(G \cap D_x \neq \emptyset) \rightarrow \exists(s,F) \in G(x \in F)$. If $(s',F') \in G$, then (s',F') and (s,F) are compatible. Eq. (3.42) implies that

$$x \cap s' \subset s \tag{3.46}$$

This implies that $d_G \cap x \subset s$, so $|d_G \cap x| < \omega$. If $y \in \mathcal{C}$, then $d_G \cap y \not\subset n \forall n \in \omega$, so $d_G \cap y$ is infinite. ■

Question 3.3.2 now can be answered by the following corollary.

Corollary 3.3.1. *Let $\mathcal{A} \subset P(\omega)$ be an almost disjoint family of power κ, where $\omega \leq \kappa \leq 2^{\omega}$. Assume* MA$(\kappa)$. *Then $\mathcal{A}$ is not maximal.*

Proof: Let $\mathcal{C} = \{\omega\}$. Then since $\mathcal{A}$ is almost disjoint it follows that $|\omega - \cup F| = \omega$ for all finite $F \subset \mathcal{A}$. Thus, Theorem 3.3.7 implies that there is an infinite $d \subset \omega$ almost disjoint from each member of $\mathcal{A}$. ■

Theorem 3.3.8. *Let $\mathcal{B} \subset P(\omega)$ be an almost disjoint family of size κ, where $\omega \leq \kappa < 2^{\omega}$. Let $\mathcal{A} \subset \mathcal{B}$. If* MA$(\kappa)$, *then there is a $d \subset \omega$ such that $\forall x \in \mathcal{A}(|x \cap d| < \omega)$ and $\forall x \in \mathcal{B} \backslash \mathcal{A}(|d \cap x| = \omega)$.*

Proof: Apply Theorem 3.3.7 with $\mathcal{C} = \mathcal{B} \backslash \mathcal{A}$. ■

Corollary 3.3.2. MA$(\kappa) \rightarrow 2^{\kappa} = 2^{\omega}$.

Proof: Fix any almost disjoint family $\mathcal{B}$ of size κ ($\mathcal{B}$ exists by Theorem 3.3.3). Define $\Phi : p(\omega) \rightarrow p(\mathcal{B})$ by $\Phi(d) = \{x \in \mathcal{B} : |x \cap d| < \omega\}$. Theorem 3.3.8 says that Φ is onto, so $2^{k} = |p(\mathcal{B})| \leq |p(\omega)| = 2^{\omega}$. ■

Corollary 3.3.3. *If* MA, *then* 2^{ω} *is regular.*

Proof: If $\kappa < 2^{\omega}$, then $2^{\kappa} = 2^{\omega}$, so by König's theorem, cf$(2^{\omega}) > \kappa$. ■

Corollary 3.3.4. MA(κ) *is equivalent to* MA(κ) *restricted to partial orders of cardinality* $\leq \kappa$.

Proof: Assume the restricted form of MA(κ), and let $(Q, \leq)$ be a ccc partial order of arbitrary cardinality, and $\mathcal{D}$ a family of at most κ many coinitial subsets of Q. We shall find a filter H in Q intersecting each $D \in \mathcal{D}$ by applying the restricted form of MA(κ) to a suitable suborder $P \subset Q$ with $|P| \leq \kappa$.

We first show that we can find $P \subset Q$ such that

(1) $|P| \leq \kappa$.

(2) $D \cap P$ is coinitial in P for each $D \in \mathcal{D}$.

(3) If $p, q \in P$, p and q are compatible in P if and only if they are compatible in Q.

For $D \in \mathcal{D}$, let $f_D : Q \to Q$ be such that

$$\forall p \in Q(f_D(p) \in D \wedge f_D(p) \leq p) \tag{3.47}$$

and let $g : Q \times Q \to Q$ be such that

$$\forall p, q \in Q(p, q \text{ compatible} \to g(p,q) \leq p \wedge g(p,q) \leq q) \tag{3.48}$$

Let $P \subset Q$ be such that $|P| \leq \kappa$ and P is closed under g and each f_D; i.e., $g(P \times P) \subset P$ and $f_D(P) \subset P$, for all $D \in \mathcal{D}$. P satisfies (3) by closure under g and (2) by closure under f_D.

Statement (3) implies that P has the ccc, so by applying the restricted MA(k) to P, let G be a filter in P such that

$$\forall D \in \mathcal{D}(G \cap D \cap P \neq \emptyset) \tag{3.49}$$

If H is the filter in Q generated by G, i.e.,

$$H = \{q \in Q : \exists p \in G(p \leq q)\} \tag{3.50}$$

then H is a filter in Q intersecting each $D \in \mathcal{D}$. ■

A tree is a partial order in the strict sense $(T, \leq)$ such that, for each $x \in T$, $\{y \in T : y < x\}$ is well-ordered by $\leq$. As usual, we abuse notation and refer to T when we mean $(T, \leq)$.

Let T be a tree. If $x \in T$, the height of x in T, denoted by $\text{ht}(x, T)$, is type $(\{y \in T : y < x\})$. For each ordinal α, the αth level of T, denoted by $\text{Lev}_\alpha(T)$, is $\{x \in T : \text{ht}(x, T) = \alpha\}$. Then $\text{ht}(T)$ is the least α such that $\text{Lev}_\alpha(T) = \emptyset$. A subtree of T is a subset $T' \subset T$ with the induced order such that

$$\forall x \in T' \forall y \in T(y < x \to y \in T') \tag{3.51}$$

Proposition 3.3.1.

(a) $\text{ht}(T) = \bigcup\{\text{ht}(x, T) + 1 : x \in T\}$.

(b) If T' is a subtree of T, then for each $x \in T'$, $\text{ht}(x, T) = \text{ht}(x, T')$.

The proof is immediate from the definitions involved.

Example 3.3.3.

(a) Let a partial order $(T, \leq)$ be defined by $\leq = \{(x, x) : x \in T\}$. Then $\text{ht}(x, T) = 0$ for all $x \in T$ and $\text{ht}(T) = 1$.

(b) Let I be any set and δ an ordinal. A tree $({}^{<\delta}I, \leq)$ is defined as follows: ${}^{<\delta}I = \bigcup\{{}^{\alpha}I : \alpha < \delta\}$, and $s \leq t$ if and only if $s \subset t$. This tree is called the complete I-ary tree of height δ. We think of elements of ${}^{\alpha}I$ as sequences of elements of I of length α. If $\alpha < \delta$, then $\mathrm{Lev}_{\alpha}({}^{<\delta}I) = {}^{\alpha}I$ and $\mathrm{ht}({}^{<\delta}I) = \delta$. When $I = 2$, we refer to ${}^{<\delta}2$ as the complete binary tree of height δ.

For any infinite cardinal κ, a κ-Suslin tree is a tree T such that $|T| = \kappa$ and every chain and every antichain of T have cardinality $< \kappa$. Suslin trees were introduced by Kurepa (1936c). We first continue with a general discussion of κ-Suslin trees and related concepts. We confine our attention to the case when κ is regular. When κ is singular, κ-Suslin trees exist but are of little interest.

For any regular κ, a κ-tree is a tree T of height κ such that $\forall\alpha < \kappa(|\mathrm{Lev}_{\alpha}(T)| < \kappa)$.

Theorem 3.3.9. *For any regular κ, every κ-Suslin tree is a κ-tree.*

Proof: $\mathrm{Lev}_{\kappa}(T) = \emptyset$, since if $x \in \mathrm{Lev}_{\kappa}(T), \{y : y < x\}$ would be a chain of cardinality κ. Thus, $\mathrm{ht}(T) \leq \kappa$. Since each $\mathrm{Lev}_{\alpha}(T)$ is an antichain, $|\mathrm{Lev}_{\alpha}(T)| < \kappa$. Since $|T| = \kappa$ and $T = \bigcup\{\mathrm{Lev}(T) : \alpha < \mathrm{ht}(T)\}$, regularity of κ implies that $\mathrm{ht}(T) = \kappa$. ■

Theorem 3.3.10 [König's Theorem]. *If T is an ω-tree, then T has an infinite chain.*

Proof: Pick $x_0 \in \mathrm{Lev}_0(T)$ such that $\{y \in T : y \geq x_0\}$ is infinite. This is possible since $\mathrm{Lev}_0(T)$ is finite, T is infinite, and every element of T is $\geq$ some element of $\mathrm{Lev}_0(T)$. By a similar argument, we may inductively pick $x_n \in \mathrm{Lev}_n(T)$ so that for each $n, x_{n+1} > x_n$ and $\{y \in T : y \geq x_{n+1}\}$ is infinite. Then $\{x_n : n \in \omega\}$ is an infinite chain in T. ■

For any regular κ, a κ-Aronszajn tree is a κ-tree such that every chain in T is of cardinality $< \kappa$. Thus, each κ-Suslin tree is a κ-Aronszajn tree, but Theorem 3.3.10 says that there are no ω-Aronszajn trees. At ω_1 the situation is different. The existence of an ω_1-Suslin tree is independent of ZFC, but there is always an ω_1-Aronszajn tree.

Theorem 3.3.11. *There is an ω_1-Aronszajn tree.*

Proof: Let

$$T = \{s \in {}^{<\omega_1}\omega : s \text{ is } 1\text{–}1\} \tag{3.52}$$

Thus, T is a subtree of ${}^{<\omega_1}\omega$. The $\mathrm{ht}(T) = \omega_1$, since for every $\alpha < \omega_1$, there is a 1–1 function from α into ω. If C were an uncountable chain in T, then $\bigcup C$ would be a 1–1 function from ω_1 into ω; thus, every chain in T is countable. Unfortunately, T is not Aronszajn since T is not an ω_1-tree; $\mathrm{Lev}_\alpha(T)$ is uncountable for $\omega \leq \alpha < \omega_1$. However, we can define a subtree of T which is Aronszajn.

If $s, t \in {}^\alpha\omega$, define $s \sim t$ if and only if the set $\{\xi < \alpha : s(\xi) \neq t(\xi)\}$ is finite. We will find s_α for each $\alpha < \omega_1$ such that

(i) $s_\alpha \in {}^\alpha\omega$ and s_α is 1–1.

(ii) $\alpha < \beta \to s_\alpha \sim \beta|\alpha$.

(iii) $\omega \backslash R(s_\alpha)$ is infinite.

Assuming such s_α may be found, let

$$T^* = \bigcup_{s<\omega_1} \{t \in \mathrm{Lev}_\alpha(T) : t \sim s_\alpha\} \tag{3.53}$$

By (ii), T^* is a subtree of T. By (i), each $s_\alpha \in T^*$ so $\mathrm{Lev}_\alpha(T) \neq \emptyset$, for each $\alpha < \omega_1$. Unlike T, T^* is an ω_1-tree since $\{t \in {}^\alpha\omega : t \sim s_\alpha\}$ is countable. Thus, T^* is an ω_1-Aronszajn tree.

We now see how to construct the sequence $\{s_\alpha\}$ by induction. Given s_α, take any $n \in \omega \backslash R(s_\alpha)$ and let $s_{\alpha+1} = s_\alpha \cup \{(\alpha, n)\}$; it is here (iii) is used. Now suppose that we have s_α for $\alpha < \gamma$, where γ is a limit ordinal. Fix α_n for $n < \omega$ so that $\alpha_0 < \alpha_1 < \alpha_2 < \cdots$ and $\bigcup\{\alpha_n : n \in \omega\} = \gamma$. Let $t_0 = s_{\alpha_0}$, and inductively define $t_n : \alpha_n \to \omega$ so that t_n is 1–1, $t_n \sim S_{\alpha_n}$, and $t_{n+1}|\alpha_n = t_n$. Let $t = \bigcup\{t_n : n \in \omega\}$. Then $t \in {}^\gamma\omega$ and t is 1–1; if we set $s_\gamma = t$, then (i) would hold for $\alpha = \gamma$ and $t(\alpha_{2n})$ and $s_\gamma(\xi) = t(\xi)$ for $\xi \notin \{\alpha_n : n \in \omega\}$. Then $\{t(\alpha_{2n+1}) : n \in \omega\} \subset (\omega \backslash R(s_\gamma))$, so (iii) holds as well. ■

A well-pruned κ-tree is a κ-tree T such that $|\mathrm{Lev}_0(T)| = 1$ and

$$\forall x \in T \forall \alpha (\mathrm{ht}(x,T) < \alpha < \kappa \to \exists y \in \mathrm{Lev}_\alpha(T)(x < y)) \tag{3.54}$$

Theorem 3.3.12. *If κ is regular and T is a κ-tree, then T has a well-pruned κ-subtree.*

Proof: Let T^1 be the set of $x \in T$ such that

$$|\{z \in T : z > x\}| = \kappa \tag{3.55}$$

Then T^1 is clearly a subtree of T. To verify Eq. (3.54) for T^1, let us fix $x \in T^1$ and α such that $\mathrm{ht}(x,T) < \alpha < \kappa$. Let $Y = \{y \in \mathrm{Lev}_\alpha(T) : x < y\}$. By definition of T^1 and the fact that each $|\mathrm{Lev}_\beta(T)| < \kappa, \{z \in T : z > x \wedge \mathrm{ht}(z,T) > \alpha\}$ has cardinality

κ, and each element of this set is above some element of Y. Since $|Y| < \kappa$, there is $y \in Y$ such that $|\{z \in T : z > y\}| = \kappa$, and this y is in T^1. A similar argument shows that $\mathrm{Lev}_0(T^1) \notin \emptyset$, so $T^1 \neq \emptyset$.

Now for every $x \in \mathrm{Lev}_0(T^1)$, $\{y \in T^1 : y \geq x\}$ is a well-pruned subtree of T.■

Theorem 3.3.13. *If κ is regular, T is a well-pruned κ-Aronszajn tree, and $x \in T$, then*

$$\forall n \in \omega \exists \alpha > \mathrm{ht}(x,T)(|\{y \in \mathrm{Lev}_\alpha(T) : y > x\}| \geq n) \tag{3.56}$$

Proof: For $n = 2$, this follows from the fact that $\{y : y > x\}$ meets all levels above x and cannot form a chain. For $n > 2$, we proceed by induction. If the theorem holds for n, fix $\alpha > \mathrm{ht}(x,T)$ and distinct $y_1, y_2, \ldots, y_n \in \mathrm{Lev}_\alpha(T)$ with each $y_i > x$. Now let $\beta > \alpha$ be such that there are distinct $z_n, z_{n+1} \in \mathrm{Lev}_\beta(T)$ with $z_n, z_{n+1} > y_n$. For $i < n$, there are $z_i \in \mathrm{Lev}_\beta(T)$ with $z_i > y_i$. Then $\{z_1, z_2, \ldots, z_{n+1}\}$ establishes the theorem for $n + 1$. ■

A topological space (Engelking, 1975) is a pair (X,T) consisting of a set X and a family T of subsets of X satisfying the following conditions:

(i) $\emptyset \in T$ and $X \in T$.

(ii) If $U \in T$ and $V \in T$, then $U \cap V \in T$.

(iii) If $\mathcal{A} \subset T$, then $\bigcup \mathcal{A} \in T$.

The set X is called a space, the elements of X are called points of the space, and the subsets of X belonging to T are called open in the space. The family T is also called a topology on X.

Let $(X, \leq)$ be a total ordering. The order topology T, or the topology induced by the ordering $\leq$, on the set X is the topology generated by the following sets, called intervals: For $a, b \in X$ satisfying $a < b$, let

$$(a,b) = \{x \in X : a < x < b\} \tag{3.57}$$

$$(\leftarrow, a) = \{x \in X : x < a\} \tag{3.58}$$

and

$$(a, \rightarrow) = \{x \in X : x > a\} \tag{3.59}$$

That is, each element in T is either a union on a finite intersection of some sets of the forms in (3.57)–(3.59).

A topological space has ccc, if each collection of pairwise disjoint open sets of the space is countable; it is separable if it has a countable subset A such that each nonempty open set intersects A.

A Suslin line is a total ordering $(X, \leq)$ such that in the order topology X is ccc but not separable. Suslin's hypothesis (SH) states that "there are no Suslin lines." SH arose in an attempt to characterize the order type of the real numbers $(R, <)$. It was well known that any total ordering $(X, \leq)$ satisfying

(a) X has no first and no last elements,

(b) X is continuous, and

(c) X is separable in the order topology

is isomorphic to $(R, <)$ (see Theorems 3.4.5 and 3.4.7). Suslin (1920) asked whether (c) may be replaced by

(c′) X is ccc in the order topology.

Clearly, under SH, (c) and (c') are equivalent, and one can show that if there is a Suslin line, then there is one satisfying (a) and (b). Thus, SH is equivalent to the statement that (a), (b), and (c') characterize the ordering $(R, <)$.

Theorem 3.3.14. *There is an ω_1-Suslin tree if and only if there is a Suslin line.*

Proof: First, let T be an ω_1-Suslin tree. By Theorem 3.3.12, we can assume that T is well-pruned. Let

$$L = \{C \subset T : C \text{ is a maximal chain in } T\} \tag{3.60}$$

If $C \in L$, there is an ordinal $h(C)$ such that C contains exactly one element from $\mathrm{Lev}_\alpha(T)$ for $\alpha < h(C)$ and no elements from $\mathrm{Lev}_\alpha(T)$ for $\alpha \geq h(C)$. Since T is Aronszajn, $h(C) < \omega_1$. Since T is well-pruned, a maximal chain cannot have a largest element, so each $h(C)$ is a limit ordinal. For $\alpha < h(C)$, let $C(\alpha)$ be the element of C on level α.

We order L as follows: Fix an arbitrary total order $<$ of T. If $C, D \in L$, $C \neq D$, let $d(C,D)$ be the least α such that $C(\alpha) \neq D(\alpha)$. Observe that $d(C,D) < \min\{h(C), h(D)\}$. Let $C \lhd D$ if and only if $C(d(C,D)) < D(d(C,D))$. We have thus defined an order on L. It is easily verified that it is indeed a total order of L. We now show that $(L, \lhd)$ is a Suslin line.

First, we show that L has the ccc. Suppose that $\{(C_\xi, D_\xi) : \xi < \omega_1\}$ is a family of disjoint nonempty open intervals. Pick $E_\xi \in (C_\xi, D_\xi)$ and α_ξ so that

$$\max\{d(C_\xi, E_\xi), d(E_\xi, D_\xi)\} < \alpha_\xi < h(E_\xi) \tag{3.61}$$

then $\{E_\xi(\alpha_\xi) : \xi < \omega_1\}$ forms an antichain in T, contradicting that T is a Suslin tree.

To show that L is not separable, it is sufficient to see that for each $\delta < \omega_1$, $\{C : h(C) < \delta\}$ is not dense in L. Fix $x \in \mathrm{Lev}_s(T)$. By Theorem 3.3.13, there exists an $\alpha > \delta$ with three distinct elements, $y, z, w \in \mathrm{Lev}_\alpha(T)$ above x. Let D, E, F be elements of L containing y, z, w, respectively. Say they are ordered $D \lhd E \lhd F$ then (D, F) is a nonempty interval, but since $x \in D \cap F$, (D, F) contains no $C \in L$ with $h(C) < \delta$.

On the other hand, suppose that we are given a Suslin line $(L, \lhd)$. We may assume that L is densely ordered by putting gaps in the set and no nonempty open subset of L is separable. Let $\mathcal{F}$ be the set of all nonempty open intervals of L; thus, elements of $\mathcal{F}$ are of the form (a, b), where a $\lhd$ b. Then $\mathcal{F}$ is partially ordered by reverse inclusion: $I \leq J$ if and only if $I \supset J$. We will define a subset $T \subset \mathcal{F}$ so that $\leq$ is a Suslin tree ordering on T.

To find T, we first find $\mathcal{F}_\mathcal{B} \subset \mathcal{F}$ for each $\beta < \omega_1$ so that for each β,

(1) the elements of $\mathcal{F}_\beta$ are pairwise disjoint,

(2) $\bigcup \mathcal{F}_\beta$ is a dense subset in L, and

(3) if $\alpha < \beta$, $I \in \mathcal{F}_\alpha$ and $\mathcal{J} \in \mathcal{J}_\beta$, then either

  (a) $I \cup \mathcal{J} = \emptyset$, or

  (b) $\mathcal{J} \subset I$ and $I \backslash \mathrm{cl}(\mathcal{J}) \neq \emptyset$

where $\mathrm{cl}(\mathcal{J})$ stands for the closure of $\mathcal{J}$, which is the smallest subset Y whose complement belongs to $\mathcal{F}$.

Assuming that this can be done, we let $T = \bigcup_\beta \mathcal{F}_\beta$. By (1)–(3), T is a tree and each $\mathcal{F}_\beta = \mathrm{Lev}_\beta(T)$. If $A \subset T$ is an antichain, then the elements of A are pairwise disjoint, so $|A| \leq \omega$. T can have no uncountable chains, since if $\{I_\xi : \xi < \omega_1\}$ were such a chain, with $\xi < \eta \to I_\xi \leq I_\eta$, then by (3)(b)

$$\xi < \eta \to (I_\eta \subset I_\xi \wedge I_\xi \backslash \mathrm{cl}(I_\eta) \neq \emptyset) \tag{3.62}$$

so $\{I_\xi \backslash \mathrm{cl}(I_{\xi+1}) : \xi < \omega_1\}$ would contradict the ccc of L. Finally, $|T| = \omega_1$, since (2) implies in particular that each $\mathcal{F}_\beta \neq \emptyset$. Thus, T is a Suslin tree.

We now construct the sets $\mathcal{F}_\beta$ by induction. $\mathcal{F}_0$ is any maximal disjoint subfamily of $\mathcal{F}$; maximality implies $\bigcup \mathcal{F}_0$ is dense. Given $\mathcal{F}_\alpha$, we define $\mathcal{F}_{\alpha+1}$ as follows: For $I \in \mathcal{F}$, let $\mathcal{K}_I$ be a maximal disjoint subfamily of

$$\{K \in \mathcal{F} : K \subset I \wedge I - \mathrm{cl}(K) \neq \emptyset\} \tag{3.63}$$

Let $\mathcal{F}_{\alpha+1} = \bigcup \{\mathcal{K}_I : I \in \mathcal{F}_\alpha\}$.

Finally, we assume γ is a limit ordinal and have defined the $\mathcal{F}_\alpha$ for each $\alpha < \gamma$ satisfying (1)–(3) for $\alpha < \beta < \gamma$. Let

$$\mathcal{K} = \{K \in \mathcal{F} : \forall\alpha < \gamma \forall I \in \mathcal{F}_\alpha [I \cap K = \emptyset \vee (K \subset I \wedge I \backslash \mathrm{cl}(K) \neq \emptyset)]\} \quad (3.64)$$

and let $\mathcal{F}_\gamma$ be a maximal disjoint subfamily of $\mathcal{K}$. Then (1) and (3) now hold for all $\alpha < \beta < \gamma$. For $\beta = \gamma$, (2) says that no $J \in \mathcal{F}$ is disjoint from all members of $\mathcal{F}_\gamma$; this will follow by maximality of $\mathcal{F}_\gamma$ if we can show that for each $J \in \mathcal{F}$, $\exists K \in \mathcal{K}(K \subset J)$. Let E be the set of all left and right endpoints of all intervals in $\bigcup_{\alpha<\gamma} \mathcal{F}_\alpha$. E is countable and J is not separable, so fix $K_1 \in \mathcal{F}$ with $K_1 \subset J$ and $K_1 \cap E = \emptyset$. If $I \in \bigcup_{\alpha<\gamma} \mathcal{F}_\alpha$, then K_1 does not contain the endpoints of I, so $I \cap K_1 = \emptyset$ or $K_1 \subset I$. Now take $K \in \mathcal{F}$ with $K \subset K_1$ and $K_1 \backslash \mathrm{cl}(K) \neq \emptyset$; then $K \subset \mathcal{F}$ and $K \in \mathcal{K}$. ■

Theorem 3.3.15. MA(ω_1) *implies that there are no* ω_1*-Suslin trees.*

Proof: Let $(T, \leq)$ be an ω_1-Suslin tree, and let $P = (T, \geq)$, the reverse order of T. Since T has no uncountable antichains, P has ccc. By Theorem 3.3.12, we may assume that T is well-pruned, in which case $D_\alpha = \{x \in T : \mathrm{ht}(x, T) > \alpha\}$ is dense in P. By MA(ω_1), there is a filter G intersecting each D_α. Then G is an uncountable chain, contradicting the fact that T is a Suslin tree. ■

3.4. Well-Founded Sets

Some questions about sets are relevant to applications of mathematics. We give two of the most important questions. A detailed discussion on the relevance between the questions and applications of mathematics will be given in Chapter 8.

First question: Is there anything which is not a set? We have declared that our axioms of set theory deal with sets — in fact, hereditary sets (Section 3.1). Furthermore, the Axiom of Extensionality has embodied in it the assertion that all things in our domain of discourse are sets. It seems likely that we have not left any interesting mathematics behind by so restricting our universe, since mathematical objects like $\mathbb{R}$, the set of all real numbers, and $\mathbb{C}$, the set of all complex numbers, are hereditary sets and have been defined explicitly within this domain in Section 3.2.

Second question: Is there an "object" x such that $x = \{x\}$? This question is independent of the existence of physical reality, and such a set x would clearly be an hereditary set. Such an object x did not occur in the construction of mathematical objects like $\mathbb{R}$ and $\mathbb{C}$. This is also an important question in systems theory (see Chapters 8 and 12).

In this section, we study the position of the Axiom of Regularity relative to the axioms in axiomatic set theory. We work in the axiom system of all axioms in ZFC except Regularity and define the class **WF** of well-founded sets by starting with $\emptyset$ and iterating the power set operation. Then we show that **WF** is closed under the other set-theoretic operations as well.

Let us define $R(\alpha)$ for each $\alpha \in \mathbf{ON}$ by transfinite recursion as follows:

(a) $R(0) = \emptyset$.

(b) $R(\alpha+1) = p(R(\alpha))$.

(c) $R(\alpha) = \bigcup_{\xi<\alpha} R(\xi)$ when α is a limit ordinal.

A set A is well-founded if there is an $\alpha \in \mathbf{ON}$ such that $A \in R(\alpha)$. Let **WF** be the class of all well-founded sets, i.e., $\mathbf{WF} = \bigcup\{R(\alpha) : \alpha \in \mathbf{ON}\}$.

Theorem 3.4.1. *For each ordinal α:*

(a) $R(\alpha)$ is transitive.

(b) $\forall \xi \leq \alpha (R(\xi) \leq R(\alpha))$.

Proof: We use transfinite induction on α: We assume that the result holds for all $\beta < \alpha$ and prove it for α.

Case 1. $\alpha = 0$. the result is trivial.

Case 2. α is a limit. Statement (b) is immediate from the definition; (a) follows from the fact that the union of transitive sets is transitive.

Case 3. $\alpha = \beta + 1$. Since $R(\beta)$ is transitive, $p(R(\beta)) = R(\alpha)$ is transitive and $R(\beta) \subset R(\alpha)$. This establishes (a) and (b) for all ordinals α. ■

Let $x \in \mathbf{WF}$, the least ordinal α with $x \in R(\alpha)$ must be a successor ordinal. The rank of x, denoted by rank(x), is the least β such that $x \in R(\beta+1)$. Hence, if $\beta = \text{rank}(x)$, then $x \subset R(\beta)$, and $x \notin R(\beta)$, and $x \notin R(\alpha)$ for all $\alpha > \beta$.

Theorem 3.4.2. *For each ordinal α,*

$$R(\alpha) = \{x \in \mathbf{WF} : \text{rank}(x) < \alpha\}$$

Proof: For $x \in \mathbf{WF}$, rank$(x) < \alpha$ if and only if $\exists \beta < \alpha (x \in R(\beta+1))$ if and only if $x \in R(\alpha)$. ■

Theorem 3.4.3. *If* $y \in \mathbf{WF}$, *then*

(a) $\forall x \in y(x \in \mathbf{WF} \wedge \operatorname{rank}(x) < \operatorname{rank}(y))$.

(b) $\operatorname{rank}(y) = \bigcup\{\operatorname{rank}(x) + 1 : x \in y\}$.

Proof: For (a), let $\alpha = \operatorname{rank}(y)$, then $y \in R(\alpha + 1) = p(R(\alpha))$. If $x \in y$, $x \in R(\alpha)$, so $\operatorname{rank}(x) < \alpha$.

For (b), let $\alpha = \bigcup\{\operatorname{rank}(x) + 1 : x \in y\}$. By (a) $\alpha \leq \operatorname{rank}(y)$. Furthermore, each $x \in y$ has rank $< \alpha$, so $y \subset R(\alpha)$. Thus, $y \in R(\alpha + 1)$, so $\operatorname{rank}(y) \leq \alpha$. ■

Theorem 3.4.3 says that the class **WF** is transitive and that we may think of the elements $y \in \mathbf{WF}$ as being constructed, by transfinite recursion, from well-founded sets of smaller rank. Therefore, the class **WF** excludes sets which are built up from themselves. More formally, there is no $x \in \mathbf{WF}$ such that $x \in x$, since from this we would have $\operatorname{rank}(x) < \operatorname{rank}(x)$. Likewise, the class **WF** excludes circularities like $x \in y \wedge y \in x$, since this would yield $\operatorname{rank}(x) < \operatorname{rank}(y) < \operatorname{rank}(x)$. It is clear that the following result holds.

Theorem 3.4.4. *Each ordinal is in* **WF** *and its rank is itself.*

Proof: What we need to show is that $\forall \alpha \in \mathbf{ON}(\alpha \in \mathbf{WF} \wedge \operatorname{rank}(\alpha) = \alpha)$. We prove this by transfinite induction on α. Assume that the theorem holds for all $\beta < \alpha$. Then, for $\beta < \alpha$, $\beta \in R(\beta + 1) \subset R(\alpha)$, so $\alpha \subset R(\alpha)$, so $\alpha \in R(\alpha + 1)$. By Theorem 3.4.3(b), $\operatorname{rank}(\alpha) = \bigcup\{\beta + 1 : \beta < \alpha\} = \alpha$. Thus the theorem holds for all ordinals α. ■

Theorem 3.4.5. $\forall \alpha \in \mathbf{ON}(R(\alpha) \cap \mathbf{ON} = \alpha)$.

The proof follows from Theorems 3.4.4 and 3.4.2.

Theorem 3.4.6.

(a) If $x \in \mathbf{WF}$, *then* $\bigcup x$, $p(x)$, *and* $\{x\} \in \mathbf{WF}$, *and the rank of these sets is less than* $\operatorname{rank}(x) + \omega$.

(b) If $x,y \in \mathbf{WF}$, *then* $x \times y$, $x \cup y$, $x \cap y$, $\{x,y\}$, *and* ${}^{y}x$ *are all in* **WF**, *and the rank of each of these sets is less than* $\max\{\operatorname{rank}(x), \operatorname{rank}(y)\} + \omega$.

Proof: For (a), let $\alpha = \operatorname{rank}(x)$. Then $x \subset R(\alpha)$, so $p(x) \subset p(R(\alpha)) = R(\alpha + 1)$, so $p(x) \in R(\alpha + 2)$. Similarly, $\{x\} \in R(\alpha + 2)$ and $\bigcup x \in R(\alpha + 1)$.

For (b), let $\alpha = \max\{\operatorname{rank}(x), \operatorname{rank}(y)\}$. As in (a), we show, e.g., $\{x,y\} \in R(\alpha + 2)$, $(x,y) \in R(\alpha + 3)$. Any ordered pair of elements of $x \cup y$ is in $R(\alpha + 3)$, so ${}^{y}x \subset R(\alpha + 3)$, so ${}^{y}x \in R(\alpha + 4)$. We leave the rest of the details for the reader. ■

Theorem 3.4.7. *The sets* $\mathbb{Z}$, $\mathbb{Q}$, $\mathbb{R}$, *and* $\mathbb{C}$ *are all in* $R(\omega+\omega)$.

The proof follows from Theorem 3.4.6 and the definitions of these sets in Section 3.2.

Theorem 3.4.8.

(a) $\forall n \in \omega(|R(n)| < \omega)$.

(b) $|R(\omega)| = \omega$.

Proof: (a) is easy by induction on n. For (b), since $\omega \subset R(\omega)$, it is sufficient to see that $R(\omega)$ is countable. To show this by induction, for each n we may identify $R(n+1)$ with ${}^{R(n)}2$ and order it lexicographically. Then the ordered sum of the sets $R(n)$, $n \in \omega$, has order type ω. ■

The cardinalities of the sets $R(\alpha)$ increase exponentially: $|R(\omega)| = \omega$, $|R(\omega+1)| = 2^{\omega}$, $|R(\omega+2)| = 2^{2^{\omega}}$, etc. Generally, let a_α be defined by transfinite recursion on α as follows:

(1) $a_0 = \omega$.

(2) $a_{\alpha+1} = 2^{a_\alpha}$.

(3) For a limit γ, $a_\gamma = \bigcup\{a_\alpha : \alpha < \gamma\}$.

Then we have the following theorem.

Theorem 3.4.9. $|R(\omega+2)| = a_\alpha$.

The proof follows from transfinite induction on α.

The following theorem is an example of showing that all mathematics takes place in the class **WF**.

Theorem 3.4.10.

(1) Every group is isomorphic to a group in **WF**.

(2) Every topological space is homeomorphic to a topological space in **WF**.

Proof: A group is an ordered pair $(G, \times)$ where $\times : G \times G \to G$. Theorem 3.4.6 therefore implies that

$$(G, \times) \in \mathbf{WF} \leftrightarrow G \in \mathbf{WF} \leftrightarrow G \subset \mathbf{WF} \tag{3.65}$$

If $(G, \times)$ is any group, let $\alpha = |G|$ and let f be a 1–1 mapping from α onto G. Define an operation $\circ$ on α by $\xi \circ \eta = f^{-1}(f(\xi) \times f(\eta))$. Then $(\alpha, \circ)$ is a group isomorphic to $(G, \times)$. The same idea can be used to prove (b). We leave the proof of (b) to the reader. ■

Generalizing the notion of well-orderings, we have the following: A relation H is well-founded on a set A if and only if

$$\forall X \subset A[X \neq \emptyset \rightarrow \exists y \in X(\neg \exists z \in X(zHy))] \tag{3.66}$$

The element y in Eq. (3.66) is called H-minimal in X. Thus, a relation H is well-founded on A if and only if every nonempty subset of A has an H-minimal element.

Theorem 3.4.11. *If $A \in$ **WF**, $\in$ is well-founded on A.*

Proof: Let X be a nonempty subset of A. Let $\alpha = \min\{\text{rank}(y) : Y \in X\}$, and pick a $y \in X$ with $\text{rank}(y) = \alpha$. Then y is $\in$-minimal in X by Theorem 3.4.3(a). ■

Theorem 3.4.12. *If A is transitive and $\in$ is well-founded on A, then $A \in$ **WF**.*

Proof: It is sufficient to show $A \subset \mathbf{WF}$, because if $A \subset \mathbf{WF}$, let $\alpha = \bigcup\{\text{rank}(y) + 1 : y \in x\}$; then $x \in R(\alpha)$, so $x \in R(\alpha + 1) \subset \mathbf{WF}$. If $A \not\subset \mathbf{WF}$, let $X = A \backslash \mathbf{WF} \neq \emptyset$ and let y be the $\in$-minimal in X. If $z \in y$, then $z \notin X$, but $z \in A$ since A is transitive; so $z \in \mathbf{WF}$. Thus, $y \subset \mathbf{WF}$, so $y \in \mathbf{WF}$. This contradicts the definition that $y \in A \backslash \mathbf{WF}$. ■

Let A be a set. By recursion on n, we define

(1) $\bigcup^0 A = A$.

(2) $\bigcup^{n+1} A = \bigcup(\bigcup^n A)$.

(3) $\text{tr cl}(A) = \bigcup\{\bigcup^n A : n \in \omega\}$.

Here tr cl is the "transitive closure," which is the least transitive set containing A as a subset.

Theorem 3.4.13.

(a) $A \subset \text{tr cl}(A)$.

(b) $\text{tr cl}(A)$ *is transitive.*

(c) If $A \subset T$ and T is transitive, then $\text{tr cl}(A) \subset T$.

(d) If A is transitive, then $\text{tr cl}(A) = A$.

(e) If $x \in A$, *then* $\text{tr cl}(x) \subset \text{tr cl}(A)$.

(f) $\text{tr cl}(A) = A \cup \bigcup\{\text{tr cl}(x) : x \in A\}$.

The proof is straightforward from the definition of transitive closure.

Theorem 3.4.14. *For any set A the following are equivalent:*

(a) $A \in \mathbf{WF}$.

(b) $\text{tr cl}(A) \in \mathbf{WF}$.

(c) $\in$ *is well-founded on* $\text{tr cl}(A)$.

Proof: (a) $\to$ (b). If $A \in \mathbf{WF}$, then by induction on n, $\bigcup^n A \in \mathbf{WF}$ since **WF** is closed under the union operation (Theorem 3.4.6). Thus, each $\bigcup^n A \subset \mathbf{WF}$, so $\text{tr cl}(A) \subset \mathbf{WF}$, so $\text{tr cl}(A) \in \mathbf{WF}$.

The proof of (b) $\to$ (c) is Theorem 3.4.11.

(c) $\to$ (a). By (c) and Theorem 3.4.12, $\text{tr cl}(A) \in \mathbf{WF}$, so $A \subset \text{tr cl}(A) \subset \mathbf{WF}$, so $A \in \mathbf{WF}$. ■

Let V be the class defined by

$$V = \{x : x = x\} \tag{3.67}$$

In other words, V is the class of all sets. If we are convinced by the previous theorems that all mathematics takes place in **WF**, it is reasonable to adopt as an axiom the statement that $V = \mathbf{WF}$. This does not mean that the two classes V and **WF** are really identical, but only that we restrict our domain of discourse to be just **WF**. In the following, we discuss some of the consequences of adding the axiom $V = \mathbf{WF}$ to the axiom system of ZFC axioms except Regularity.

Theorem 3.4.15. *The following are equivalent:*

(a) The axiom of regularity.

(b) $\forall A(\in$ *is well-founded on* $A)$.

(c) $V = \mathbf{WF}$.

Proof: (a) $\leftrightarrow$ (b) is immediate from the definition of well-founded relations. For (b) $\to$ (c), (b) implies that for any A, $\in$ is well-founded on $\text{tr cl}(A)$, so $A \in \mathbf{WF}$. For (c) $\to$ (b), apply Theorem 3.4.11. ■

Unlike the other axioms of ZFC, Regularity has no application in ordinary mathematics, since accepting it is equivalent to restricting our attention to **WF**, where all mathematics takes place. In Chapters 5 and Chapter 9, we will see how the Axiom of Regularity can be used to study level structures of general systems.

Since Regularity is equivalent to $V = \mathbf{WF} = \bigcup\{R(\alpha) : \alpha \in \mathbf{ON}\}$, it gives us a picture of all sets being created by an iterative process, starting from nothing.

Theorem 3.4.16. *A is an ordinal if and only if A is transitive and totally ordered by $\in$.*

Proof: To see that $\in$ well-orders A, let X be any nonempty subset of A. Then Regularity implies there exists an $\in$-minimal element in X. ■

3.5. References for Further Study

There are many good books in the area of axiomatic set theory, such as (Drake, 1974; Engelking, 1975; Fraenkel et al., 1973; Jech, 1978; Kleene, 1952; Kreisel and Krivine, 1967; Kunen, 1980; Kuratowski and Mostowski, 1976; Kurepa, 1936c; Martin and Solovay, 1970; Quigley, 1970; Rudin, 1977; Shoenfield, 1975; Suslin, 1920).

CHAPTER 4

Centralizability and Tests of Applications

The concept of centralized systems is written in the language of set theory in order to take advantage of the rigorous mathematical reasoning. The concept of centralizable systems is introduced. Applications of the concept in sociology, concerning the existence of factions in human society, public issues of contention, and importance level of problems versus media, are listed. Two real-life examples are given to illustrate the results obtained in this chapter. At the end some open questions are posed.

4.1. Introduction

Hall and Fagen (1956) introduced the concept of centralized systems, where a centralized system is a system in which one object or a subsystem plays a dominant role in the system operation. The leading part can be thought of as the center of the system, since a small change in it would affect the entire system, causing considerable changes. Lin (1988a) applied the concept of centralized systems to the study of some phenomena in sociology. Several interesting results were obtained, including the argument on why there must be a few people in each community who dominate others. In this chapter we look at the overall study of the concept of centralized systems, related concepts, and some applications.

In the rest of this section, all necessary concepts of systems theory and terminology of set theory will be listed in order for the reader to follow the discussions.

A system is an ordered pair of sets, $S = (M,R)$, such that M is the set of all objects of S, and R is a set of some relations defined on M. The sets M and R are called the object set and the relation set of S, respectively. Here, let $r \in R$ be a relation of S. Then r is defined as follows: There exists an ordinal number $n = n(r)$, a function of r, called the length of the relation r, such that $r \subseteq M^n$, where

$$M^n = \underbrace{M \times M \times \cdots \times M}_{n \text{ times}}$$
$$= \{f : n \to M \text{ is a mapping}\}$$

is the Cartesian product of n copies of M.

A system $S = (M,R)$ is trivial (resp., discrete) if $M = R = \emptyset$ (resp., $M \neq \emptyset$ and either $R = \emptyset$ or $R = \{\emptyset\}$). Given two systems $S_i = (M_i,R_i)$, $i = 1,2$, S_1 is a partial system of S_2 if either (1) $M_1 = M_2$ and $R_1 \subseteq R_2$ or (2) $M_1 \underset{\neq}{\subset} M_2$ and there exists a subset $R' \subseteq R_2$ such that $R_1 = R'|M_1 \equiv \{f : f$ is a relation on M_1 and there exists $g \in R'$ such that f is the restriction of g on $M_1\}$. In symbols, $R'|M_1 = \{f : g \in R'(f = g|M_1)\}$, where $g|M_1 \equiv g \cap M_1^{n(g)}$. The system S_1 is a subsystem of S_2 if $M_1 \subseteq M_2$ and for each relation $r_1 \in R_1$ there exists a relation $r_2 \in R_2$ such that $r_1 \subseteq r_2|M_1$.

A system $S = (M,R)$ has n levels, where n is a fixed whole number, if (1) each object $S_1 = (M_1,R_1)$ in M is a system, called the first-level object system; (2) if $S_{n-1} = (M_{n-1},R_{n-1})$ is an $(n-1)$th-level object system, then each object $S_n = (M_n,R_n) \in M_{n-1}$ is a system, called the nth-level object system of S.

Let $S_i = (M_i,R_i)$, $i = 1,2$, be two systems and $h : M_1 \to M_2$ a mapping. By transfinite induction, two classes $\widehat{M}_i$, $i = 1,2$, and a class mapping $\widehat{h} : \widehat{M}_1 \to \widehat{M}_2$ can be defined with the properties

$$\widehat{M_i} = \bigcup_{n \in \text{Ord}} M_i^n, \quad i = 1,2$$

and for each $x = (x_0, \ldots, x_\alpha, \ldots) \in \widehat{M}_1$,

$$\widehat{h}(x) = (h(x_0), \ldots, h(x_\alpha), \ldots) \in \widehat{M_2}$$

where Ord is the class of all ordinals. For each relation $r \in R_1$, $\widehat{h}(r) = \{\widehat{h}(x) : x \in r\}$ is a relation on M_2 with length $n(r)$. Without confusion, h will be used to indicate the class mapping $\widehat{h}$, and h is a mapping from the system S_1 into the system S_2, denoted by $h : S_1 \to S_2$. When $h : M_1 \to M_2$ is surjective, injective, or bijective, the mapping $h : S_1 \to S_2$ is also surjective, injective, or bijective, respectively.

The systems S_i are similar if there exists a bijection $h : S_1 \to S_2$ such that $h(R_1) = \{h(r) : r \in R_1\} = R_2$. The mapping h is called a similarity mapping from S_1 onto S_2. Evidently, if h is a similarity mapping from S_1 onto S_2, the inverse h^{-1} is a similarity mapping from S_2 onto S_1. A mapping $h : S_1 \to S_2$ is termed a homomorphism from S_1 into S_2, if $h(R_1) \subseteq R_2$.

Note that a theorem followed by (ZFC) means that the theorem is true only under the assumption that the ZFC axioms are true, where ZFC stands for the Zermelo–Fränkel system with the axiom of choice. For set X, $|X|$ denotes the cardinality of X. A cardinal number $\alpha = \aleph_\lambda$ is called regular if for any sequence of ordinal numbers $\{\alpha_\beta < \aleph_\lambda : \beta < \lambda\}$, the limit of the sequence is less than $\aleph_\lambda$. For details and other terminology of set theory, see (Kuratowski and Mostowski, 1976).

4.2. Centralized Systems and Centralizability

To use the mathematical reasoning to study centralized systems, Lin (submitted) redefined the concept in the language of set theory as follows.

Definition 4.2.1. A system $S = (M, R)$ is called a centralized system if each object in S is a system and there exists a nontrivial system $C = (M_C, R_C)$ such that for any distinct elements x and $y \in M$, say $x = (M_x, R_x)$ and $y = (M_y, R_y)$, then $M_C = M_x \cap M_y$ and $R_C \subseteq R_x|M_C \cap R_y|M_C$; the system C is called a center of S.

Theorem 4.2.1 [ZFC (Lin and Ma, 1993)]. *Let κ be an arbitrary infinite cardinality and $\theta > \kappa$ a regular cardinality such that for any ordinal number $\alpha < \theta, |\{\{f : \lambda \to \alpha\} : \lambda < \kappa\}| < \theta$. Assume that $S = (M, R)$ is a system satisfying $|M| \geq \theta$ and each object $m \in M$ is a system with $m = (M_m, R_m)$ and $|M_m| < \kappa$. If there exists an object contained in at least θ objects in M, there then exists a partial system $S' = (M', R')$ of S such that S' forms a centralized system and $|M'| \geq \theta$.*

This result is a restatement of the well-known Δ-lemma in axiomatic set theory (Kunen, 1980). The following question, posed by Dr. Robert Beaudoin, is still open.

Question 4.2.1. A system $S = (M, R)$ is strongly centralized if each object in S is a system and there is a nondiscrete system $C = (M_C, R_C)$ such that for any distinct elements x and $y \in M$, say $x = (M_x, R_x)$ and $y = (M_y, R_y), M_C = M_x \cap M_y$ and $R_C = R_x|M_C \cap R_y|M_C$. Give conditions under which a given system has a partial system which is strongly centralized.

A system S_0 is n-level homomorphic to a system A, where n is a fixed natural number, if there exists a mapping $h_{S_0} : S_0 \to A$, called an n-level homomorphism, satisfying the following:

(1) The systems S_0 and A have no nonsystem kth-level objects, for each $k < n$.

(2) For each object S_1 in S_0, there exists a homomorphism h_{S_1} from the object system S_1 into the object system $h_{S_0}(S_1)$.

(3) For each $i < n$ and each ith-level object S_i of S_0, there exist level object systems S_k, for $k = 0, 1, \ldots, i-1$, and homomorphisms h_{S_k}, $k = 1, 2, \ldots, i$, such that S_k is an object of the object system S_{k-1} and h_{S_k} is a homomorphism from S_k into $h_{S_{k-1}}(S_k)$, for $k = 1, 2, \ldots, i$.

Definition 4.2.2 [Lin et al. (1998)]. A system S is centralizable if it is 1-level homomorphic to a centralized system S_C under a homomorphism $h : S \to S_C$ such that for each object m in S the object systems m and $h(m)$ are similar. Each center of S_C is also called a center of S.

Theorem 4.2.2 [Lin et al. (1998)]. *A system $S = (M,R)$ with two levels is centralizable if and only if there exists a nontrivial system $C = (M_C, R_C)$ such that C is embeddable in each object of S — that is, C is similar to a partial system of S_1.*

Similar to the existence theorem of centralized systems (i.e., Theorem 4.2.1), we have the following result.

Theorem 4.2.3 [Lin et al. (1998)]. *Let κ and θ be cardinalities satisfying the conditions in Theorem 4.2.1. Assume that S is a system with an object set of cardinality $\geq \theta$ and each object in S is a nontrivial system with object set of cardinality $< \kappa$. There then exists a partial system S' of S such that the object set of S' is of cardinality $\geq \theta$ and S' forms a centralizable system.*

For detailed proofs of these theorems, please refer to Chapter 9.

4.3. Several Tests of Applications

In this section we show how the concept of centralizability has been applied in several research areas.

4.3.1. A Societal Phenomenon

Sociology concerns the environment in which we live and the problems encountered in human society. In (Lin, 1988a) the following social phenomenon was studied in the light of centralizability: the dominance of one person or a small group of persons over their family or neighbors. For example, in a relationship between two individuals, one of them must dominate the other. This is true in relationships between husband and wife, brothers, sisters, or even friends. To make any progress, let us first write a special case of Theorem 4.2.1.

Theorem 4.3.1 [ZFC]. *Suppose that $S = (M,R)$ is a system such that $|M| \geq c$, where c is the cardinality of the set of all real numbers and that each object in S is a system with finite object set. If there exists such an element that belongs to at least c objects in M, there then exists a partial system B of S with an object set of cardinality $\geq c$, and B forms a centralized system.*

Suppose that A is a collection of people and $[A]^{<\omega}$ is the collection of all finite subcollections of people in A. Then for any finite subcollection $x \in [A]^{<\omega}$, three possibilities exist:

(1) There exists exactly one relation among the people in x.

(2) There exists more than one relation among the people in x.

(3) There does not exist any relation among the people in x.

If situation 1 occurs, we construct a system $S_x = (x, R_x)$ as follows: in the relation set R_x there exists only one element which describes the relation among the people in the finite set x. If situation 2 occurs, we construct a collection of systems $\{S_x^i : i \in I_x\}$, where I_x is an index set that may depend on the finite subcollection x such that for any $i \in I_x$ the relation set of the system S_x^i has cardinality 1 and for any fixed relation f among the people in x there exists exactly one $i \in I_x$ such that $S_x^i = (x, \{f\})$. Finally, if situation 3 occurs, we construct a system $S_x = (x, \emptyset)$, where the relation set is empty.

We now consider the system $(M, \emptyset)$ with empty relation set, where $M = \{S_x : x \in [A]^{<\omega}$, there exists at most one relation among the people in $x\} \cup \{S_x^i : i \in I_x, x \in [A]^{<\omega}$, there exists more than one relation among the people in $x\}$. Theorem 4.3.1 says that if (1) the cardinality $|M|$ is greater than or equal to c and (2) there exists at least one person who is an object in at least c elements in M, then there exists a subcollection $M^* \subseteq M$ such that $(M^*, \emptyset)$ forms a centralized system and $|M^*| \geq c$.

Based upon the construction of the system $(M, \emptyset)$, it follows that condition (1) means that there exists a complicated network of relations among the people in A. Condition (2) implies that there exists at least one person in A who has enough relations with the other people in A. However, when we study problems with people, how can we enumerate the relations that exist among these people? The difficulty comes from the notion of relation. To apply Theorem 4.3.1 correctly, we must rewrite each relation between people in the language of set theory. That is, each relation must be written in such a way that it consists of the following basic blocks only: "x is a set," "$x \in y$," "$x = y$," "and," "or," "if ... then," "if and only if." For details see (Kuratowski and Mostowski, 1976). For the time being, the question is still open.

However, no matter how complicated the relations among the people in a real community can be, Theorem 4.3.1 says that a person or a subcollection of people $x \in [A]^{<\omega}$ who want to be a center in the collection A must possess the following background: (3) the people in x have many transverse and longitudinal relations with the others in A. Thus, if we suppose that in any social society there are uncountably many relations in the sense of set theory among the people, then from the point of view of centralizability, we can understand why there always exist factions in human society, namely because there is more than one center.

4.3.2. Public Issues of Contention

If history is seen as a moving picture, each moment of the moving picture shows a few public issues attracting major public attention. To prove this point, Vierthaler

(1993) used the *Reader's Guide* to investigate shifts and changes in American public concern for social problems. The research uses data available in the series of 52 volumes (1990–92) to consider historic variations in the series of volumes. Then Dr. Vierthaler looked at the topics identified as special "social problems" in the 92-year series to analyze the degree and types of "connectiveness" that exist among the specific social problems as a way to study changes in "American social problems consciousness." Besides the data supporting the observation that at any moment in history some public issues seem to attract more attention than others, in (Lin and Vierthaler, 1998) a systems theoretical evidence was given to show that this observation is a historical truth. In this subsection we briefly describe how the systems theoretical evidence was developed.

In the discussion of the previous subsection, we replace possibilities in 1–3 by the following:

(1′) There exists exactly one social issue connnecting the people in x.

(2′) There exists more than one social issue connnecting the people in x.

(3′) There does not exist any social issue connnecting the people in x.

where a social issue connecting the people in x means that the people in x are either affected by or attracted to a social issue. Now we consider the system $(M, \emptyset)$ accordingly, and Theorem 4.3.1 will be applied. Note that when we study the collection of all social issues connecting the people at a fixed historical moment, there is no straightforward method to enumerate all social issues, both minor and major, because there are so many that we do not know how to enumerate them. As a result, social scientists use several methods to obtain indicators. Most recognized are social survey polls that repeat the question from 1935–1971, asking respondents: "What is the most important problem facing our country?" Other researchers use documents, such as the *Readers' Guide*, a record of popular magazine articles about "social problems" from 1900–1991, or further original sources (Krippendorff, 1980) such as front page newspaper stories and the study of other texts over time to identify public issues of importance. So, a consequence of Theorem 4.3.1 is that because there always exist opposite central social issues, which divide the public, there will always exist opposite political parties in human society. In addition, different "positions" are possible on the same issue. hence, the basis to unify or divide depends on peoples' interest and their ideas or positions toward how to best resolve a social issue. Thus, a public issue is a matter of importance to the public precisely because it addresses a matter of concern that is controversial — whether over the goals or the means considered plausible to attain the goal.

4.3.3. Public Issues: Importance Level of Problems versus Media

One school's opinion says that there are "real" social issues at any given time in history which derive from major problems in the social structure that affects people, thus attracting public attention, and that the media just reports to the public truthfully about all aspects of these social issues. A different school's opinion says that there is no such thing as major social issues at each given moment in history which attract public attention, because the mass media manufactures issues by exaggerated propaganda or by accentuating extreme ideological positions. For details, see (Lauer, 1976; Spector and Kitsuse, 1977).

In (Lin and Vierthaler, 1998), the concept of systems was used to show that these two opposite theories are actually the two opposite sides of reality. They are unified in the name of systems. Let us see how to achieve this end.

If reality is considered as truth, we then need to understand the meaning of truth first. In general, "truth" is a fuzzy concept. For example, the laws of nature are the truths about the world. A mathematical truth is a true statement in mathematics. Then what should be the general meaning of truth? Klir (1985) introduced a philosophical definition of systems. He says that a system is what is distinguished as a system. Based upon this understanding, it can be seen that for any system $S=(M,R)$ and each relation $r \in R$, r is a true relation in S connecting some objects in the set M. So, r can be an S-truth. From Russell's paradox (Kuratowski and Mostowski, 1976), we have the following theorem.

Theorem 4.3.2 [Lin and Ma (1993)]. *There is no system whose object set consists of all systems.*

This result means that we cannot consider a relation which is true with respect to all systems as its objects, and that for each given statement, if it is a truth, it then implies that there exists a system in which the given statement describes a relation. One interesting consequence of this understanding explains why there always exist different and opposite interpretations about each social issue. The reason is because for a given social issue, there always exist systems S_1 and S_2 such that the social issue is looked at differently in S_1 than in S_2. That is, the concept of "reality" is relative. The concept of truth was first studied in the name of systems in (Lin, 1990b).

Let (M,R) be a system such that the object set M is defined as in Section 4.3.2 and R is a set of relations between the object systems in M. It is then reasonable to believe that this system describes all aspects or all "realities" about the collection of all social issues existing at a fixed moment of time. Then the fact that (M,R) is a system implies that M is a pile of isolated social issues and the relation set R describes how the historical moment (M,R) can be studied as an entity of continuity, connection, and movement. As pointed out in (Lin, 1990b), Klir's definition of systems means the following.

Axiom 4.3.1. That a system A possesses a property B means that there exists a system C, different from system A, such that C recognizes that A possesses property B.

This axiom implies that consciousness and matter coexist. That is, the importance level of social issues catches the media's interest; in turn, the media makes the issues more accessible to the public and so more public attention is created. Another interesting consequence of this axiom is that the media could do whatever possible to promote the importance level of a social issue. However, if the issue does not have many transverse and longitudinal connections with other social issues, no matter how hard the media tries, there will be a "plateau" over which the importance level or the level of public attention will never go. This might be the reason why during the 1992 U.S. presidential campaign, the percentage of supporters for then-president George Bush rose dramatically during the final days and then began to stall. Consequently, public support for Bush never reached a high level nor went beyond the observed level, which was lower than desired by that side of campaign.

In the next two sections, two examples are used to elucidate the results obtained in this chapter.

4.4. Growth of the Polish School of Mathematics

A good book about the developing history of the Polish School of Mathematics is Kuratowski (1980), which is the source of the story used here. Briefly, in 1911 there were four distinguished professors of mathematics in the only two Polish universities: J. Puzyna, W. Sierpinski, S. Zaremba, and K. Zorawski. They had no common interests in the field of mathematics, since each worked in a different field: Puzyna in analytic functions, Sierpinski in number theory and set theory, Zaremba in differential equations, and Zorawski in differential geometry. As a consequence of the discussion in the preceding sections, they could not have students in common because there did not exist enough relations between them; this means that there was no possibility of organizing a mathematical school.

For that to happen, there had to be a common set of problems on which a number of people were engaged so that the people would be working as a whole community. At the same time, no matter how small the group of people was at the beginning, it had to contain sufficiently bright and active people to attract the interest of other mathematical adepts by its topics; that is, the people in the group should be active enough to set up relations with other mathematicians outside.

A few years later, two exceptionally talented and creative young mathematicians, Z. Janiszewski and S. Mazurkiewicz, were working in almost the same field, topology. Janiszewski received his doctor's degree in Paris in 1912, Mazurkiewicz in Lwow in the same year.

In early 1918, some people spoke about a fairly strong Warsaw center of set theory, topology, and their applications under the direction of Professors Janiszewski, Mazurkiewicz, and Sierpinski. Since the arrival of Janiszewski and Mazurkiewicz, a community of the three mathematicians with common interests had been formed. The development of the hard-working community of mathematicians linked by common scientific interests was one of the essential factors leading to the creation of the Polish School of Mathematics. In the next two years the school was created and developed exactly as described by Janiszewski in the keynote address of the school.

In 1918, in the first volume of a publication titles *Polish Science, Its Needs, Organization and Development*, Janiszewski published the article, "On the needs of mathematics in Poland." He began with the assumption that Polish mathematicians could not afford "to be just the recipients or customers of foreign centers" but "to win an individual position for Polish mathematics." One of the principal means suggested by him for attaining that end was the concentration of scientific staff in a relatively narrow field of mathematics but one in which Polish mathematicians had common interests and — what was more important — one in which they had achievements that counted on a world scale. Though mathematicians need no laboratories or expensive and sophisticated auxiliary equipment for their work, they do need a proper atmosphere; this proper atmosphere can be created only by the cultivation of common topics. For research workers, collaborators are almost indispensable, for in isolation they will in most cases be lost. The causes are not only psychological, such as lack of incentive, but an isolated researcher knows much less than those who work as a team. Only the results of research, the finished ripe ideas can reach an isolated researcher and then only when they appear in print, often several years after their conception. The isolated researcher does not know how and when they have been obtained, is far from those forges or melting pots where mathematics is produced, comes late, and must inevitably lag behind.

Because of the concentration of scientific staff, Janiszewski suggested the establishment of a periodical devoted exclusively to the fields of mathematics connected with set theory and foundations of mathematics. Such a periodical, if published in a language known abroad, would serve a double goal: It would present the achievements of Polish mathematicians to the world of learning, while attracting foreign authors with similar interests.

Such a journal, entitled *Fundamenta Mathematicae*, was founded, and the first volume appeared in 1920. That day was thought to be the inauguration of the Polish School of Mathematics. Although *Fundamenta* was conceived as an international journal, the first volume deliberately contained papers only by Polish authors. It was something of an introduction to the world of the newly risen school of mathematics.

Fundamenta was a journal limited to just one field of mathematics. Publication in languages known abroad made the accomplishments of Polish mathematicians accessible to the scientific world on a large scale, and it was an indispensable condition for attracting the works of foreign mathematicians for publication in the journal and thus giving it an international character. In contrast to the prevalent practice today, where there exist strictly specialized mathematical journals, in those days it was a novelty and met with undisguised skepticism on the part of many mathematicians.

The part played by *Fundamenta* in the development of fields represented in it is not limited to the publication of papers. A considerable role has also been played by the problems sections, which have been maintained since the first volume. Some problems became classics; others resulted in the appearance of many valuable papers. Proof of this may be found by inspecting the new edition of the first volume, which appeared in 1937. The editors had hit upon the very happy idea of presenting in a new edition the actual state of problems discussed in the first edition of that volume (i.e., 17 years earlier), both in individual papers and in the problem section. The purpose of recalling the original problems was used to show the world of learning that *Fundamenta* was a journal that published important problems.

In 1935, *Fundamenta* celebrated its twenty-fifth volume. A jubilee volume of twice the normal size was published, and the most distinguished scholars working in the fields represented in the journal were invited to publish their works in it. Thus, the papers in the jubilee volume showed the world that *Fundamenta* was a prestigious journal of the fields represented in it and showed the position of the Polish School of Mathematics.

According to Tamarkin (Tamarkin, 1936), under the masterful guidance of its editors, *Fundamenta Mathematicae* immediately developed into a unique periodical that attracted international recognition and cooperation and whose history became the history of development of the modern theory of functions and point sets. From this brief description, we can see how the idea of centralized systems was used extremely successfully in the establishment of a scientific center.

4.5. The Appearance of Nicolas Bourbaki

We can read about Nicolas Bourbaki, a French School of Mathematics, in many articles, such as (Cartan, 1980; Halmos, 1957). In this section we describe some stories useful to systems analysis.

Nicolas Bourbaki is the author of a wide-ranging textbook on mathematics written in French. Although the first volume appeared in 1939, the work is by no means complete; 21 volumes totaling more than 3000 pages have been published to date. Bourbaki has also written articles for mathematical journals, among them the *Archiv der Mathematik*. In addition, he runs a seminar (the Séminaire Bourbaki) in Paris at which noted professors and researchers from all over France and from neighboring countries come together three times a year for a three-day mathematical conference held in the Institut Henri Poincaré. Nearly 150 papers covering a broad range of mathematical topics have been delivered in this seminar and then to mathematical circles around the world.

Since the first notes, reviews, and papers with Nicolas Bourbaki's name appeared in some journals in the middle 1930s, his courage and resourcefulness in using the axiomatic method to develop the whole of mathematics, the especially ingenious terminology, and the clean and economical organization of ideas and style of presentation have fascinated many mathematicians. They had been asking: Who is Nicolas Bourbaki? How was the scientific research group formed? How are they working? Why did they choose this name? In short, Nicolas Bourbaki appeared in the world as if he came from a dim mist.

In about 1946, in an article for the *Encyclopedia Britannica*, R. P. Boas, then-executive editor of *Mathematical Reviews*, printed his opinion that Bourbaki was simply the pseudonym for a group of French mathematicians. The publishers of *Britannica* soon found themselves in an acutely embarrassing position, for they received a scathing letter signed by Nicolas Bourbaki in which he declared that he was not about to allow anyone to question his right to exist. And to avenge himself on Boas, Bourbaki began to circulate the rumor that the mathematician Boas did not exist and that the initials B.O.A.S. were simply a pseudonym for a group of editors of *Mathematical Reviews*.

In his work, Bourbaki introduced many new and important concepts. Because there do not exist general rules in mathematics by which to judge what kinds of problems are important and what others are not of interest, mathematicians are extremely interested in the criteria on which Bourbaki decided what to do. Bourbaki divided the class of all mathematical problems into several categories (Dieudonne, 1982, p. 2):

(I) Stillborn problems (e.g., the determination of Fermat primes or the irrationality of Euler's constant)

(II) Problems without issue

(III) Problems that beget a method

(IV) Problems that belong to an active and fertile general theory

(V) Theories in decline for the time being

(VI) Theories in a state of dilution

The majority of Bourbaki's topics belong to categories IV and III. The study of problems in category IV ultimately (and perhaps only after a long time) reveals the existence of unexpected underlying structures that not only illuminate the original questions but also provide powerful general methods for elucidating a host of problems in other areas of mathematics. For the problems in category III, an examination of the techniques used to solve the original problems enables one to apply them (perhaps by making them considerably more complicated) to similar or more difficult problems without necessarily feeling that one really understands why they work.

Mathematics has changed radically in the past 20 years (a development in which Bourbaki may have had a hand). There may be some concepts among the fundamentals in Bourbaki's textbook that have already become outdated. After Bourbaki finishes the first part of his work, he may feel obligated to start all over again. Now people are watching Bourbaki's behavior with some further questions: What is the working direction of Bourbaki with newcomers? Will he still restrict himself to the mathematical foundation? What are his plans now?

In comparing the appearance of Nicolas Bourbaki and the growth of the Polish School of Mathematics, we can observe some similarities and differences. In the present situation, where there are uncountably many schools of thought in the world of learning, how can scientists in the coming generations choose appropriate methods that will enable them to tower above the forest of scientific schools? This is an actual problem. But no matter how many thousands of methods exist to achieve superiority in science, according to the study in this chapter, one point will never change — that is, that the establishment of a network of relations is essential to achieving the desired superiority successfully.

4.6. Some Related Problems and a Few Final Words

All applications of the concept of centralizability, discussed in this chapter, depend on the assumption that all axioms in ZFC are true. Furthermore, the question of how to enumerate human relations in terms of the basic blocks of set theory is still unknown. The assumption and the question need to be addressed in order to make this piece of research more reliable. The study of the feasibility of the assumption will have impact on the entire research of applied mathematics, since in applications of mathematics no one in the known literature has ever tried to address whether the setting of an application really makes the axioms in ZFC true. As is well known, if some of the axioms were not true in an application setting, then the truthfulness of the mathematical theorems applied would be in doubt. So, the study for the feasibility of this assumption will no doubt lead to an active research area in applied mathematics.

As for the question, it has been open in sociology for years. So far, no one seems to have developed a convincing method to enumerate human relations.

Compared with applications of systems theory in sociology, done earlier by many scholars, the study in this chapter is different in the following ways: First, each application is based on rigorous logical reasoning. Because of this feature, the problems mentioned appeared. Second, any cap appearing in the logical reasoning is pointed out so that the applicability of the current approach of systems theory can be checked; at the same time, the concept of feedback can play an important role here. That is, if for some reason the approach needs to be modified, systems theory will accordingly be enriched. Consequently, more and deeper insights in related areas can be obtained.

CHAPTER 5

A Theoretical Foundation for the Laws of Conservation

Finite divisibility of general systems is shown. Based upon this result, Lavoisier's Axiom, introduced in 1789, is verified by using the concept of fundamental particles, where a fundamental particle is one which can no longer be divided into smaller ones. The conservation principle of ancient atomists is recast and proved mathematically, based on which the law of conservation of matter–energy can be seen clearly. A brief history of atoms, elements, and laws of conservation is given, and some important open questions are posed. This chapter is based on (Lin, 1996).

5.1. Introduction

The law of conservation of matter–energy, one of the greatest achievements of all time, and systems, a current fashionable concept in modern science and technology, are joined under the name of general systems theory. Research in the foundation of mathematics, named set theory, has found its way through general systems theory into the realm of scientific activities related to the search for order of the natural world in which we live. The goal of this chapter is to establish a rigorous theoretical frame-work for the part of the world of learning on the laws of conservation. Hopefully, this study will provide an early casting of a stone to attract many beautiful and useful gems in future research along the line described here.

In the rest of this section, we briefly list some basic terminology and concepts to fully understand the discussion in Section 5.2.

According to Cantor, the creator of set theory, the word "set" means a collection of objects (or elements). However, such a view is untenable, as in certain cases the intuitive concept has been proved to be unreliable. So, in axiomatic set theory the theory is based on a system of axioms (usually on the system of ZFC axioms), from which all theorems are obtained by deduction. "Set" is one of the two primitive notions of the theory and is not examined directly about its meaning.

The ZFC axiom system is the Zermelo–Fränkel system with the axiom of choice. For details, refer to (Kunen, 1980; Kuratowski and Mostowski, 1976). For convenience, we state the Axiom of Regularity, one of the ZFC axioms: If A is a nonempty set of sets, then there exists a set X in A such that

$$X \cap A = \emptyset$$

All mathematical results, introduced in Section 5.2, are based on the assumption that the system of ZFC axioms is consistent, denoted by Theorem (ZFC). That is, the theory derived from ZFC axioms does not contain contradictory statements.

Let x and y be two sets. The set $\{\{x\},\{x,y\}\}$ is called an ordered pair, denoted by (x,y). The first term of (x,y) is x, and the second term is y. For details, see (Kuratowski and Mostowski, 1976).

The concept of systems, used in the following discussion, was first introduced in (Lin, 1987). It generalizes those concepts introduced by Whitehead (1978), Tarski (1954–1955), Hall and Fagen (1956), Mesarovic and Takahara (1975), and Bunge (1979). The definition reads as follows: S is a (general) system, if S equals an ordered pair (M,R) of sets, where M is the set of objects and R is a set of some relations on M. The sets M and R are called the object set and the relation set of S, respectively. (For those readers who are sophisticated enough in mathematics, a relation r in R implies that there exists an ordinal number $n = n(r)$, a function of r, such that r is a subset of the Cartesian product of n copies of the set M.)

5.2. Brief History of Atoms, Elements, and Laws of Conservation

The countless ideas in thousands of volumes in a library are arrangements of 26 letters of the alphabet, 10 numerals, and a few punctuation marks. Because of this and similar observations, the great thinkers throughout the history have been wondering whether the rich profusion from objects around us are formed of a basic alphabet of nature itself in various combinations? If so, what would be the characteristics of nature's "building blocks"?

The ancient Greeks were intrigued by the diverse, changing, temporary character of objects around them. But they were confident that the universe is one, and they looked for something unifying and eternal in the variety and flux of things. In the search for order, the one-element Ionians (624–500 B.C.) had a naturalistic and materialistic bent. They sought causes and explanations in terms of the eternal working of things themselves rather than in any divine, mythological, or supernatural intervention. Looking for a single basic reality, each Ionian believed that all things have their origin in a single knowable element: water, air, fire, or some indeterminate, nebulous substance.

The four-element philosophers, including Empedocles (490–435 B.C.), Pythagoras and his followers (from 5th century B.C. on), Philolaus (480 B.C.–?), Plato (427–347 B.C.) and Aristotle (384–322 B.C.), claimed that there is not one basic element but four: earth, water, air, and fire. The Egyptians had long recognized four elements with qualities of male or female. For example, earth is male when it has the form of boulders and crags; female when it is cultivable land. When air is windy, it is male; when cloudy or sluggish, female. The Chinese, as early as the 12th century B.C., had five elements, the previous four and wood, together with five virtues, tastes, colors, tones, and seasons.

Among his physical ideas of matter, Aristotle introduced a fifth element. This element cannot be "generated, corrupted, or transformed." This pure, eternal, ethereal substance makes up the heavens and all their unchangeable objects. Beneath the lunar sphere, the many products of the terrestrial four elements ceaselessly transform. He pointed out that, although all substances of the four elements are individually "generated, corrupted, and transformed, the universe as a whole is ungenerated and indestructible." Thus, in him a conservation of matter concept can be found.

Anaxagoras (510–428 B.C.) in his "seeds" idea was unwilling to submerge the tremendous varieties in things into any common denominator. He preferred to accept the immediate diversity of things as is. With his philosophy, every object is infinitely divisible. No matter how far an object is divided, what is left would have characteristics of the original substance. That brings us to the very interesting question of the transformation of matter. If all substances are derived from unique seeds of themselves, how can one substance develop into another? The contribution of Anaxagoras' concept of seeds was its refinement, its idea of taking substances as they are and breaking them down minutely in order to know more about them. One main weakness of the concept is passing down the complexities of a large-scale object to unseen miniatures of itself. Since these miniatures were infinitely divisible, a fundamental unit was lacking.

The Leucippus–Democritus atom (500–55 B.C.) combined features of the Ionian single element, Anaxagoras' seeds, and Empedocles' four elements and yet was an improvement over all of them. "Atom" means "not divisible" in Greek. This term was intentionally chosen by Democritus to emphasize a particle so small that it could no longer be divided. To Leucippus and Democritus, the universe originally and basically consisted entirely of atoms and a "void" in which atoms moved. Eternally, the atom is indestructible and unchanging. Atoms are all of the same substance, but by their various sizes and shapes, they can be used to explain the large variety of objects they compose. Individual atoms, solid, eternal, and indestructible, always maintain their identity in uniting or separating. It is the union and separation of atoms that is temporary and that results in the transformation of objects. The result is the principle of conservation of matter: Matter is neither created nor destroyed but is transformed. However, after the atomists, the idea of Plato and Aristotle on the nature of matter might seem almost an anticlimax.

Without the introduction of experimentation or the refinements of measurement, the Greek atomic theory, as it turned out, did not have a chance against the authority of Plato and particularly that of Aristotle. However, experimental techniques and knowledge had not yet reached the point where they could prevent the atomists from being eclipsed by Aristotle — and experimental techniques and knowledge are social accumulations. That is why we need to develop the fundamental theory and understanding about the existence and feasibility of the modern concept of elements and particles.

Men search for order through a few unifying principles, but nature has a way of seeping through categories set for it. Meanwhile, men gain detailed knowledge of their surroundings and develop more sophisticated concepts, techniques, and unifying principles. Experiments, becoming respectable in the Scientific Revolution, eventually showed that earth, water, air, and fire can be resolved into simpler substances and that the Aristotelian elements are not elements after all, or that they are more appropriate as states of matter that apply to every substance.

Bacon and Galileo had accepted the Greek idea of atoms. D. Sennert, a German physician, had applied atomic theory to specific natural processes. P. Gassendi, influential French mathematician and philosopher, developed an extensive non-mathematical atomic theory. Boyle, a founder of the British Royal Society, was among the first serious experimenters of the Scientific Revolution to use atoms to explain specific experiments. His contemporary Newton used a corpuscular theory to explain his investigations of light.

To ancient atomists, the total count of atoms in the universe was constant. The total amount of matter, therefore, remained the same regardless of changes. The atomists had a conservation principle of matter. In 1789, Lavoisier abstracted from his laboratory observations a law of conservation of matter: All matter existing within any closed system constitutes the universe for a specific experiment, where a "closed" system is one in which outside matter does not enter and inside matter does not leave. According to William Dampier, in 1929, "Because of its practical use, and for its own intrinsic interest, the principle of conservation of energy may be regarded as one of the great achievements of the human mind." Let us now take a brief look at the development of the conservation law of energy.

Heat of friction was the Achilles' heel of the caloric theory. Rubbing two blocks of wood together, makes the surfaces warmer. Why? From where does this heat of friction come? The calorists theorized that heat fluid is squeezed out from the interior of the blocks. That is, caloric is neither created nor destroyed; it is merely forced to the surface by friction.

Count Rumford (1753–1814) denied the calorists' theory, as he supervised the boring of cannon. Sir Humphrey Davy (1778–1829) corroborated Rumford by rubbing two pieces of ice together. Ship's Surgeon J. R. Mayer (1814–1878) speculated that the calorists were wrong while observing venous blood color in different climates. And J. P. Joule (1818–1889) argued against it as his paddles vigorously stirred water in calorimeters. All these men claimed evidence that heat of friction is heat produced by motion, that heat is a form of motion (energy), not a form of matter. Conversely, Watt's steam engines, invented in 1763, transformed heat into mechanical energy. The ancient Chinese invented gunpowder and used it in rockets. Europeans of the Middle Ages used the powder to propel musket shot and cannonballs. Actually, a gun is an engine that converts the heat and expansive force of an explosion into mechanical motion. So, by Joule's time, conservation of energy as a broad principle was in the air.

Some limited aspects of the broad principle appeared as follows: Newton's law of action and reaction led to conservation of momentum. Huygens proposed a conservation of kinetic energy for elastic collisions. Work concepts and swinging pendulums led to a principle of conservation of mechanical energy and to efforts toward perpetual-motion machines. Black took seven-league steps with his conservation of heat principle. Experiments by Rumford and Davy pointed to heat as a form of energy, and Joule established a quantitative connection between mechanical energy and heat through persistent, precise techniques.

What remained was for someone with enough insight, interest, and imagination to put the pieces together and present the whole picture. This was what the German physician J. R. Mayer (1814–1878) did, later reinforced by the technical language, mathematical rigor, and pointed applications of his countryman H. von Helmholtz (1821–1894), a famous biophysicist.

Chemistry was based on the law of conservation of matter. Physics was based on the law of conservation of energy. It has been shown however that matter transforms into energy, and energy into matter. Now hen what? Where is the constancy of matter or energy? What happens to the indispensable "=" signs in the equations of physicist and chemist? What happens to physics and chemistry as "exact" sciences? Einstein combined the two conservation laws into one, a law of the conservation of matter–energy: The total amount of both matter and energy is always the same.

For a more detailed and comprehensive history, please refer to Perlman's work (Perlman, 1970).

5.3. A Mathematical Foundation for the Laws of Conservation

From the discussion in the previous section, it can be seen that all these laws of conservation were established based on experiments or intuitive interpretations of some related experimental results or simple observations. In this section some mathematical results of general systems theory are introduced and used to establish a theoretical foundation for the laws of conservation. The models or understanding of the laws of conservation used in this section may not be those currently accepted by research physicists or chemists, but they are surely some of the major steps along the path leading to the current ones. By doing so purposely, I hope that a greater audience can be reached.

A system $S_n = (M_n, R_n)$ is an nth-level object system (Lin and Ma, 1987) of a system $S_0 = (M_0, R_0)$ if there exist systems $S_i = (M_i, R_i)$, for $i = 1, 2, \ldots, n-1$, such that the system S_i is an object in $M_{i-1}, 0 < i < n+1$. Each element in M_n is called an nth-level object of S_0. A chain of object systems of a system S is a sequence $\{S_i : i < \alpha\}$, for some ordinal number α, of different-level object systems of S, such that for each pair $i,j < \alpha$ with $i < j$, there exists an integer $n = n(i,j)$, which depends on the ordinal numbers i and j, satisfying that the system S_j is an nth-level object system of S_i. Then we can prove the following result.

Theorem 5.3.1 [ZFC]. *Suppose that S is a system. Then each chain of object systems of S must be finite.*

Proof: For the sake of completeness of this chapter, a proof of this result is given. It is not intended for general audiences but for a few readers who have the adequate mathematical background. The theorem will be proven by contradiction.

Suppose that there exists a chain of object systems of infinite length, such as $\{S_i = (M_i, R_i) : i = 1, 2, 3, \ldots\}$, satisfying that S_i is an object of the system $S_{i-1}, i = 1, 2, 3, \ldots$. Now define a set X as follows:

$$X = \{M_i : i = 1, 2, \ldots\} \cup \{S_i : i = 1, 2, \ldots\} \cup \{\{M_i\} : i = 1, 2, \ldots\}$$

From the Axiom of Regularity (Kunen, 1980), it follows that there exists a set Y in X such that

$$Y \cap X = \emptyset \tag{5.1}$$

There are now three possibilities: (1) $Y = M_i$, for some $i = 1, 2, \ldots$; (ii) $Y = S_i$, for some $i = 1, 2, \ldots$; and (iii) $Y = \{M_i\}$, for some $i = 1, 2, \ldots$. If (i) holds, $S_{i+1} \in Y \cap X \neq \emptyset$. This contradicts Eq. (5.1). So, the possibility (i) cannot be true. If (ii) holds, $Y = S_i = (M_i, R_i) = \{\{M_i\}, \{M_i, R_i\}\}$. That is, $\{M_i\} \in Y \cap X \neq \emptyset$. This contradicts Eq. (5.1). Hence, (ii) cannot be correct. If (iii) holds, $M_i \in Y \cap X \neq \emptyset$, again a contradiction of Eq. (5.1). Therefore, (iii) cannot be true. These contradictions show that the assumption of a chain of object systems of infinite length cannot be correct. ■

As introduced by Klir (1985), a system is what is distinguished as a system. We then can do the following general systems modeling: For each chosen matter, real situation problem, or environment, such as a chemical reaction process, a system, describing the matter of interest, can always be defined. For example, let $S = (M, R)$ be a systems representation of a chemical reaction process, such that M stands for the set of all substances used in the reaction and R is the set of all relations between the substances in M. It can be seen that S represents the chemical reaction of interest. Changes of S represents changes of the objects in M and changes of relations in R. Based upon this understanding and by virtue of Theorem 5.3.1, it follows that we have proven theoretically Lavoisier's Axiom, given in 1789.

Axiom 5.3.1 [Lavoisier's Axiom (Perlman, 1970, p. 414)].

> In all the operations of art and nature, nothing is created! An equal quantity of matter exists both before and after the experiment ... and nothing takes place beyond changes and modifications in the combination of the elements. Upon this principle, the whole art of performing chemical experiments depends.

In brief, this conservation law says that all matter existing within any closed system constitutes the universe for a specific experiment. That is, if $S = (M,R)$ is the systems representation of a chemical reaction defined as before, then a chemical reaction between the substances in M is just simply a change and modification of some relations between the fundamental particles. Here, a fundamental particle is a particle which can no longer be divided into smaller ones. The existence of fundamental particles is shown by Theorem 5.3.1, since if a particle A can still be divided into smaller particles, A then can be studied as a system, based upon Klir's definition of systems. Theorem 5.3.1 says that each chain of object systems must stop; i.e., each particle must be finitely divisible.

Based upon this understanding, the conservation principle of ancient atomists (Perlman, 1970, p. 414), which states that the total count of atoms in the universe is constant; therefore, the total amount of matter remains the same regardless of changes, can be restated as follows:

Law [Conservation of Fundamental Particles]. The total count of basic particles in the universe is constant; therefore, the total amount of matter remains the same regardless of changes and modifications.

A brief reasoning for the feasibility of this law is the following: There are three basic known changes in the universe: physical, chemical, and nuclear. A change in which a substance entirely retains its original composition is a physical change. A change in which a substance loses its original composition and one or more new substances are formed is a chemical change. A change where nuclei, which are unstable, shoot out "chunks" of mass and energy, and other changes, while take place in the nucleus of an atom, are called nuclear reactions. None of these changes is a change at the fundamental particle level. Rather each is a recombination of some higher level systems of the fundamental particles. At the same time, the following theorem guarantees that for each given system $S = (M,R)$, the set of fundamental particles is uniquely determined. Hence, if the physical universe is studied as a system, the set of basic particles is eternal, indestructible, and unchangeable. It, in turn, proves the feasibility of the Law of Conservation of Fundamental Particles.

Theorem 5.3.2 [ZFC]. *For each system $S = (M,R)$, the set of all fundamental objects are unique.*

Proof: Again, the proof given here is for completeness of this chapter. Let

$$M_1 = \{m \in M : m \text{ is not a system}\}$$

and

$$M_1^* = \bigcup\{M_x : x = (M_x, R_x) \in M - M_1\}$$

By mathematical induction, assume that all sets M_i and M_i^* have been defined for each $i < n+1$ for a natural number n satisfying

$$M_i = \{m \in M_{i-1}^* : m \text{ is not a system}\}$$

and

$$M_i^* = \bigcup\{M_x : x = (M_x, R_x) \in M_{i-1}^* - M_i\}$$

Then M_{n+1} and M_{n+1}^* can be defined as follows:

$$M_{n+1} = \{m \in M_n^* : m \text{ is not a system}\}$$

and

$$M_{n+1}^* = \bigcup\{M_x : x = (M_x, R_x) \in M_n^* - M_{n+1}\}$$

Theorem 5.3.1 guarantees that there is a natural number n such that M_j and M_j^* equal $\emptyset$, for all $j > n+1$. Thus, the set

$$M^* = \bigcup\{M_i : i = 1, 2, 3, \ldots\}$$

consists of all fundamental objects of the system S and is unique, since each set M_i is uniquely defined. ■

Let us discuss some impacts of what has been done on the Law of Conservation of Matter–Energy. The law states that the total amount of matter and energy is always the same. Historically, this law was developed based upon the new development of science and technology: Energy and matter are two different forms of the same thing. That is, matter can be transformed into energy, and energy into matter. Each matter consists of fundamental particles, and so does energy. Since energy can be in different forms, such as light, it confirms that each fundamental particle has no volume or size. Actually, Theorem 5.3.1 also proves this fact, because if a particle has certain a volume, it would be possible to chop it up into smaller pieces. That is, the particle can be considered a system; it is not a fundamental particle. As a consequence of the Law of Conservation of Basic Particles, the Law of Conservation of Matter–Energy becomes clear and obvious.

5.4. A Few Final Words

In the previous mathematical analysis of the laws of conservation, many important questions are left open. For example, it is necessary to develop conservation equations in terms of fundamental particles and to write them in modern symbolic form. Only with the indispensable "=" signs in the equation can the theory, developed on the law of conservation of fundamental particles, be made into one "exact science" with the capacity of prediction.

It took several thousand years to develop the Law of Conservation of Matter–Energy. The law has been considered as one of the greatest syntheses of all time, comparable to those of Ptolemy and Newton. Without the "=" signs of the conservation laws, certainly modern physics, chemistry, and technology could hardly exist. Not only this, the great idea, contained in the conservation law, contradicts the common notion that all things must change. This idea is still developing. The full range of its implications is not yet clear, but surely the consequence will be revolutionary.

CHAPTER 6

A Mathematics of Computability that Speaks the Language of Levels

In this chapter we use the modern systems theory to retrace the history of important and interesting philosophical problems. Based on the discussion, we show that the multilevel structure of nature can be approached by using the general systems theory approach initiated by Mesarovic in the early 1960s. Two difficulties appearing in modern physics are listed and studied in terms of a new mathematical theory. This theory reflects the characteristic of the multilevel nature. Some elementary properties of the new theory are listed. Some important and fruitful leads for future research are posed in the final section.

6.1. Introduction

Since the concept of systems was first formally introduced by von Bertalanffy in the 1920s, the meaning of system has been getting wider and wider. As the application of the systems concept has become more widespread, the need to establish the theoretical foundation of all systems theories becomes evident. To meet the challenge of forming a unified foundation of the theory of systems, Mesarovic and Takahara introduced a concept of general systems, based on Cantor's set theory, in the 1960s. Since the theory of general systems has been established on set theory, one shortfall of this "theoretical foundation" is that it is difficult or impossible to quantify any subject matter of research. In this chapter recent results in general systems theory are used to show the need for a mathematical theory of multilevels, and, based on Wang's work (1985; 1991), one is introduced. This theory is used to study two problems in modern physics. Hopefully, this work will make up the computational deficiency of the current theory of general systems.

6.2. Multilevel Structure of Nature

Even though the major part of classical and modern mathematics deals with subject matters of one level, such as problems on the real number line, the plane, the n-dimensional Euclidean spaces, etc., human minds have been searching for the organic structure of multilevels of the universe since at least the time of the one-element Ionians (624–500 B.C.). For example, in the search for order, Ionians had a naturalistic and materialistic bent. They sought causes and explanations in terms of the eternal working of things themselves rather than in any divine, mythological, or supernatural intervention. Looking for a single basic reality, each Ionian believed that all things have their origin in a single knowable element: water, air, fire, or some indeterminate, nebulous substance. For more on this, see (Perlman, 1970). This section shows that general systems theory, based on Cantor's set theory, first introduced formally by Mesarovic in the early 1960s, can be seen as a new research approach to the multilevel structure of nature.

Mesarovic's definition of a general system is as follows (Mesarovic and Takahara, 1975):

A system S is a relation on nonempty sets V_i:

$$S \subseteq \prod \{V_i : i \in I\}$$

where I is an index set.

This structure of general systems is multilevel, since elements in the nonempty sets V_i could be systems also. Using this idea inductively, we see that an element S_1 of the system S could be a system, an element S_2 of system S_1 could be a system, ... Can this process go on forever? If the answer to the question is "yes," then "the world would be infinitely divisible"?! If the answer to the question is "no," does that mean the world is made up of fundamental elements?

The idea that the world is infinitely divisible can be found in Anaxagoras' "seeds" philosophy (510–428 B.C.). He was unwilling to submerge the tremendous varieties of things into any common denominator, and preferred to accept the immediate diversity of things as is. With his philosophy, every object is infinitely divisible. No matter how far an object is divided, what is left would have characteristics of the original substance. The opposite idea — that the world is made up of fundamental particles — appeared from 500 to 55 B.C. The Leucippus–Democritus atom combined features of the Ionians' single element, Anaxagoras' "seeds," and some thoughts of other schools, and yet was an improvement over all of them. "Atom" means "not divisible" in Greek. This term was intentionally chosen by Democritus to emphasize a particle so small that it could no longer be divided. To Leucippus and Democritus, the universe originally and basically consisted entirely of "atoms" and a "void" in which atoms move.

In the search for an answer as to which of the two opposite ideas could be right, Ma and Lin (1987), introduced the following definition of general systems: *S is a (general) system, if S is an ordered pair (M,R) of sets, where R is a set of some relations defined on M. Each element in M is called an object of S, and M and R are called the object set and the relation set of S, respectively.* For each relation $r \in R$, r is defined as follows: there exists an ordinal number $n = n(r)$, depending on r, such that $r \subseteq M^n$, that can be either finite or infinite, and called the length of r. Assume the length of the empty relation $\emptyset$ is 0; i.e., $n(\emptyset) = 0$. This new definition of general systems generalizes that of Mesarovic. Many important structures of systems can now be studied that otherwise could not have been with the one-relation approach. We note that the concept of general systems was defined in more general terms by many other authors as well. For example, Wu and Klir independently defined the concepts of pansystems and systems at a much higher level. For more details see (Wu, 1994; Klir, 1985).

Two systems $S_i = (M_i, R_i)$, $i = 1,2$, are identical, if $M_1 = M_2$ and $R_1 = R_2$. System S_1 is a partial system of system S_2 if $M_1 \subseteq M_2$ and for each relation $r_1 \in R_1$ there exists a relation $r_2 \in R_2$ such that $r_1 = r_2|M_1 = r_2 \cap (M_1)^{n(r_2)}$. System S_1 is a subsystem of S_2, if $M_1 \subseteq M_2$ and for each relation $r_1 \in R_1$ there exists a relation $r_2 \in R_2$ such that $r_1 \subseteq r_2|M_1$. A system $S_n = (M_n, R_n)$ is an nth-level object system of a system $S_0 = (M_0, R_0)$ if there exist systems $S_i = (M_i, R_i)$, $i = 1, 2, \ldots, n-1$, such that S_i is an object in M_{i-1}, $1 \leq i \leq n$. Each element in M_n is called an nth-level object of S_0.

Based on these notations, the following mathematical results (Lin and Ma, 1993) can be proved, where (ZFC) means that we assume all the axioms in the ZFC axiom system (Kuratowski and Mostowski, 1976) are true.

Theorem 6.2.1 [Russell's paradox]. *There is no system such that the object set of the system consists of all systems.*

Theorem 6.2.2 [ZFC]. *Let $S_0 = (M_0, R_0)$ be a system and $S_n = (M_n, R_n)$ an nth-level object system of S_0. Then it is impossible for S_0 to be a subsystem of S_n for each integer $n > 0$.*

A chain of object systems of a system S is a sequence $\{S_i : i < \alpha\}$ for some ordinal number α, of different-level object systems of the system S, such that for each pair $i,j < \alpha$ with $i < j$, there exists an integer $n = n(i,j)$ such that the system S_j is an nth-level object system of the system S_i.

Theorem 6.2.3 [ZFC]. *Suppose that S is a system. Then each chain of object systems of S must be finite.*

Theorem 6.2.4 [ZFC]. *For each system S there exists exactly one set $M(S)$, consisting of all basic objects in S, where a basic object in S is a level object of S which is no longer a system.*

For detailed treatment of these theorems, please refer to Chapter 9.

As introduced by Klir (Klir, 1985), a system is what is distinguished as a system. We can then do the following general systems modeling: For each chosen matter, real situation problem, or environment, such as a chemical reaction process, a system, describing the matter of interest, can always be defined. For example, let $S = (M, R)$ be a systems representation of a chemical reaction process, such that M stands for the set of all substances used in the reaction and R is the set of all relations between the substances in M. It can be seen that S represents the chemical reaction of interest. Based upon this modeling and Theorems 6.2.3 and 6.2.4, it follows that we have argued theoretically that the world is finitely divisible.

In (Lin, 1996) or in Chapter 5 the same methodology was used to show that a theoretical foundation can be established for the Law of Conservation of Matter–Energy.

6.3. Some Hypotheses in Modern Physics

Without using a mathematics of multilevels, some hypotheses in modern science seem to be contradictory. However, the theories, established on the hypotheses, have been winning victories one after another. Einstein's relativity and Dirac's quantum mechanics, for example, are two of these theories. In this section two such hypotheses are listed to further support the need for a mathematics of multilevels.

6.3.1. The Rest Mass of a Photon

According to Einstein, the inertial mass m of a moving particle is related to its rest mass m_0 as follows:

$$m = \frac{m_0}{\sqrt{1 - v^2/c^2}} \tag{6.1}$$

where c stands for the speed of light and v is the particle speed. As for the photon, since its speed is $v = c$, the rest mass m_0 should be zero, or from Eq. (6.1) $m = \infty$, contradicting observations. However, even it is assumed that $m_0 = 0$, the fallacy involved in the calculation still remains in Eq. (6.1) because in classical mathematics on the real number line, $0 \div 0$ is undetermined.

6.3.2. Dirac's δ Function

In Dirac's well-known work (1958), a function, later named Dirac's δ function, was introduced as follows: $\delta : \mathbb{R} \to \mathbb{R}$, where $\mathbb{R}$ is the set of all real numbers, satisfies the conditions:

$$\begin{aligned} \delta(x) &= 0 \quad \text{whenever } x \neq 0 \\ \delta(0) &= \infty \\ \int_{-\infty}^{\infty} f(x)\delta(x)\,dx &= f(0) \end{aligned} \tag{6.2}$$

where $f(x)$ is a function on $\mathbb{R}$ with compact support and is infinitely differentiable.

This function has played an important role in quantum mechanics. However, in classical mathematics, including modern measure theory, it is impossible to have such a function, since each function $f(x)$ satisfying the first condition must possess the property that $\int_{-\infty}^{\infty} f(x)\delta(x)\,dx = 0$.

Because of the magnificent role that Dirac's δ function had played in the success of quantum mechanics, the international mathematics community became desperate to come up with a feasible explanation for the strange δ function. In 1950, Schwartz introduced the concept of distributions, based on which the δ function was understood as a linear functional defined on the "elementary space" consisting of real-valued functions with some special mathematical properties. For details, see (Richards and Youn, 1990).

6.4. A Non-Archimedean Number Field

In this section a number field, called a generalized number system (**GNS**), is introduced. The concept of the **GNS** is based upon (Wang, 1964), which successfully solved an open problem in general topology. Independently, Laugwitz (1961; 1962; 1968) studied the concept of generalized power series. The concept of the **GNS** was first introduced in the early 1980s by Wang (1985; 1991) without reference to Laugwitz's work and Robinson's nonstandard analysis. The **GNS** is really a non-Archimedean number field and has found many successful applications in modern physics. Algebraically, the **GNS** is a subfield of Laugwitz's generalized power series.

Historically speaking, nonstandard analysis, established in the early 1960s by Robinson (Davis, 1977), first introduced the infinitesimal and infinitely large numbers formally and legally, based on mathematical logic. This theory broke through the limitation of the classical concept of real numbers, where infinities are usually considered ideal objects. With nonstandard analysis, many classical theories, such as mathematical analysis, algebra, norm spaces, some classical mechanics problems, etc., can be greatly simplified and become more intuitive. However, the shortfall of Robinson's nonstandard analysis is that infinitesimals and infinities can only be shown to exist and have no concrete expressions of any tangible form. Based on the argument of the existence of infinitesimals and infinities given in Robinson's theory, the **GNS** has the computational advantage, since each number here, whether ordinary real, infinitesimal, or an infinity, has an exact expression in terms of the real numbers in the classical sense.

A generalized number x is a function from $\mathbb{Z}$ to $\mathbb{R}$, where $\mathbb{Z}$ is the set of all integers and $\mathbb{R}$ is the set of all real numbers, such that there exists a number $k = k(x) \in \mathbb{Z}$, called the level index of x, satisfying

$$x(t) = 0 \text{ for all } t < k, \quad \text{and} \quad x(k(x)) \neq 0 \tag{6.3}$$

Then x can be expressed as a generalized power series as follows:

$$\begin{aligned} x &= \sum_{t \geq k(x)} x(t) \times \mathbf{1}_{(t)} \\ &\equiv \sum_{t \geq k(x)} x(t)\mathbf{1}_{(t)} \\ &\equiv (\dots, 0, \underset{\substack{\uparrow \\ k(x)\text{th level}}}{x(k(x))}, x(k(x)+1), \dots) \\ &= (\dots, 0, x_{k(x)}, x_{k(x)+1}, \dots) \end{aligned} \tag{6.4}$$

where $\mathbf{1}_{(t)}$ stands for the function from $\mathbb{Z}$ to $\mathbb{R}$ such that $\mathbf{1}_{(t)}(s) = 0$ if $s \neq t$, and $\mathbf{1}_{(t)}(t) = 1$ and $x(t) \equiv x_t$. Intuitively, the right-hand side of Eq. (6.4) is the "graph" of the function $x : \mathbb{Z} \to \mathbb{R}$. The set of all generalized numbers will be written as **GNS**.

We now define operations of generalized numbers as follows:

(1) **Addition and subtraction.:**
Let x, y be arbitrary from **GNS**. Define

$$x \pm y = \sum_t [x(t) \pm y(t)] \tag{6.5}$$

(2) **Ordering.:**
The order relation between x and $y \in$ **GNS** is the lexicographical ordering; i.e., $x < y$ iff there is an integer k_0 such that $x(t) = y(t)$ if $t < k_0$ and $x(k_0) < y(k_0)$.

(3) **Scalar multiplication.:**
Let $c \in \mathbb{R}$ and $x \in$ **GNS** define

$$cx = \sum_t cx(t) \times \mathbf{1}_{(t)} \tag{6.6}$$

Because of the ordering relation defined in 2, we can now think of **GNS** as a system of numbers of different level; and the function $\mathbf{1}_{(t)}$ will be called the "unit element" of the tth level.

(4) **Multiplication.:**
Let $x, y \in \mathbf{GNS}$. Then the product $x \times y$ or xy is defined by

$$x \times y = \sum_t \sum_{m+n=t} [x(m) \times y(n)] \times \mathbf{1}_{(t)} \tag{6.7}$$

Specifically, $\mathbf{1}_{(t)} \times \mathbf{1}_{(s)} = \mathbf{1}_{(t+s)}$.

(5) **Division.:**
For $x, y \in \mathbf{GNS}$, if there exists a $z \in \mathbf{GNS}$ such that

$$y \times z = x \tag{6.8}$$

then the generalized number z is called the quotient of x divided by y, denoted by $x \div y$ or x/y.

Theorem 6.4.1. *If $y \neq 0 \equiv \sum_t 0 \times 1_{(t)}$, the zero function from $\mathbb{Z}$ to $\mathbb{R}$, then the quotient $x \div y$ always uniquely exists.*

Proof: Assume that the known and the unknown generalized numbers x, y, and z are

$$y = \sum_{m \geq k(y)} y(m)\mathbf{1}_{(m)}$$

$$x = \sum_{m \geq k(x)} x(m)\mathbf{1}_{(m)}$$

and

$$z = \sum_{m \geq k(z)} z(m)\mathbf{1}_{(m)}$$

We need to find z such that

$$\sum_{u+v=s} y(u) \times z(v) = x(s) \tag{6.9}$$

with $k(y) \leq u \leq s - k(z)$ and $k(z) \leq v \leq s - k(y)$. This second inequality implies that we can define $k(z) = k(x) - k(y)$. Hence, replacing s by $k(x)$ in Eq. (6.9), we have

$$y(k(y)) \times z(k(x) - k(y)) = x(k(x))$$

That is,

$$y(k(y)) \times z(k(z)) = x(k(x))$$

Therefore, $z(k(z)) = x(k(x))/y(k(y))$. When s is replaced by $k(x)+1$ in Eq. (6.9), we have

$$y(k(y)) \times z(k(z)+1) + y(k(y)+1) \times z(k(z)) = x(k(x)+1)$$

Therefore,

$$z(k(z)+1) = \{x(k(x)+1) - y(k(y)+1) \times z(k(z))\}/y(k(y))$$

By induction, we can continue this process and define each $z(m)$, for $m \geq k(z) = k(x) - k(y)$. Thus, the generalized number z is defined based on the given x and y.

The uniqueness of z can be shown by observing that the level index of z cannot be less than $k(x) - k(y)$. To this end, assume that $k(z) = k(x) - k(y) - n$, for some nonzero whole number n. Then replacing s by $k(x) - n$, in Eq. (6.9), we obtain

$$0 \neq y(k(y)) \times z(k(z)) = x(k(x) - n) = 0$$

contradiction. Therefore, the whole number n must be zero. That is, the generalized number z must be unique. ■

Theorem 6.4.2. *For each $x \in$ **GNS** such that $x(k(x)) \geq 0$ and $k(x)$ is an even number, there exists a unique $y \in$ **GNS** such that $y^2 = y \times y = x$. This generalized number y is called the square root of x, denoted $\sqrt{y}$.*

Proof: Assume the unknown generalized number y is

$$y = \sum_{m \geq k(y)} y(m)\mathbf{1}_{(m)}$$

Then the equation $y \times y = x$ is equivalent to that for any integer s,

$$\sum_{m+n=s} y(m)y(n) = x(s) \tag{6.10}$$

where $k(y) \leq m,\ n \leq s - k(y)$. That is, $2k(y) \leq k(x)$. We can then define $k(y) = k(x)/2$. When $s = k(x)$, Eq. (6.10) becomes

$$y(k(y)) \times y(k(y)) = x(k(x))$$

Since $x(k(x)) \geq 0$, $y(k(y)) = \sqrt{x(k(x))}$. When $s = k(x)+1$, Eq. (6.10) becomes

$$y(k(y)) \times y(k(y)+1) + y(k(y)+1) \times y(k(y)) = x(k(x)+1)$$

So, $y(k(y)+1) = x(k(x)+1)/2y(k(y))$.

By applying induction on the index $k(y)+i$, the generalized number y can be well defined.

To show the uniqueness of y, it suffices to see that $k(y)$ cannot be less than $k(x)/2$, as defined. To this end, assume $k(y) = k(x)/2 - n$ for some nonzero whole number n. Replacing s by $k(x) - 2n$, changes Eq. (6.10) to

$$0 \neq y(k(y)) \times y(k(y)) = x(k(x) - n) = 0$$

contradiction. That is, the definition of the generalized number y is unique. ∎

The proof of Theorem 6.4.2 can be generalized to obtain 6.4.3.

Theorem 6.4.3. *For each natural number $n \geq 2$ and each $x \in$ **GNS** such that $k(x)$ is a multiple of n, then*

*(1) If n is even and $x(k(x)) \geq 0$, there exists a unique $y \in$ **GNS** such that*

$$y^n \equiv \underbrace{y \times y \times \cdots \times y}_{n \text{ times}} = x$$

*(2) If n is odd, there exists a unique $y \in$ **GNS** such that $y^n = x$.*

The generalized number y in (1) or (2) will be called the nth root of x and denoted by $\sqrt[n]{x}$ or $x^{1/n}$.

Since $y \times y = \sum_s \sum_{m+n=s} y(m)y(n)\mathbf{1}_{(s)}$, $k(y^2) = 2k(y)$. In general, $k(y^n) = nk(y)$, $k(x^1x^2 \cdots x^n) = k(x^1) + k(x^2) + \cdots + k(x^n)$, where $x^i \in$ **GNS**, $i = 1, 2, \ldots, n$, and $k(x^1/x^2) = k(x^1) - k(x^2)$. That is, if $k(x)$ is not a multiple of n, there is no $y \in$ **GNS** such that $y^n = x$. In this sense, then Theorem 6.4.3 is the best we can get.

To conclude this section, let us observe the following facts.

Theorem 6.4.4. (**GNS**$, +, \times, <$) *forms an ordered non-Archimedean field. If $\mathbf{I}_{(k)} = \{x = x_k \times \mathbf{1}_{(k)} : x_k \in \mathbb{R}\}$ and k and h are integers satisfying $k < h$, then the generalized numbers contained in $\mathbf{I}_{(h)}$ are infinitesimals, compared with those contained in $\mathbf{I}_{(k)}$; conversely, nonzero generalized numbers in $\mathbf{I}_{(k)}$ are infinities compared with those contained in $\mathbf{I}_{(h)}$.*

6.5. Applications

In history, "laws" of the nature have been frequently expressed and treated with the rigor and beauty of mathematical symbols. For the most part, this application has been very successful. However, in the past 40 some years, theoretical physicists have been faced with difficulties in mathematical calculation. According to Wang (1991), such difficulties arise mainly because the modern world of physics is multilevel in character, whereas applicable mathematics still remains at the stage of single level where the only infinities are denoted by $-\infty$ or $+\infty$, and are considered in our macroscopic world as ideal objects and treated with a different set of rules, compared with those of ordinary real numbers. Now, in **GNS**, the field $\mathbb{R}$ is isomorphic to $\mathbf{I}_{(0)}$, which represents measurable quantities within the macroscopic world. Also, all numbers in **GNS**, whether infinities or infinitesimals, are treated with the same set of rules. In this section we show how the difficulties listed in Section 6.3 can be resolved with **GNS**.

6.5.1. The Hypothesis on the Rest Mass of a Photon

The difficulty in the calculation of $m = 0 \div 0$, discussed in Section 6.3, can be overcome as follows. Suppose that the rest mass $\mathbf{m}_0$ of a photon is

$$\mathbf{m}_0 = (\dots, 0, \dots, \underset{\substack{\uparrow \\ \text{0th level}}}{0}, m_1, \dots)$$

and that the inertial mass $\mathbf{m}$ in Eq. (6.1) is a finite real number; that is, $\mathbf{m} \in \mathbf{I}_{(0)}$ has the representation

$$\mathbf{m} = (\dots, 0, \underset{\substack{\uparrow \\ \text{0th level}}}{m}, 0, \dots)$$

Now the velocity of a photon is

$$\mathbf{v} = (\dots, 0, \dots, 0, \underset{\substack{\uparrow \\ \text{0th level}}}{c}, 0, v_2, v_3, \dots)$$

where $v_2 < 0$, since c is assumed by Einstein to be the maximum speed in the universe. Equation (6.1) can then be written as

$$\mathbf{m} = \frac{\mathbf{m}_0}{\sqrt{\mathbf{1}_{(0)} - \mathbf{v}^2/\mathbf{c}^2}}$$

where $\mathbf{c}$ is understood in the same way as $\mathbf{m} \in \mathbf{I}_{(0)}$. By substituting the **GNS** values of $\mathbf{m}$, $\mathbf{m}_0$, $\mathbf{1}_{(0)}$, $\mathbf{v}$, and $\mathbf{c}$, into this equation, and by comparing the leading nonzero terms (i.e., the coefficients of $\mathbf{1}_{(2)}$), we obtain $(-2c)m^2 v_2 = m_1^2$. That gives the relation between v_2 and m_1.

In this argument, the existence of the square root $\sqrt{\mathbf{1}_{(0)} - \mathbf{v}^2/\mathbf{c}^2}$ can be seen as follows:

$$\mathbf{v}^2 = \sum_{t\geq 0} \sum_{i+j=t} \mathbf{v}(i)\mathbf{v}(j)\mathbf{1}_{(t)}$$
$$= (\dots, 0, \underset{\substack{\uparrow \\ \text{0th level}}}{c^2}, 0, 2cv_2, 2cv_3, \dots)$$

and

$$\mathbf{v}^2/\mathbf{c}^2 = (\dots, 0, \underset{\substack{\uparrow \\ \text{0th level}}}{1}, 0, 2v_2, \dots)$$

So

$$k(\mathbf{1}_{(0)} - \mathbf{v}^2/\mathbf{c}^2) = 2$$

and

$$\mathbf{v}^2/\mathbf{c}^2(k(\mathbf{1}_{(0)} - \mathbf{v}^2/\mathbf{c}^2)) = -2\mathbf{v}_2 > 0$$

From Theorem 6.4.2, it follows that $\sqrt{\mathbf{1}_{(0)} - \mathbf{v}^2/\mathbf{c}^2}$ exists.

6.5.2. A New Look at Dirac's δ Function

We show that, as a real-valued function is a function from a subset of $\mathbf{I}_{(0)}$ to a subset of $\mathbf{I}_{(0)}$, Dirac's δ function can be considered as a function from **GNS** into **GNS**. Therefore, it is a more natural representation. At the same time, the integral sign in Eq. (6.2) is no longer a symbol without much mathematical meaning. For convenience, boldfaced letters will be used for **GNS**-related concepts and lightface letters for real-valued concepts. For example, $f(x)$ is a real-valued function, and $\mathbf{f}(\mathbf{x})$ is a function from a subset of **GNS** to a subset of **GNS**.

Without loss of generality, let $\mathbf{y} = \mathbf{f}(\mathbf{x})$ be a function from **GNS** to **GNS**. We define the integral of the function inductively as follows.

Let $E = \{\mathbf{x} \in \mathbf{GNS} : \mathbf{f}(\mathbf{x}) \neq \mathbf{0}\}$.

Case 1: Suppose that for each $\mathbf{x} \in E$, $x_{-m} = 0$ for all $m \in \mathbb{N}$, the set of all natural numbers, and that $\mathbf{y} = (\dots, y_{-m}, \dots, y_0, \dots, y_n, \dots)$, where, in general, $y_k = y_k(\mathbf{x})$ is a function of all x_m, $m \in \mathbb{N} \cup \{0\}$.

Step 1: Define a subset $H^{(0)} \subseteq \mathbb{R}$ as follows: $x_0 \in H^{(0)}$ iff

$$y_{-n} = 0 \text{ if } n > 0 \quad \text{and} \quad y_0 = y_0(x_0) \text{ a function of } x_0 \tag{6.11}$$

and let

$$f^{(0)}(x_0) = \begin{cases} y_0(x_0) & \text{if } x_0 \in H^{(0)} \\ \infty & \text{otherwise} \end{cases} \tag{6.12}$$

If the real-valued function $f^{(0)}$ is Lebesgue integrable on $\mathbb{R}$, then $\mathbb{R} - H^{(0)}$ has Lebesgue measure zero, denoted $\int_{\mathbb{R}} f^{(0)}(x_0)\,dx_0 = a^{(0)}$.

Step n: Assume the real numbers $a^{(0)}, a^{(1)}, \ldots, a^{(n-1)}$ have all been defined. For fixed $x_0, \ldots, x_{n-1}$, define a subset $H^{(n)} \subseteq \mathbb{R}$ as follows: $x_n \in H^{(n)}$ iff

$$y_{-m} = 0 \text{ if } m > n \quad \text{and} \quad y_{-n} = y_{-n}(x_n) \text{ a function of } x_n \tag{6.13}$$

and let

$$f^{(n)}_{(x_0,\ldots,x_{n-1})}(x_n) = \begin{cases} y_{-n}(x_n) & \text{if } x_n \in H^{(n)} \\ \infty & \text{otherwise} \end{cases} \tag{6.14}$$

If the real-valued function $f^{(n)}_{(x_0,\ldots,x_{n-1})}$ is Lebesgue integrable on $\mathbb{R}$, denote the integral on $\mathbb{R}$ by $a^{(n)}(x_0, \ldots, x_{n-1})$. If $\sum_{x_0,\ldots,x_{n-1}} a^{(n)}(x_0, \ldots, x_{n-1})$ is meaningful — that is, if there are only up to countably many nonzero summands and the summation is absolutely convergent — then the sum is denoted $a^{(n)}$.

Definition 6.5.1 [Wang (1985)]. If $\sum_{n=1}^{\infty} a^{(n)}$ is convergent and equals a, then the function $\mathbf{f}(\mathbf{x})$ is GNL-integrable and we write

$$\int_{\mathbf{GNS}} \mathbf{f}(\mathbf{x})\,d\mathbf{x} = a$$

Case 2: Partition the set E as follows: $E = \bigcup\{E_k : k \in \mathbb{Z}, k \geq 0\}$, where $E_k = \{\mathbf{x} \in \mathbf{GNS} : k(\mathbf{x}) = -k\}$. Consider the right-shifting of E_k, $E_k^* = \{\mathbf{x} = \mathbf{x}^* \times \mathbf{1}_{(k)} : \mathbf{x}^* \in E_k\}$, and the left-shifting of $\mathbf{f}(\mathbf{x})$:

$$\mathbf{f}^{(k)}(\mathbf{x}) = \begin{cases} \mathbf{f}(\mathbf{x} \times \mathbf{1}_{(-k)}) \times \mathbf{1}_{(-k)} & \text{if } \mathbf{x} \in E_k^* \\ 0 & \text{otherwise} \end{cases}$$

Using the method developed in Case 1, denote $a_k = \int_{\mathbf{GNS}} \mathbf{f}^{(k)}(\mathbf{x})d\mathbf{x}$.

Definition 6.5.2 [Wang (1985)]. If $\sum_k a_k$ is convergent, then the function $\mathbf{f}(\mathbf{x})$ is GNL-integrable and we write

$$\int_{\mathbf{GNS}} \mathbf{f}(\mathbf{x})d\mathbf{x} = \sum_k a_k$$

We are now ready to look at Dirac's δ function. Let $f:\mathbb{R}\to\mathbb{R}$ be infinitely differentiable and define $\mathbf{f}:\mathbf{GNS}\to\mathbf{GNS}$ as follows:

$$\begin{aligned}\mathbf{f}(\mathbf{x}) &= f(x_0) + f'(x_0)[\sum_{m=1}^{\infty} x_m \mathbf{1}_{(m)}] + \\ &+\cdots+\frac{1}{n!}f^{(n)}(x_0)[\sum_{m=1}^{\infty} x_m\mathbf{1}_{(m)}]^n+\cdots \\ &= \sum_{n=1}^{\infty}\frac{1}{n!}f^{(n)}(x_0)[\sum_{m=1}^{\infty} x_m\mathbf{1}_{(m)}]^n\end{aligned}$$

This function is called the **GNS** function induced by $f(x)$. Then it can be shown that 6.5.1 holds.

Theorem 6.5.1 [Wang (1985)]. *Let f and $\mathbf{f}$ be defined as before. Then there is a* GNL-*integrable function* $\mathbf{g}:\mathbf{GNS}\to\mathbf{GNS}$ *such that*

$$\int_{\mathbf{GNS}}\mathbf{f}(\mathbf{x})\mathbf{g}(\mathbf{x})d\mathbf{x} = \mathbf{f}(\mathbf{0})$$

In particular, let $g:\mathbb{R}\to\mathbb{R}$ be arbitrary and infinitely differentiable with bounded support: that is, $\mathrm{Sup}(g)=\{x\in\mathbb{R}: g(x)\neq 0\}$ *is bounded in $\mathbb{R}$, and $\int_{\mathbb{R}} g(x)\,dx = 1$. Then the* GNL-*integrable function* $\mathbf{g}:\mathbf{GNS}\to\mathbf{GNS}$ *can be defined*

$$\mathbf{g}(\mathbf{x}) = \begin{cases} g(x_1)\times\mathbf{1}_{(-1)} & \text{if } \mathbf{x}=(\ldots,0,x_1,x_2,\ldots) \\ 0 & \text{otherwise}\end{cases}$$

That is, $g(x)$ is a Dirac δ function according to Eq. (6.2).

Proof: Pick a real-valued function $g:\mathbb{R}\to\mathbb{R}$ such that g is infinitely differentiable with bounded support and satisfies

$$\int_{-\infty}^{\infty} g(t)\,dt = 1$$

Define

$$\mathbf{g}(\mathbf{x}) = \begin{cases} g(x_1)\mathbf{1}_{(-1)} & \text{for } x=(\ldots,0,x_1,x_2,\ldots) \\ 0 & \text{otherwise}\end{cases}$$

Then

$$\mathbf{f}(\mathbf{x})\mathbf{g}(\mathbf{x}) = \begin{cases} \sum_{n=0}^{\infty}[\frac{1}{n!}f^{(n)}(0)g(x_1) & (\sum_{m=1}^{\infty}x_m\mathbf{1}_{(m)})^n]\mathbf{1}_{(-1)} \\ & \text{for } \mathbf{x}=(\ldots,0,x_1,x_2,\ldots) \\ 0 & \text{otherwise}\end{cases}$$

Since $(\mathbf{f}(\mathbf{x})\mathbf{g}(\mathbf{x}))_{-n} = 0$, for each $n = 0, 1, 2, \ldots$, and $\int_{-\infty}^{\infty} g(t)\,dt = 1$, it follows that

$$\int_{\mathbf{GNS}} \mathbf{f}(\mathbf{x})\mathbf{g}(\mathbf{x})\,d\mathbf{x} = \mathbf{f}(\mathbf{0}) \int_{-\infty}^{\infty} g(t)\,dt = \mathbf{f}(\mathbf{0})$$

This completes the proof. ■

For a more rigorous treatment of **GNS**, please refer to Chapter 11

6.6. Some Final Words

If the ordered field **GNS** is substituted for ℝ, we can see that the theory and applications of **GNS** will have a great future, since ℝ has been applied in almost all areas of human knowledge and been challenged numerous times. As shown in (Wang, 1985; Wang, 1991), many studies can be carried out along the lines of classical research, such as mathematical analysis, distribution theory, quantum mechanics, etc., on **GNS**. Not only this, we also expect that when one is studying topics from various disciplines with **GNS**, some open problems in relevant fields might be answered, and some new fields of research might be opened up. Most importantly, we hope that the study of **GNS** will furnish a convenient tool for a computational general systems theory.

CHAPTER 7

Bellman's Principle of Optimality and Its Generalizations

In dynamic programming Bellman's principle of optimality is extremely important. The principle says (Bellman, 1957) that each subpolicy of an optimum policy must itself be an optimum policy with regard to the initial and the terminal states of the subpolicy. Here, the terms "policy," "subpolicy," "optimum policy," and "state" are are primitive and are not given specific meanings. In each application, these terms are understood in certain ways comparable to the situation of interest. For example, in a weighted graph, an optimum path from a vertex v_1 to vertex v_2 is the one with the smallest sum of weights of the edges along the path. In this case, a "policy" means a path, a "subpolicy" a subpath, an "optimum policy" stands for a path with the smallest weight, and "the initial and terminal states" for the beginning and ending vertices v_1 and v_2, respectively. The principle implies that for each chosen optimum path from v_1 to v_2, each subpath must be an optimum path connecting the initial and terminal vertices of the subpath.

7.1. A Brief Historical Note

During the last half century or so, Bellman's principle has become the cornerstone in the study of the shortest-path problem, inventory problem, traveling-salesman problem, equipment-replacement models, the resource allocation problems, etc. For details see Dreyfus and Law (1977). However, since the early 1970s, a number of practical optimization problems have been put forward in management science, economics, and many other fields. One feature common to all these problems is that the performance criteria are not real numbers, and consequently, the optimum options are no longer maxima or minima of some set of real numbers. Many dynamic programming models with nonscalar-valued performance criteria have been established. See (Bellman and Zadeh, 1970; Mitten, 1974; Wu, 1980; Wu, 1981a; Furukawa, 1980; Henig, 1983; Baldwin and Pilsworth, 1982).

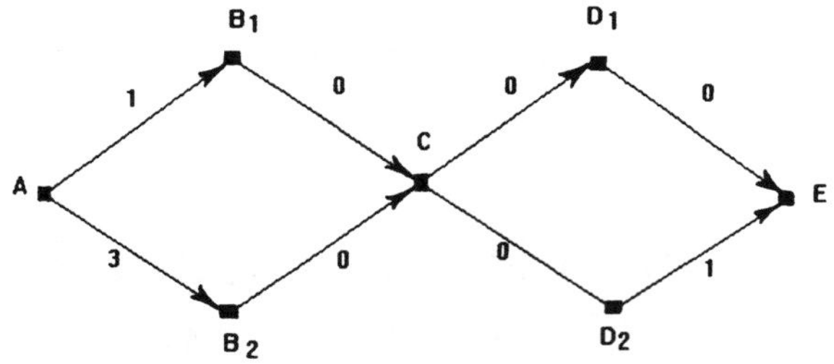

Figure 7.1. A subpolicy of the optimum policy is no longer optimum.

Since the application of Bellman's principle is no longer restricted to classical cases, scholars began to question how far the validity of the principle can be carried. Hu (1982) constructed an example to show that the principle does not hold in general. In fact, it was shown by Qin (1991), that if the goal set is strong antisymmetric, then the sum of local optimization orders is less than or equal to the global optimization order. Therefore, the following natural question arises.

Question 7.1.1 [Hu (1982)]. Under what characterizations can a given problem be solved by dynamic programming?

At the same time, as many scholars have realized, it is the fundamental equation approach, not the principle of optimality, that makes up the theoretical foundation of dynamic programming. The optimum principle gives one a property of the optimum policy, while the fundamental equation approach offers one an effective method to obtain the desired optimal policy. Since the validity of Bellman's principle is limited, is the fundamental equation approach also limited? Again, Hu's counterexample shows that the fundamental equation approach does not always work. Thus, the following questions are natural:

Question 7.1.2 [Wu and Wu (1986)]. In what range does the fundamental equation approach hold? How can its validity be proved? What relations are there, if any, between the fundamental equation approach and Bellman's Principle of Optimality?

To seek the answer to these questions, several approaches of systems analysis are presented.

7.2. Hu's Counterexample and Analysis

Let us first look at Hu's counterexample.

Example 7.2.1. Consider the network in Fig. 7.1. Define an optimum path as one where the sum of all arc lengths (mod4) is the minimum. Then the policy $A \to B_2 \to C \to D_2 \to E$ is optimum, but its subpolicy $B_2 \to C \to D_2 \to E$ is not optimum, since $B_2 \to C \to D_1 \to E$ is, which has weight 0.

Now let $W = \{0,1,2,3,\dots\}$ be the set of all possible weights, $\theta_1 =$ minimum: $p(W) \to W$ be the optimum option, which is a mapping from the power set $p(W)$ of W, the set of all subsets of W, into W such that

$$\theta_1(X) \in X \quad \text{for any } \emptyset \neq X \in p(W) \tag{7.1}$$

and let $\theta_2 = \{+\} \circ (\text{mod}4) : W^2 \to W$ be the binary relation on W such that

$$x\theta_2 y = \theta_2(x,y) = (x+y)(\text{mod}4) \quad \text{for any } x,y \in W \tag{7.2}$$

where $\circ$ stands for the ordinary composition of mappings.

If G is a nonempty set, representing the set of all vertices, then each network or graph or weighted graph is simply an element $g \in W \uparrow G^2$, where $W \uparrow G^2$ is the set of all mappings from $G^2 = G \times G$ into W. For any a_0, $a_n \in G$, define

$$\begin{aligned} g^{\langle n\rangle}(a_0,a_n) &= \theta_1\{(\cdots((g(a_0,x_1)\theta_2 g(x_1,x_2)\theta_2 g(x_2,x_3))\theta_2\cdots)\theta_2 g(x_{n-1},a_n)) : \\ &\qquad x_1,x_2,\dots,x_{n-1} \in G\} \\ &= \theta_1\{\omega(a_0x_1\cdots x_{n-1}a_n) : x_1,x_2,\dots,x_{n-1} \in G\} \end{aligned}$$

where $\omega(y_0,\dots y_n) = (\cdots((g(y_0,y_1)\theta_2 g(y_1,y_2))\theta_2\cdots)\theta_2 g(y_{n-1},y_n))$.

Then Bellman's principle of optimality can be written as follows.

Theorem 7.2.1 [Bellman's Principle of Optimality]. *For any natural number n, elements $a_0,a_1,a_2,\dots,a_n \in G$, if*

$$g^{\langle n\rangle}(a_0,a_n) = \omega(a_0a_1\cdots a_{n-1}a_n) \neq z \tag{7.3}$$

where z is the zero in W, then for any $0 \le i \le j \le n$,

$$g^{\langle i-j\rangle}(a_i,a_j) = \omega(a_ia_{i+1}\cdots a_{j-1}a_j)$$

For any $g \in W \uparrow G^2$, a natural number n, and $a,b \in G$, the fundamental equation on $(g;n,a,b)$ is given by the following functional equations:

$$\begin{aligned} f^{\langle 1\rangle}(x,b) &= g(x,b) \\ f^{\langle m\rangle}(x,b) &= \theta_1\{g(x,y)\theta_2 f^{\langle m-1\rangle}(y,b) : y \in G\} \end{aligned}$$

where $2 \le m \le n$. Applying these equations to Hu's counterexample, one obtains: $f^{\langle 4\rangle}(A,E) = 1$, with $A \to B_1 \to C \to D_1 \to E$ being its policy. However, it can be easily seen that $g^{\langle 4\rangle}(A,E) = 0$ and $A \to B_2 \to C \to D_2 \to E$ is the global optimum policy with weight 0. That is, in this example, $f^{\langle 4\rangle}(A,E) \neq g^{\langle 4\rangle}(A,E)$. Hence, we ask when will the weight obtained from the fundamental equation approach be the same as the global optimum weight?

7.3. Generalized Principle of Optimality

By formalizing the discussion in the previous section, Wu and Wu (1984a) obtained one sufficient and three necessary conditions for Bellman's principle to hold. In this section, all results are based on (Wu and Wu, 1984a).

In general, let W and G represent two nonempty sets, $\theta_1 : p(W) \to W$ be the optimum option satisfying Eq. (7.1), and θ_2 be a mapping from W^2 into W. For any element $g \in W \uparrow G^2$ and any $a_0, a_n \in G$, define

$$\begin{aligned} g^{\langle n\rangle}(a_0,a_n) &= \theta_1\{(\cdots((g(a_0,x_1)\theta_2 g(x_1,x_2))\theta_2 g(x_2,x_3)\theta_2\cdots)\theta_2 g(x_{n-1},a_n)) : \\ &\qquad x_1,x_2,\ldots,x_{n-1} \in G\} \\ &= \theta_1\{\omega(a_0x_1\cdots x_{n-1}a_n) : x_1,x_2,\ldots,x_{n-1} \in G\} \end{aligned}$$

where $\omega(y_0,\ldots,y_n) = (\cdots((g(y_0,y_1)\theta_2 g(y_1,y_2))\theta_2\cdots)\theta_2 g(y_{n-1},y_n))$.

An element $z \in W$ is called a zero element, or simply zero if

$$z\theta_2 w = w\theta_2 z = w \quad \text{for each } w \in W$$

and

$$\theta_1(X \cup \{z\}) = \theta_1(X) \quad \text{for each } \emptyset \neq X \in p(W)$$

The optimum option θ_1 is X-optimum-preserving for $X \in p(W)$ if

$$\theta_1(X) = x \text{ and } \theta_1(Y) = x \quad \text{for any } x \in Y \subset X$$

The option θ_1 satisfies the optimum-preserving law if θ_1 is X-optimum-preserving for all $\emptyset \neq X \in p(W)$.

The relation θ_2 is distributive over θ_1 if

$$\theta_1(X)\theta_2 x = \theta_1\{w\theta_2 x : w \in X\}$$

for all $\emptyset \neq X \in p(W)$ and any $x \in W$, and

$$x\theta_2\theta_1(X) = \theta_1\{x\theta_2 w : w \in X\}$$

for all $\emptyset \neq X \in p(W)$ and $x \in W$. The relation θ_2 is associative if for any $w_1, w_2, w_3 \in W$,

$$(w_1\theta_2 w_2)\theta_2 w_3 = w_1\theta_2(w_2\theta_2 w_3)$$

The relation θ_2 satisfies the cancellation law if for any $w_1, w_2, w_3 \in W$,

$$w_1\theta_2 w_2 = w_1\theta_2 w_3 \text{ implies } w_2 = w_3$$

and

$$w_2\theta_2 w_1 = w_3\theta_2 w_1 \text{ implies } w_2 = w_3$$

whenever w_1 is not the zero element in W.

Theorem 7.3.1 [Generalized Principle of Optimality (Wu and Wu, 1984a)]. *Assume that θ_1 satisfies the optimum-preserving law and that θ_2 is associative, cancellative and distributive over θ_1. Then in the system $(W, G; \theta_1, \theta_2)$, Bellman's Principle of Optimality (Theorem 7.2.1) holds.*

Proof: Suppose that $g \in W \uparrow G^2$, n is a natural number, $a_0, a_1, \ldots, a_n \in G$, and $g^{\langle n \rangle}(a_0, a_n) = \omega(a_0 a_1 \cdots a_{n-1} a_n) \neq z$. For any $0 \leq i \leq j \leq n$, since

$$\theta_1\{\omega(a_0 x_1 x_2 \cdots x_{n-1} a_n) : x_1, x_2, \ldots, x_{n-1} \in G\} = \omega(a_0 a_1 \cdots a_n)$$

and

$$\{\omega(a_0 a_1 \cdots a_i x_{i+1} x_{i+2} \cdots x_{j-1} a_j a_{j+1} \cdots a_n) : x_{i+1}, x_{i+2}, \ldots, x_{j-1} \in G\} \\ \subset \{\omega(a_0 x_1 x_2 \cdots x_{n-1} a_n) : x_1, x_2, \ldots, x_{n-1} \in G\}$$

and

$$\omega(a_0 a_1 \cdots a_n) \in \{\omega(a_0 a_1 \cdots a_i x_{i+1} x_{i+2} \cdots x_{j-1} a_j a_{j+1} \cdots a_n) : \\ x_{i+1}, x_{i+2}, \ldots, x_{j-1} \in G\}$$

then by the optimum-preserving law of θ_1, it follows that

$$\theta_1\{\omega(a_0 a_1 \cdots a_i x_{i+1} x_{i+2} \cdots x_{j-1} a_j a_{j+1} \cdots a_n) : x_{i+1}, x_{i+2}, \ldots, x_{j-1} \in G\} \\ = \omega(a_0 a_1 \cdots a_n)$$

By the associative law and the distributive law over θ_1 of θ_2, one has

$$\begin{aligned} &\omega(a_0 a_1 \cdots a_i)\theta_2(\theta_1\{\omega(a_i x_{i+1} \cdots x_{j-1} a_j) : \\ &\qquad x_{i+1}, x_{i+2}, \ldots, x_{j-1} \in G\})\theta_2\omega(a_j a_{j+1} \cdots a_n) \\ &\quad = \theta_1\{\omega(a_0 a_1 \cdots a_i)\theta_2\omega(a_i x_{i+1} \cdots x_{j-1} a_j)\theta_2\omega(a_j a_{j+1} \cdots a_n) : \\ &\qquad x_{i+1}, x_{i+2}, \ldots, x_{j-1} \in G\} \\ &\quad = \theta_1\{\omega(a_0 a_1 \cdots a_i x_{i+1} \cdots x_{j-1} a_j \cdots a_n) : x_{i+1}, x_{i+2}, \ldots, x_{j-1} \in G\} \\ &\quad = \omega(a_0 a_1 \cdots a_n) \\ &\quad = \omega(a_0 a_1 \cdots a_i)\theta_2\omega(a_i a_{i+1} \cdots a_j)\theta_2\omega(a_j a_{j+1} \cdots a_n) \end{aligned}$$

Since $\omega(a_0 a_1 \cdots a_n) \neq z$, one has that $\omega(a_0 a_1 \cdots a_i) \neq z$ and $\omega(a_j a_{j+1} \cdots a_n) \neq z$. Therefore, by the cancellation law of θ_2,

$$\theta_1\{\omega(a_i x_{i+1} \cdots x_{j-1} a_j) : x_{i+1}, \ldots, x_{j-1} \in G\} = \omega(a_i a_{i+1} \cdots a_j)$$

That is, $g^{\langle j-i \rangle}(a_i, a_j) = \omega(a_i a_{i+1} \cdots a_j)$. ■

The generalized principle of optimality is a sufficient condition of the conservation from a global optimization to local ones. In the following section, it is shown that the conditions in the generalized optimality principle are also necessary except for the condition of associativity.

7.4. Some Partial Answers and a Conjecture

Theorem 7.4.1 [Wu and Wu (1984a)]. *Assume that* θ_2 *is associative,* n *a natural number, and* $a_0, a_1, \ldots, a_n \in G$ *satisfy Eq. (7.3). If the Generalized Principle of Optimality (Theorem 7.3.1) holds for a chosen* $g \in W \uparrow G^2$, *then* θ_2 *possesses the following property of cancellation: for any* $0 \le i < j < k \le n$ *and any* $x_{i+1}, \ldots, x_{j-1}, x_{j+1}, \ldots, x_{k-1} \in G$,

$$\omega(a_i a_{i+1} \cdots a_j)\theta_2 \omega(a_j a_{j+1} \cdots a_k) = \omega(a_i a_{i+1} \cdots a_j)\theta_2 \omega(a_j x_{j+1} \cdots x_{k-1} a_k) \quad (7.4)$$

implies

$$\omega(a_j a_{j+1} \cdots a_k) = \omega(a_j x_{j+1} \cdots x_{k-1} a_k)$$

and

$$\omega(a_i a_{i+1} \cdots a_j)\theta_2 \omega(a_j a_{j+1} \cdots a_k) = \omega(a_i x_{i+1} \cdots x_{j-1} a_j)\theta_2 \omega(a_j a_{j+1} \cdots a_k)$$

implies

$$\omega(a_i a_{i+1} \cdots a_j) = \omega(a_i x_{i+1} \cdots x_{j-1} a_j)$$

Proof: Since $g^{\langle n\rangle}(a_0, a_n) = \omega(a_0 a_1 \cdots a_{n-1} a_n) \neq z$, and from the assumption that the Generalized Principle of Optimality holds, it follows that

$$g^{\langle k-j\rangle}(a_j, a_k) = \omega(a_j a_{j+1} \cdots a_k) \quad (7.5)$$

By Eq. (7.4) and the associative law of θ_2, it follows that

$$\begin{aligned}\omega(a_0 a_1 \cdots a_{n-1} a_n) &= \omega(a_0 a_1 \cdots a_i)\theta_2(\omega(a_i a_{i+1} \cdots a_j)\theta_2 \omega(a_j a_{j+1} \cdots a_k))\\ &\quad \theta_2 \omega(a_k a_{k+1} \cdots a_n)\\ &= \omega(a_0 a_1 \cdots a_i)\theta_2(\omega(a_i a_{i+1} \cdots a_j)\theta_2 \omega(a_j x_{j+1} \cdots x_{k-1} a_k))\theta_2\\ &\quad \omega(a_k a_{k+1} \cdots a_n)\\ &= \omega(a_0 a_1 \cdots a_j x_{j+1} x_{j+2} \cdots x_{k-1} a_k a_{k+1} \cdots a_n)\end{aligned}$$

That is,

$$\begin{aligned}g^{\langle n\rangle}(a_0, a_n) &= \omega(a_0 a_1 \cdots a_j x_{j+1} x_{j+2} \cdots x_{k-1} a_k a_{k+1} \cdots a_n)\\ &\neq z\end{aligned}$$

From the assumption that the Generalized Principle of Optimality holds, it follows that

$$g^{\langle k-j\rangle}(a_j, a_k) = \omega(a_j x_{j+1} \cdots x_{k-1} a_k) \quad (7.6)$$

By Eqs. (7.5) and (7.6), it follows that

$$\omega(a_j a_{j+1} \cdots a_k) = \omega(a_j x_{j+1} x_{j+2} \cdots x_{k-1} a_k)$$

The proof of the other equation is similar and is omitted. ■

Theorem 7.4.2 [Wu and Wu (1984a)]. *Assume that θ_1 satisfies the optimum-preserving law, θ_2 is associative, n os a natural number, and $a_0, a_1, \ldots, a_n \in G$ satisfy Eq. (7.3). If the Generalized Principle of Optimality (Theorem 7.3.1) holds, then θ_2 possesses the following distributivity over θ_1 : for $1 \leq i \leq n-1$,*

$$\begin{aligned} &\theta_1\{\omega(a_0 x_1 \cdots x_{i-1} a_i) : x_1, \ldots, x_{i-1} \in G\} \theta_2 \omega(a_i a_{i+1} \cdots a_n) \\ &\qquad = \theta_1\{\omega(a_0 x_1 \cdots x_{i-1} a_i) \theta_2 \omega(a_i a_{i+1} \cdots a_n) : x_1, \ldots, x_{i-1} \in G\} \end{aligned}$$

and

$$\begin{aligned} &\omega(a_0 a_1 \cdots a_j) \theta_2 \theta_1\{\omega(a_i x_{i+1} \cdots x_{n-1} a_n) : x_{i+1}, \ldots, x_{n-1} \in G\} \\ &\qquad = \theta_1\{\omega(a_0 a_1 \cdots a_j) \theta_2 \omega(a_i x_{i+1} \cdots x_{n-1} a_n) : x_{i+1}, \ldots, x_{n-1} \in G\} \end{aligned}$$

Proof: Since it is assumed that the Generalized Principle of Optimality holds, it follows that

$$g^{\langle i \rangle}(a_0, a_i) = \omega(a_0 a_1 \cdots a_i) \tag{7.7}$$

The assumption that θ_1 satisfies the optimum-preserving law implies that

$$\begin{aligned} &\theta_1\{\omega(a_0 x_1 \cdots x_{n-1} a_n) : x_1, x_2, \ldots, x_{n-1} \in G\} \\ &= \theta_1\{\omega(a_0 x_1 \cdots x_{i-1} a_i a_{i+1} \cdots a_n) : x_1, x_2, \ldots, x_{i-1} \in G\} \end{aligned} \tag{7.8}$$

Then, from Eqs. (7.7) and (7.8) and the associative law of θ_2, it follows that

$$\begin{aligned} &(\theta_1\{\omega(a_0 x_1 x_2 \cdots x_{i-1} a_i) : x_1, x_2, \ldots, x_{i-1} \in G\}) \theta_2 \omega(a_i a_{i+1} \cdots a_n) \\ &= g^{\langle i \rangle}(a_0, a_i) \theta_2 \omega(a_i a_{i+1} \cdots a_n) \\ &= \omega(a_0 a_1 \cdots a_i) \theta_2 \omega(a_i a_{i+1} \cdots a_n) \\ &= \omega(a_0 a_1 \cdots a_n) \\ &= g^{\langle n \rangle}(a_0, a_n) \\ &= \theta_1\{\omega(a_0 x_1 x_2 \cdots x_{n-1} a_n) : x_1, x_2, \ldots, x_{n-1} \in G\} \\ &= \theta_1\{\omega(a_0 x_1 \cdots x_{i-1} a_i a_{i+1} \cdots a_n) : x_1, x_2 \cdots, x_{i-1} \in G\} \\ &= \theta_1\{\omega(a_0 x_1 \cdots x_{i-1} a_i) \theta_2 \omega(a_i a_{i+1} \cdots a_n) : x_1, x_2, \ldots, x_{i-1} \in G\} \end{aligned}$$

just what is desired. The proof of the other identity is similar and is omitted. ■

Theorem 7.4.3 [Wu and Wu (1984a)]. *Assume that θ_2 is associative and distributive over θ_1 and n is a natural number, and $a_0, a_1, \ldots, a_n \in G$ satisfy Eq. (7.3). If the Generalized Principle of Optimality (Theorem 7.3.1) holds, then θ_1 possesses the following optimum-preserving property: for each $1 \leq i \leq n-1$,*

$$\begin{aligned}
&\theta_1\{\omega(a_0x_1\cdots x_{n-1}a_n) : x_1, x_2, \ldots, x_{n-1} \in G\} \\
&= \theta_1\{\omega(a_0x_1\cdots x_{i-1}a_ia_{i+1}\cdots a_n) : x_1, x_2, \ldots, x_{i-1} \in G\} \\
&= \theta_1\{\omega(a_0a_1\cdots a_ix_{i+1}\cdots x_{n-1}a_n) : x_{i+1}, \ldots, x_{n-1} \in G\}
\end{aligned}$$

Proof: Since the Generalized Principle of Optimality is assumed to be true,

$$g^{\langle i\rangle}(a_0, a_i) = \omega(a_0a_1\cdots a_i)$$

Therefore, by associativity and the distributive law over θ_1 of θ_2, it follows that

$$\begin{aligned}
&\theta_1\{\omega(a_0x_1\cdots x_{n-1}a_n) : x_1, x_2, \ldots, x_{n-1} \in G\} \\
&= g^{\langle n\rangle}(a_0, a_n) \\
&= \omega(a_0a_1\cdots a_n) \\
&= \omega(a_0a_1\cdots a_i)\theta_2\omega(a_ia_{i+1}\cdots a_n) \\
&= (g^{\langle i\rangle}(a_0, a_i))\theta_2\omega(a_ia_{i+1}\cdots a_n) \\
&= (\theta_1\{\omega(a_0x_1\cdots x_{i-1}a_i) : x_1, x_2, \ldots, x_{i-1} \in G\})\theta_2\omega(a_ia_{i+1}\cdots a_n) \\
&= \theta_1\{\omega(a_0x_1\cdots x_{i-1}a_i)\theta_2\omega(a_ia_{i+1}\cdots a_n) : x_1, x_2, \ldots, x_{i-1} \in G\} \\
&= \theta_1\{\omega(a_0x_1\cdots x_{i-1}a_ia_{i+1}\cdots a_n) : x_1, x_2, \ldots, x_{i-1} \in G\}
\end{aligned}$$

The proof of the other identity is similar and is omitted. ■

We conclude with the following conjecture.

Conjecture 7.4.1. *In the system $(W, G; \theta_1, \theta_2)$, the Generalized Principle of Optimality holds iff there is an ordering relation $\leq$ on W such that*

(i) for any $x, y \in W$, one and only one of the following must be true:

$$x\langle y, \quad x = y, \quad x\rangle y$$

(ii) the optimum option θ_1 is defined by $\theta_1 =$ superior or $\theta_1 =$ inferior under the ordering relation $\leq$

(iii) the ordering relation $\leq$ and the relation θ_2 are consistent. That is, $x \leq y$ iff $x\theta_2w \leq y\theta_2w$ and $w\theta_2x \leq w\theta_2y$ for any $w \in W$.

7.5. Generalization of Fundamental Equations

In the previous two sections, Bellman's Principle of Optimality has been generalized and three necessary conditions given. However, one should keep in mind that the optimum principle only states a property of the optimum policy, while the fundamental equation approach offers an effective method for one to obtain the desired optimal policy. In this section, a sufficient condition and two necessary conditions are established under which the fundamental equation approach holds. That is, under the conditions, the weight computed from the fundamental equation equals that of a global optimum. The presentation here is based on Shouzhi Wu and Xuemou Wu (1986).

For a given system $(W, G; \theta_1, \theta_2)$, defined in the same way as in the previous section, for any $g \in W \uparrow G^2$, a natural number n, and $a, b \in G$, the following functional equations are called a fundamental equation on $(g; n, a, b)$, or simply a fundamental equation:

$$f^{\langle 1\rangle}(x,b) = g(x,b)$$
$$f^{\langle m\rangle}(x,b) = \theta_1\{g(x,y)\theta_2 f^{\langle m-1\rangle}(y,b) : y \in G\} \tag{7.9}$$

for $2 \le m \le n$. Here, the mappings θ_1 and θ_2 express the recursiveness of operation, which in turn shows that Eq. (7.9) is a generalization of the fundamental equations in dynamic programming (Bellman, 1957).

Theorem 7.5.1 [Wu and Wu (1986)]. *In the system $(W, G; \theta_1, \theta_2)$, if θ_1 satisfies the optimum-preserving law, and θ_2 the associative and distributive laws over θ_1, then for any $g \in W \uparrow G^2$, $n \in \mathbb{N}$, and $a, b \in G$, the following is true:*

$$g^{\langle n\rangle}(a,b) = f^{\langle n\rangle}(a,b)$$

Proof: For any $g \in W \uparrow G^2$, $n \in \mathbb{N}$, and $a, b \in G$, mathematical induction will be used on the index m to show that

$$g^{\langle m\rangle}(a,b) = f^{\langle m\rangle}(a,b)$$

When $m = 1$, it is obvious that

$$g^{\langle 1\rangle}(a,b) = g(a,b) = f^{\langle 1\rangle}(a,b)$$

Suppose that $g^{\langle m\rangle}(a,b) = f^{\langle m\rangle}(a,b)$ for some $m \in \mathbb{N}$ and any a and $b \in G$. It is now sufficient to show that $g^{\langle m+1\rangle}(a,b) = f^{\langle m+1\rangle}(a,b)$. Since θ_2 satisfies the associative and distributive laws, for any $x \in G$, one has

$$\begin{aligned} g(a,x)\theta_2 f^{\langle m\rangle}(x,b) &= g(a,x)\theta_2 g^{\langle m\rangle}(x,b) \\ &= g(a,x)\theta_2\theta_1\{\omega(xy_1y_2\cdots y_{m-1}b) : y_1, y_2, \ldots, y_{m-1} \in G\} \\ &= \theta_1\{g(a,x)\theta_2\omega(xy_1y_2\cdots y_{m-1}b) : y_1, y_2, \ldots, y_{m-1} \in G\} \\ &= \theta_1\{\omega(axy_1y_2\cdots y_{m-1}b) : y_1, y_2, \ldots, y_{m-1} \in G\} \end{aligned}$$

Therefore,

$$f^{\langle m+1\rangle}(a,b) = \theta_1\{g(a,x)\theta_2 f^{\langle m\rangle}(x,b) : x \in G\}$$
$$= \theta_1\{\omega(axy_1y_2\cdots y_{m-1}b) : y_1,y_2,\ldots,y_{m-1} \in G\}$$

Claim. If θ_1 satisfies the optimum-preserving law, then for any $g \in W \uparrow G^2$, $n \in \mathbb{N}$ and a and $b \in G$, the following equation holds:

$$g^{\langle n\rangle}(a,b) = \theta_1\{\theta_1\{\omega(ax_1x_2\cdots x_{n-1}b) : x_2,\ldots,x_{n-1} \in G\} : x_1 \in G\}$$

Now the assumption that θ_1 satisfies the optimum-preserving law and the Claim guarantee that

$$g^{\langle m+1\rangle}(a,b) = \theta_1\{\theta_1\{\omega(axy_1y_2\cdots y_{m-1}b) : y_1,y_2,\ldots,y_{m-1} \in G\} : x \in G\}$$

Thus,

$$g^{\langle m+1\rangle}(a,b) = f^{\langle m+1\rangle}(a,b)$$

which is what we desired.

To finish the proof of the theorem, let us see why the claim is true.

Proof of claim: First note that for any $x_1 \in G$, the following is true:

$$\{\omega(ax_1x_2\cdots x_{n-1}b) : x_2,\ldots,x_{n-1} \in G\}$$
$$\subset \{\omega(ax_1x_2\cdots x_{n-1}b) : x_1,x_2,\ldots,x_{n-1} \in G\}$$

Therefore, for any $x_1 \in G$, one has

$$\theta_1\{\omega(ax_1x_2\cdots x_{n-1}b) : x_2,\ldots,x_{n-1} \in G\}$$
$$\in \{\omega(ax_1x_2\cdots x_{n-1}b) : x_1,x_2,\ldots,x_{n-1} \in G\}$$

Hence, it follows that

$$\{\theta_1\{\omega(ax_1x_2\cdots x_{n-1}b) : x_2,\ldots,x_{n-1} \in G\} : x_1 \in G\}$$
$$\subset \{\omega(ax_1x_2\cdots x_{n-1}b) : x_1,x_2,\ldots,x_{n-1} \in G\} \quad (7.10)$$

Assume that $g^{\langle n\rangle}(a,b) = \omega(ac_1c_2\cdots c_{n-1}b)$ for some $c_1,c_2,\ldots,c_{n-1} \in G$. Since

$$\{\omega(ac_1x_2\cdots x_{n-1}b) : x_2,\ldots,x_{n-1} \in G\}$$
$$\subset \{\omega(ax_1x_2\cdots x_{n-1}b) : x_1,x_2,\ldots,x_{n-1} \in G\}$$

and

$$\omega(ac_1c_2\cdots c_{n-1}b) \in \{\omega(ac_1x_2\cdots x_{n-1}b) : x_2,\ldots,x_{n-1} \in G\}$$

and θ_1 satisfies the optimum-preserving law, it follows that

$$\theta_1\{\omega(ac_1x_2\cdots x_{n-1}b):x_2,\ldots,x_{n-1}\in G\}=\omega(ac_1c_2\cdots c_{n-1}b)$$

This equation implies at once that

$$\omega(ac_1c_2\cdots c_{n-1}b)\in\{\theta_1\{\omega(ax_1x_2\cdots x_{n-1}b):x_2,\ldots,x_{n-1}\in G\}:x_1\in G\} \tag{7.11}$$

Now, Eqs. (7.10) and (7.11) and the assumption that θ_1 satisfies the optimum-preserving law imply that

$$\begin{aligned}\theta_1\{\theta_1\{\omega(ax_1x_2\cdots x_{n-1}b):x_2,\ldots,x_{n-1}\in G\}:x_1\in G\}\}&=\omega(ac_1c_2\cdots c_{n-1}b)\\&=g^{\langle n\rangle}(a,b)\end{aligned}$$

This ends the proof of the theorem. ■

The following results show that the conditions in Theorem 7.5.1 are also necessary.

Theorem 7.5.2 [Wu and Wu (1986)]. *In the system $(W,G;\theta_1,\theta_2)$ with θ_1 satisfying the optimum-preserving law and θ_2 satisfying the associative law, if for any $g\in W\uparrow G^2$, $n\in\mathbb{N}$, and $a,b\in G$, it is true that*

$$g^{\langle n\rangle}(a,b)=f^{\langle n\rangle}(a,b)$$

then θ_2 possesses the following distribution over θ_1 : for any $g\in W\uparrow G^2,n\in\mathbb{N}$, and $a,b\in G$,

$$\begin{aligned}&\theta_1\{g(a,x_1)\theta_2\theta_1\{\omega(x_1x_2\cdots x_{n-1}b):x_2,x_3,\ldots,x_{n-1}\in G\}:x_1\in G\}\\&\quad=\theta_1\{\theta_1\{g(a,x_1)\theta_2\omega(x_1x_2\cdots x_{n-1}b):x_2,x_3,\ldots,\cdots,x_{n-1}\in G\}:x_1\in G\}\end{aligned}$$

Proof: For any $g\in W\uparrow G^2,n\in\mathbb{N}$, and $a,b\in G$, and since θ_1 satisfies the optimum-preserving law, from the Claim it follows that

$$g^{\langle n\rangle}(a,b)=\theta_1\{\theta_1\{\omega(ax_1x_2\cdots x_{n-1}b):x_2,x_3,\ldots,x_{n-1}\in G\}:x_1\in G\}$$

From the assumption that θ_2 satisfies the associative law, it follows that

$$\omega(ax_1x_2\cdots x_{n-1}b)=g(a,x_1)\theta_2\omega(x_1x_2\cdots x_{n-1}b)$$

Therefore,

$$g^{\langle n\rangle}(a,b)=\theta_1\{\theta_1\{g(a,x_1)\theta_2\omega(x_1x_2\cdots x_{n-1}b):x_2,x_3,\ldots,x_{n-1}\in G\}:x_1\in G\} \tag{7.12}$$

Now, from the assumption that for any $g \in W \uparrow G^2$, $n \in \mathbb{N}$, and $a,b \in G$, the following is true: $g^{\langle n \rangle}(a,b) = f^{\langle n \rangle}(a,b)$. Thus,

$$\begin{aligned} g^{\langle n \rangle}(a,b) &= \theta_1 \{g(a,x_1)\theta_2 g^{\langle n-1 \rangle}(x_1,b) : x_1 \in G\} \\ &= \theta_1 \{g(a,x_1)\theta_2 \theta_1 \{\omega(x_1 x_2 \cdots x_{n-1} b) : \\ &\quad x_2,\dots,x_{n-1} \in G\} : x_1 \in G\} \end{aligned} \tag{7.13}$$

By Eqs. (7.12) and (7.13), it finally follows that

$$\begin{aligned} &\theta_1 \{g(a,x_1)\theta_2 \theta_1 \{\omega(x_1 x_2 \cdots x_{n-1} b) : x_2,\dots,x_{n-1} \in G\} : x_1 \in G\} \\ &= \theta_1 \{\theta_1 \{g(a,x_1)\theta_2 \omega(x_1 x_2 \cdots x_{n-1} b) : x_2,\dots,x_{n-1} \in G\} : x_1 \in G\} \end{aligned}$$

This ends the proof of the theorem. ■

Theorem 7.5.3 [Wu and Wu (1986)]. *In the system $(W,G;\theta_1,\theta_2)$ with θ_2 satisfying the associative law and the distributive law over θ_1, if for any $g \in W \uparrow G^2$, $n \in \mathbb{N}$, and $a,b \in G$, it is true that*

$$g^{\langle n \rangle}(a,b) = f^{\langle n \rangle}(a,b)$$

then θ_1 possesses the following optimum-preserving property: for any $g \in W \uparrow G^2, n \in \mathbb{N}$ and a and $b \in G$, the following is true:

$$\begin{aligned} &\theta_1 \{\theta_1 \{\omega(a x_1 x_2 \cdots x_{n-1} b) : x_2,\dots,x_{n-1} \in G\} : x_1 \in G\} \\ &= \theta_1 \{\omega(a x_1 x_2 \cdots x_{n-1} b) : x_1,x_2,\dots,x_{n-1} \in G\} \end{aligned}$$

Proof: Since for any $g \in W \uparrow G^2, n \in \mathbb{N}$, and $a,b \in G$, it is true that $g^{\langle n \rangle}(a,b) = f^{\langle n \rangle}(a,b)$, and θ_2 satisfies the associative law and the distributive law over θ_1, it then follows that

$$\begin{aligned} &\theta_1 \{\omega(a x_1 x_2 \cdots x_{n-1} b) : x_1,x_2,\dots,x_{n-1} \in G\} \\ &= g^{\langle n \rangle}(a,b) \\ &= \theta_1 \{g(a,x_1)\theta_2 g^{\langle n-1 \rangle}(x_1,b) : x_1 \in G\} \\ &= \theta_1 \{g(a,x_1)\theta_2 \theta_1 \{\omega(x_1 x_2 \cdots x_{n-1} b) : x_2,\dots,x_{n-1} \in G\} : x_1 \in G\} \\ &= \theta_1 \{\theta_1 \{g(a,x_1)\theta_2 \omega(x_1 x_2 \cdots x_{n-1} b) : x_2,\dots,x_{n-1} \in G\} : x_1 \in G\} \\ &= \theta_1 \{\theta_1 \{\omega(a x_1 x_2 \cdots x_{n-1} b) : x_2,\dots,\cdots,x_{n-1} \in G\} : x_1 \in G\} \end{aligned}$$

as desired. ■

7.6. Operation Epitome Principle

This section presents a different approach to systems analysis, which is due to Qin (1991). With this approach, Bellman's Principle of Optimality once again is generalized. The generalized result is called the "operation epitome principle."

For an N-stage sequential decision process, let X_i be the set of all states of the ith stage, $i = 0, 1, 2, \ldots, N$, where $X_0, X_1, \ldots, X_N$ are some fixed sets. Let $E_i = X_{i-1} \times X_N$, $i = 1, 2, \ldots, N$, $E = \prod\{E_i : i = 1, 2, \ldots, N\}$, $W \subseteq \mathbb{R}$, and define a binary relation $g \subseteq W^2$ by the following: for any $x, y \in W^2$, $(x, y) \in g$ iff $x \leq y$. Further, $h_{ij} : E_i \times X_j \to W$ is the gain (function) of decisions. Intuitively speaking, the set E consists of all available policies, and the set W is made up of all the criteria used in the decision-making process. The binary relation $g \subseteq W^2$ establishes the comparison between criteria.

Define a mapping $F : \bigcup_{i=1}^{N} W_i \to W$ so that the following conditions are satisfied:

(1) $F(w) = w$ for each $w \in W$.

(2) For each $(w_1, w_2, \ldots, w_i) \in W^i$, $i \leq N$, and $1 \leq k \leq i$,

$$F(w_1, w_2, \ldots, w_i) = F(F(w_1, w_2, \ldots, w_k), F(w_{k+1}, w_{k+2}, \ldots, w_i))$$

(3) For each fixed $w \in W$, the mappings $F(\cdot, w)$ and $F(w, \cdot) : W \to W$ are strictly order-preserving. That is, for any $x, y \in W$, $x < y$ iff $F(x, w) < F(y, w)$, and $x < y$ iff $F(w, x) < F(w, y)$.

For example, it can be shown that the addition operation of real numbers can be the mapping F. In addition, if W contains all positive real numbers, then the multiplication operation can also be the mapping F.

The gain of the policy $e_{1-N} = (e_1, e_2, \ldots, e_N) \in E$ is measured by the function $f : E \to W$, defined by

$$f(e_{1-N}) = F(h_{1N}(e_1), h_{2N}(e_2), \ldots, h_{NN}(e_N))$$

For a subpolicy $e_{i-j} = (e_i, e_{i+1}, \ldots, e_j)$ of $e_{1-N} = (e_1, e_2, \ldots, e_N)$ with initial state $x_i \in X_i$ and terminal state $x_j \in X_j$, the gain of e_{i-j} is defined as follows:

$$E_k^* = X_{k-1} \times X_j,\ k = i+1, i+2, \ldots, j$$
$$E_{i-j} = \prod\{E_k^* : k = i+1, i+2, \ldots, j\}$$

and

$$f(e_{i-j}) = F(h_{ij}(e_i), h_{i+1,j}(e_{i+1}), \ldots, h_{jj}(e_j))$$

Now, for each policy $e \in E$, the stationary set (of the policy e) is defined

$$\begin{aligned}\mathrm{STY}(e) &\equiv \{w \in f(E) : w \neq f(e) \text{ and } (f(e), w) \in g\} \\ &= g(f(e)) \cap f(E) - \{f(e)\}\end{aligned}$$

If $\mathrm{STY}(e) = \emptyset$, then $f(e)$ is called a generalized extreme value of $f(E)$. Then, Bellman's Principle of Optimality can be written as follows.

Theorem 7.6.1 [Bellman's Principle of Optimality]. *For any policy $e_{1-N} = (e_1, e_2, \ldots, e_N) \in E$, if the equation $\mathrm{STY}(e_{1-N}) = \emptyset$ holds in the system $(W, E; f, g)$, then for any natural numbers i and j satisfying $0 \leq i \leq j \leq N$, the subpolicy $e_{i-j} = (e_i, e_{i+1}, \ldots, e_j)$ of the policy e_{1-N} satisfies the condition that the equation $\mathrm{STY}(e_{i-j}) = \emptyset$ holds in the system $(W, E_{i-j}; f, g)$.*

To prove this new version of Bellman's principle, we first formalize the previous discussion.

Consider the system $(W,E;f,g)$ such that E and W are nonempty sets, $f : E \to W$ is a mapping and $g \subseteq W^2$ is a binary relation. This system is called the optimization model of the general systems. The set W consists of all the criteria used in an optimization problem of interest, and E is the set of all possible policies. These sets are called the criteria set and the policy set of $(W, E : f, g)$, respectively. Intuitively speaking, for each chosen criterion $e \in E$, there exists exactly one policy $w \in W$ and the rule matching the criterion e with the policy w is described by the function $f : E \to W$. The binary relation $g \subseteq W^2$ defines how to compare criteria used.

For any subset $X \subseteq W$, define

$$g * X = \{w \in X : \text{ for any} x \in X,\ x \neq w \text{ implies } (w,x) \notin g\} \tag{7.14}$$

If $X = f(E)$ and $w \in X - g * X$, then there exists at least one $x \in X = f(E)$ such that $x \neq w$ and $(w,x) \in g$. Let $e \in E$ be such that $w = f(e) \in f(E) - g * f(E)$. Then the element $w \in W$ is called a generalized extreme value of the set $X = f(E)$, and the element $e \in E$ is a generalized optimal solution (or policy) of the model $(W,E;f,g)$.

Lemma 7.6.1 [Qin (1991)]. *If $X \subseteq W$ is a nonempty subset, then*

$$g * X = \{x \in X : X \cap (g \cup I)(x) = \{x\}\}$$

where I is the diagonal of the Cartesian product X^2, defined by $I = \{(x,x) : x \in X\}$.

Proof: For each element $w \in g * X$, by the definition of $g * X$ it follows that, for any $x \in X$, if $x \neq w$, then $(w,x) \notin g$. That is, $w \in X$ and $w \in (g \cup I)(w) =$

$g(w) \cup \{w\}$. If $x \in X \cap (g \cup I)(w)$ and $x \neq w$, then $x \in g(w)$. That is, $(w,x) \in g$, contradiction. So, $X \cap (g \cup I)(w) = \{w\}$. This proves that

$$g * X \subseteq \{x \in X : X \cap (g \cup I)(x) = \{x\}\}$$

Conversely, if for $w \in X$, $X \cap (g \cup I)(w) = \{w\}$, then for any $x \in X, x \neq w$ must imply $(w,x) \notin g$, since otherwise, $x \in X \cap (g \cup I)(w) = \{w\}$, which is impossible.■

Now, for each policy $e \in E$ the stationary set (of the policy e) is defined

$$\begin{aligned} \text{STY}(e) &\equiv \{w \in f(E) : w \neq f(e) \text{ and } (f(e), w) \in g\} \\ &= g(f(e)) \cap f(E) - \{f(e)\} \end{aligned}$$

If $\text{STY}(e) = \emptyset$, then $f(e)$ is called a generalized extreme value of $f(E)$; that is, $f(e) \in g * f(E)$.

A mapping $F : X \to W$, where $X \subset W$, is called strictly order-preserving if the following property holds: for any $x, y \in X$,

$$F(x) \neq F(y),\ (F(x), F(y)) \in g \text{ if and only if } x \neq y,\ (x,y) \in g$$

Theorem 7.6.2 [Operation Epitome Principle (Qin, 1991)]. *Let W and $E = E_1 \times E_2$ be nonempty sets, $g \subseteq E^2$ a binary relation, and $h_i : E_i \to W$, $i = 1,2$, mappings. If a mapping $F : W \cup W^2 \to W$ satisfies*

(i) $F(w) = w$, for each $w \in W$.

(ii) For each fixed w, the mappings $F(w,\cdot)$ and $F(\cdot,w) : W \to W$ are strictly order-preserving, then a mapping $f : E \cup E_1 \cup E_2 \to W$ can be defined by

$$f(e) = F(h_1(e_1), h_2(e_2)) \text{ and } f(e_i) = F(h_i(e_i)), \qquad i = 1,2$$

for each $e = (e_1, e_2) \in E$ such that

$$|\text{STY}(e_i)| \leq |\text{STY}(e)|$$

where $|X|$ is the cardinality of the set X.

Proof: For any $e' \in E_2$, define the mapping $p_1 : E_1 \to E$, which depends on fixed e', by

$$p_1(e_1) = (e_1, e') \quad \text{for each } e_1 \in E_1$$

For any $e' \in E_1$ define the mapping $p_2 : E_2 \to E$, which depends on fixed e', by

$$p_2(e_2) = (e', e_2) \quad \text{for each } e_2 \in E_2$$

Now the desired mapping $f : E \cup E_1 \cup E_2 \to W$ is induced by

$$f(e) = F(h_1(e_1), h_2(e_2)) \quad \text{for any } e = (e_1, e_2) \in E = E_1 \times E_2$$

and

$$f(e_i) = F(h_i(e_i)) \quad \text{for any } e_i \in E_i,\ i = 1,2$$

Since the mapping $F : W \cup W^2 \to W$ is strictly order-preserving, one has that, for each $e = (e_1, e_2) \in E = E_1 \times E_2$,

$$|\mathrm{STY}(e_i)| \leq |\{w \in f(p_i(E_i)) : w \neq f(e) \text{ and } (f(e), w) \in g\}| \tag{7.15}$$

On the other hand, it is easy to see that

$$\{w \in f(p_i(E_i)) : w \neq f(e) \text{ and } (f(e), w) \in g\} \subseteq \mathrm{STY}(e) \tag{7.16}$$

Equations (7.15) and (7.16) imply the desired inequality. ■

Now, Bellman's Principle of Optimality (Theorem 7.6.1) is clear.

7.7. Fundamental Equations of Generalized Systems

For an optimization model $(W, E; f, g)$ of general systems and a natural number $n \in N$, the relation $g^n \subseteq W^2$ is defined

$$g^n = \{(x,y) \in W^2 : \exists w_i \in W \text{ such that } (x, w_1), (w_{i-1}, w_i), (w_{n-1}, y) \in g,\\ i = 2, 3, \ldots, n-1\}$$

A binary relation $g \subseteq W^2$ is strongly antisymmetric (Qin, 1991), if for any natural numbers $m, n \in \mathbb{N}$, there do not exist distinct elements $x, y \in W$ such that the conditions $(x,y) \in g^m$ and $(y,x) \in g^n$ hold simultaneously. A binary relation $g \subseteq W^2$ is strictly strongly antisymmetric (Qin, 1991), if there is no infinite sequence $\{w_i\}_{i=0}^{\infty}$ such that $w_i \neq w_{i+1}$ and $(w_i, w_{i+1}) \in g$ for each $i = 0, 1, 2, \ldots$

It can be seen that the concept of strong antisymmetry is a generalization of the concept of partial ordering relations. To this end, recall that a relation $g \subseteq W^2$ is called a partial ordering (Kuratowski and Mostowski, 1976), if the following conditions are satisfied:

(i) (Antireflexivity) $(x,x) \notin g$ for any $x \in W$.

(ii) (Antisymmetry) If $(x,y) \in g$ and $x \neq y$, then $(y,x) \notin g$ for $x, y \in W$.

(iii) (Transitivity) If (x,y) and $(y,z) \in g$, then $(x,z) \in g$ for $x, y, z \in g$.

First, if a relation $g \subseteq W^2$ is a partial ordering, g must be strongly antisymmetric. Suppose that this statement is not true. That is, the relation $g \subseteq W^2$ is a partial ordering but not a strongly antisymmetric relation. Hence, there exist $w_1, w_2 \in W$ such that $w_1 \neq w_2$, and natural numbers $m, n \in \mathbb{N}$ such that $(w_1, w_2) \in g^m$ and $(w_2, w_1) \in g^n$. From the transitivity of g, it follows that $(w_1, w_2) \in g$, $(w_2, w_1) \in g$, and $w_1 \neq w_2$. This last condition contradicts the antisymmetry of g. Second, the following example shows that a strongly antisymmetric relation $g \subseteq W^2$ may not be a partial ordering relation.

Example 7.7.1. Let $W = \mathbb{R}$, and for any $a, b \in W$, $(a,b) \in g \subseteq W^2$, iff $b+1 > a > b$. It can be shown that $(a,b) \in g^m$, for $m \in \mathbb{N}$, iff $b+m \geq b+1 > a > b$; and that $(b,a) \in g^n$, for $n \in \mathbb{N}$, iff $a+m \geq a+1 > b > a$. Therefore, g is strongly antisymmetric. However, it is not a partial ordering relation on W, since the transitivity relation is not satisfied.

Theorem 7.7.1 [Qin (1991)]. *If a set W is finite, then a binary relation $g \subseteq W^2$ is strictly strongly antisymmetric iff the relation g is strongly antisymmetric.*

Proof: Necessity. Suppose that $g \subseteq W^2$ is strictly strongly antisymmetric but not strongly antisymmetric. From the definition of strong antisymmetry it follows that there exist $w_1, w_2 \in W$ and $m, n \in \mathbb{N}$ such that $w_1 \neq w_2$, $(w_1, w_2) \in g^m$, and $(w_2, w_1) \in g^n$. Thus, there exist elements w_{1i}, $i = 1, \ldots, m-1$, and w_{2j}, $j = 1, 2, \ldots, n-1$, such that

$$(w_1, w_{11}), (w_{1i}, w_{1i+1}), (w_{1m-1}, w_2) \in g \quad \text{for } i = 1, \ldots, m-1$$

and

$$(w_2, w_{21}), (w_{2j}, w_{2j+1}), (w_{2n-1}, w_1) \in g \quad \text{for } j = 1, \ldots, n-1$$

That is, there exists a finite sequence $\{x_i\}_{i=1}^{n+m} \subseteq W$ such that $x_1 = w_1$, $x_{m+1} = w_2$, $x_i \neq x_{i+1}$, for $i = 1, 2, \ldots, m+n-1$ and $(x_i, x_{i+1}) \in g$, for $i = 1, 2, \ldots, m+n-1$ Now an infinite sequence $\{y_i\}_{i=1}^{\infty} \subseteq W$ can be defined as follows:

$$y_i = x_i \quad \text{if } i \leq m+n$$

and

$$y_i = x_k \quad \text{if } i \geq m+n \text{ and } k = i \ (\text{mod}(m+n))$$

It is easy to observe that $y_i \neq y_{i+1}$ and $(y_i, y_{i+1}) \in g$ for each $i \in \mathbb{N}$, which contradicts the assumption that g is strictly strongly antisymmetric.

Sufficiency. Suppose W is finite and $g \subseteq W^2$ is strongly antisymmetric. Assume that $g \subseteq W^2$ is not strictly strongly antisymmetric. There then exists an infinite sequence $\{y_i\}_{i=0}^{\infty} \subseteq W$ such that $y_{i-1} \neq y_i$ and $(y_{i-1}, y_i) \in g$ for each $i \in \mathbb{N}$.

It is true that the sequence $\{y_i\}_{i=0}^{\infty} \subseteq W$ satisfies the condition that if $(y_q, y_p) \in g$, then $p > q$. If not, then $q \geq p$. Hence, there exists a subsequence $y_p, y_{p+1}, y_{p+2}, \dots, y_q$ such that

$$(y_p, y_{p+1}), (y_{p+1}, y_{p+2}), \dots, (y_{q-1}, y_q), (y_q, y_p) \in g$$

Now take the finite sequence $w_0, w_1, w_2, \dots, w_n$, for $n = q - p + 1$, such that $w_k = y_{k+p}$, if $0 < k < n - 1$, and $w_n = y_p$. It can be seen that $(w_{i-1}, w_i) \in g$, for $i = 1, 2, \dots, n$, and $w_0 = w_n$, which contradicts the fact that g is strongly antisymmetric. Therefore, $p > q$.

To complete the proof, we show that the elements o $\{y_i\}_{i=0}^{\infty} \subseteq W$ differ from one another, contradicting that W is finite. The contradiction thus implies that $g \subseteq W^2$ must be strictly strongly antisymmetric. To this end, suppose that the elements of $\{y_i\}_{i=0}^{\infty} \subseteq W$ do not differ. There then exist $p, q \in \mathbb{N}$ such that $p < q$, $y_p = y_q$, and $(y_p, y_{p+1}) \in g$. Hence, $(y_q, y_{p+1}) \in g$ so that $q < p + 1$; i.e., $q \leq p$, contradicting the fact that $p < q$. Therefore, $\{y_i\}_{i=0}^{\infty} \subseteq W$ is an infinite set, which contradicts the assumption. This ends the proof of the theorem. ■

The proof of the previous theorem actually says that strictly strong antisymmetry implies strong antisymmetry, and that when the set W is finite the converse implication is also true.

Question 7.7.1. Construct an example to show that the concept of strong antisymmetry is different from that of strictly strong symmetry.

For any subset $X \subseteq E$, define

$$X^* = \{e \in X : f(e) \in g * f(X)\} \tag{7.17}$$

and

$$W^* = g * W \tag{7.18}$$

Lemma 7.7.1 [Qin (1991)]. *For any* $E_1 \subseteq E, f(E_1^*) = g * f(E_1)$.

Proof: For any $w \in f(E_1^*)$, there exists $e \in E_1^*$ such that $w = f(e)$. Equation (7.17) implies that $w = f(e) \in g * f(E_1)$, based on which Eq. (7.14) implies that $w = f(e) \in f(E_1)$ and that for any $x \in f(E_1)$, $x \neq w$ implies $(w,x) \notin g$. The last condition means that $w \in g * f(E_1)$. Therefore, we have shown that $f(E_1^*) \subseteq g * f(E_1)$.

Conversely, for any $w \in g * f(E_1)$, Eq. (7.14) implies that $w = f(e) \in f(E_1)$, for some $e \in E_1$, and that for any $x \in f(E_1), x \neq w$ implies $(w,x) \notin g$. This condition implies that $w \in g * f(E_1^*)$. That is, we have shown that $f(E_1^*) \supseteq g * f(E_1)$. ■

Lemma 7.7.2 [Qin (1991)]. *For any element $x \in W$ and subset $X \subseteq W$, define a subset $X(x) \subseteq W$ by*

$$X(x) = X \cap (g \cup I)(x)$$

Then

$$g * X(x) = (g * X) \cap ((g \cup I)(x))$$

where I is the diagonal of the Cartesian product X^2, defined by $I = \{(x,x) : x \in X\}$.

Proof: For each $w \in g * X(x)$, $w \in (g \cup I)(x)$ and for any $w' \in X(x)$, $w' \neq w$, $(w,w') \notin g$. To show that $w \in (g * X) \cap (g \cup I)(x)$, one need only prove that $w \in g * X$. If $w \notin g * X$, there exists $w'' \in X$ such that $w \neq w''$ and $(w,w'') \in g$; i.e., $w'' \in X(x)$. Therefore, there exists $w'' \in X(x)$ such that $(w,w'') \in g$, which contradicts $w \in g * X(x)$, so $w \in g * X$. Thus, we have obtained that

$$g * X(x) \subseteq (g * X) \cap ((g \cup I)(x))$$

Conversely, for an element $w \in (g * X) \cap ((g \cup I)(x))$, $w \in g * X$ and $w \in (g \cup I)(x)$. It follows that $w \in X$, $w \in (g \cup I)(x)$, and that for any $y \in (g \cup I)(x)$, $w \neq y$ implies $y \notin X$. It is clear that

$$w \in X \cap (g \cup I)(x) = X(x) \subseteq W$$

Therefore, for any $y \in (g \cup I)(x), y \neq w$ implies $y \notin X$. Hence, $w = X \cap (g \cup I)(w)$; i.e., $w \in g * X(x)$. The following inclusion relation is thus true:

$$g * X(x) \supseteq (g * X) \cap ((g \cup I)(x))$$

■

Lemma 7.7.3 [Qin (1991)]. *If g is a strictly strongly antisymmetric relation on the set W, and $\emptyset \neq X \subseteq W$, then*

$$g * X \neq \emptyset$$

Proof: The proof is by contradiction. Assume that there exists a subset $\emptyset \neq X \subseteq W$ such that $g * X = \emptyset$, and let $x \in X$, then $x \notin g * X = \emptyset$. From Eq. (7.14) it follows that there exists $x_1 \in X$ such that $x \neq x_1$, $(x,x_1) \in g$, and $x_1 \notin g * X$, that there exists $x_2 \in X$ such that $x_1 \neq x_2$, $(x_1,x_2) \in g$, and $x_2 \notin g * X$, and so on. Therefore, an infinite sequence $x_1, x_2, x_3, \ldots$ has been obtained, satisfying $(x_i, x_{i+1}) \in g$, for each $i = 1,2,3,\ldots$ This condition contradicts the assumption g is strictly strong antisymmetric. Thus, the proof is complete. ■

Theorem 7.7.2 [Qin (1991)]. *Let g be a strictly strongly antisymmetric relation on the set W, and $G \subseteq p(W)$, the power set of W. Then*

$$g*(\bigcup_{D\in G} D) = g*(\bigcup_{D\in G} g*D)$$

Proof: For an element $w \in g*(\bigcup_{D\in G} D)$, it follows from Lemma 7.6.1 that

$$\begin{aligned} w &= (\bigcup_{D\in G} D)\cap(g\cup I)(w) \\ &= \bigcup_{D\in G} \{D\cap(g\cup I)(w)\} \end{aligned}$$

Partition the subset $G = \bigcup\{G_1, G_2\}$ such that $G = G_1 \cup G_2$ and $G_1 \cap G_2 = \emptyset$, satisfying, for any $X \in G_1$ and $Y \in G_2$,

$$X\cap(g\cup I)(w) = w \text{ and } Y\cap(g\cup I)(w) = \emptyset$$

Therefore, for any $X \in G_1$, $w \in g*X$, and for any $w \neq y \in g(w)$, one has that $y \in X$. Hence $y \notin g*X$. That is,

$$(g*X)\cap(g\cup I)(w) = w$$

On the other hand, for any $Y \in G_2$, $Y\cap(g\cup I)(w) = \emptyset$, it can be seen that for any $y \notin (g\cup I)(w), y \notin Y$, and so $y \notin g*Y$. That is, for any $Y \in G_2$,

$$(g*Y)\cap(g\cup I)(w) = \emptyset$$

Thus,

$$\begin{aligned} w &= \bigcup\{(g*D)\cap(g\cup I)(w) : D \in G\} \\ &= (\bigcup_{D\in G} (g*D))\cap(g\cup I)(w) \end{aligned}$$

which implies that $w \in g*(\bigcup_{D\in G} g*D)$. It has thus been shown that

$$g*(\bigcup_{D\in G} D) \subseteq g*(\bigcup_{D\in G} g*D) \tag{7.19}$$

Conversely, for an element $w \in g*(\bigcup_{D\in G} g*D)$,

$$w = (\bigcup_{D\in G} g*D)(g\cup I)(w)$$

Partition the set $G = \bigcup\{G_1, G_2\}$ such that $G = G_1 \cup G_2$ and $G_1 \cap G_2 = \emptyset$, satisfying, for any $X \in G_1$, and $Y \in G_2$,

$$(g*X)\cap(g\cup I)(w) = w \text{ and } (g*Y)\cap(g\cup I)(w) = \emptyset$$

Hence, for any $X \in G_1$, $w \in g*X$; i.e., $w = X \cap (g \cup I)(w)$. It follows from Lemmas 7.7.1 and 7.7.3 that for any $Y \in G_2$,

$$Y \cap (g \cup I)(w) = \emptyset$$

Then one obtains that $(\bigcup_{D \in G} D) \cap (g \cup I)(w) = w$, which in turn implies that $w \in g*(\bigcup_{D \in G} D)$. That is,

$$g*(\bigcup_{D \in G} D) \supseteq g*(\bigcup_{D \in G} g*D) \tag{7.20}$$

Combining Eqs. (7.19) and (7.20), proved the theorem. ■

Theorem 7.7.3 [Qin (1991)]. *In the optimization model $(W,E;f,g)$ of general systems, if $F: W \to W$ is a strictly order-preserving mapping and X is a subset of W, then*

$$F(g*X) = g*F(X)$$

Proof: For an element $w \in F(g*X)$, there exists $y \in g*X$ such that $w = F(y) \in F(X)$ and such that for any $z \neq y$, $(y,z) \in g$, so $z \notin X$. It suffices to show that there is no $F(w_2) \notin F(X)$ satisfying $w = F(w_2)$ and $F(w_2) \neq F(y)$, such that $(F(y),F(w_2)) \in g$. Suppose there is such an element $F(w_2) \in F(X)$. It then follows that $(y,w) \in g$ and $y \neq w_2$, which contradicts the assumption that $y \in g*X$ and implies that $w \in g*F(X)$. From the arbitrariness of the element w, it follows that $F(g*X) \subseteq g*F(X)$.

Conversely, for any element $w \in g*F(X)$, there exists an element $y \in X$ such that $w = F(y)$. It suffices to show that there is no $z \in X$ such that $z \neq y$ and $(y,z) \in g$. Suppose this condition is not satisfied. Then from the strictly order-preserving properties of F, it follows that $F(y) \neq F(z)$ and $(F(y),F(z)) \in g$. This contradicts the assumption that $w \in g*F(X)$. Hence, $y \in g*X$; i.e., $w \in F(g*X)$. Again, the arbitrariness of the element w guarantees

$$F(g*X) \supseteq g*F(X)$$

■

Let W, X, and T be nonempty sets with $g \subseteq W^2$ given. Define

$$E = \bigcup_{T' \subseteq T} X \uparrow T'$$

A mapping $f: E \to W$ is called p-resolving, if there exists a mapping $F: W \cup W^2 \to W$ satisfying

(i) $F(w) = w$ for each $w \in W$.

(ii) For any $e = (e_1, e_2) \in E, f(e) = F(f(e_1), f(e_2))$.

(iii) For any $w \in W$, the mappings $F(w, \cdot)$ and $F(\cdot, w) : W \to W$ are strictly order-preserving.

Theorem 7.7.4 [Qin (1991)]. *If the optimization model $(W, E; f, g)$ of general systems satisfies*

(i) $E = \bigcup_{T' \subseteq T} X \uparrow T'$, for some given nonempty sets X and T,

(ii) $f : E \to W$ is p-resolving,

(iii) $T' = T_1' \cup T_2'$, $E_1 = X \uparrow T_1'$, $E_2 = X \uparrow T_2'$ and $E' = E_1 \times E_2$,

then

$$g * f(E') = g * \cup_{e_1 \in E_1} F(f(e_1), g * f(E_2)) \tag{7.21}$$

Proof: It follows from Theorem 7.7.2 that

$$\begin{aligned} g * f(E') &= g * (\bigcup_{e \in E'} f(e)) \\ &= g * \cup_{(e_1, e_2) \in E_1 \times E_2} F(f(e_1), f(e_2)) \\ &= g * [\bigcup_{e_1 \in E_1} (\bigcup_{e_2 \in E_2} F(f(e_1), f(e_2)))] \\ &= g * \{\bigcup_{e_1 \in E_1} [g * \cup_{e_2 \in E_2} F(f(e_1), f(e_2))]\} \end{aligned}$$

For fixed $e_1 \in E_1$, it follows from Theorem 7.7.3 that

$$\begin{aligned} g * \cup_{e_2 \in E_2} F(f(e_1), f(e_2)) &= g * [F(f(e_1), f(E_2))] \\ &= F(f(e_1), g * f(E_2)) \end{aligned}$$

Therefore, Eq. (7.21) is valid. ■

By Theorem 7.7.1 it follows that Theorem 7.7.4 holds with W strongly anti-symmetric if W is finite.

Let $(W, E; f, g)$ be a given optimization model of general systems with the sets W and E nonempty. The optimal solutions of $(W, E; f, g)$ are called the zeroth-order optimization solutions of the system $(W, E; f, g)$. The first-order optimization solutions of the system $(W, E; f, g)$ are the optimal solutions of the subsystem of $(W, E; f, g)$ obtained by removing all zeroth-order optimization solutions from the original system $(W, E; f, g)$. Inductively, the nth-order optimization solutions of $(W,E;f,g)$ are the optimal solutions of the subsystem obtained by removing all zeroth-order, first-order, ..., $(n-1)$st-order optimization solutions from the original system $(W,E;f,g)$. If $e \in E$ is an nth-order optimization solution of $(W,E;f,g)$, the number n, which is a function of the optimization solution e and is denoted by $n = Oo(e)$, is called the optimization order of the solution e.

Define

$$W_0 = W \qquad \text{and} \qquad W_i = W_{i-1} - W_{i-1}^* \quad \text{for each } i \in \mathbb{N}$$

Each element $e \in W_i^*$ is called an ith-order generalized extreme value of W.

Theorem 7.7.5 [Qin (1988)]. *If the optimization model $(W,E;f,g)$ of general systems satisfies the conditions*

(i) W and E are nonempty,

(ii) $f : E \to W$ is p-resolving; and

(iii) $g \subseteq W^2$ is strongly antisymmetric,

then the sum of the optimization orders of all subpolicies is less than or equal to the optimization order of the policy itself.

The proof is left to the reader.

7.8. Applications

(i) In Hu's counterexample it can be shown that θ_1 satisfies the optimum-preserving law, and that θ_2 is associative but not distributive over θ_1. Therefore, Theorem 7.4.2 says that Bellman's Principle of Optimality does not hold. At the same time, by Theorem 7.5.2, it follows that the fundamental equations will not give the global optimum solution. For details one has that

$$f^{\langle 4\rangle}(A,E) = 1$$

with $A \to B_1 \to C \to D_1 \to E$ its policy, and

$$g^{\langle 4\rangle}(A,E) = 0$$

with $A \to B_2 \to C \to D_2 \to E$ the global optimum policy. Here,

$$f^{\langle 4\rangle}(A,E) = 1 \neq 0 = g^{\langle 4\rangle}(A,E)$$

(ii) Consider the network in Fig. 7.2. Let θ_1 = maximum and $\theta_2 = \wedge$. That is, for any $x,y \in W$, $\theta_2(x,y) = \min\{x,y\}$. Then $A \to B_1 \to C_2 \to D$ is an optimum

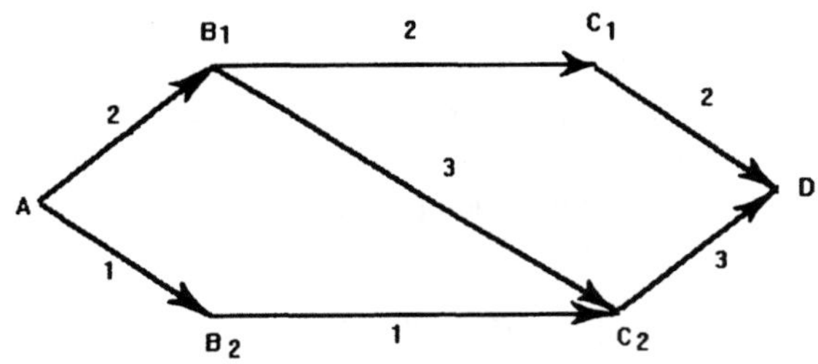

Figure 7.2. The principle of optimality does not hold.

path with weight 2, but its subpath $B_1 \to C_2 \to D$ is not optimum, since its weight is 3 and the weight of $B_1 \to C_1 \to D$ is 2. It can be shown that θ_2 is associative but does not satisfy a cancellation property. From Theorem 7.4.1 it follows that the Generalized Principle does not hold in this case.

(iii) Consider the network in Fig. 7.2 again with $\theta_1 =$ subminimum and $\theta_2 = +$. That is, θ_2 is the regular addition, and θ_1 is defined as

$$\begin{aligned}\theta_1(X) &= \min\{X - \{\min\{X\}\}\} \\ &= \text{the second minimum in } X\end{aligned}$$

for each $X \in p(W)$ with $|X| > 1$. Then $A \to B_1 \to C_1 \to D$ is the optimum path with weight 6, while its subpolicy $B_1 \to C_1 \to D$ with weight 4 is not optimum. It can be seen that θ_2 is associative and distributive over θ_1, but θ_1 does not possess an optimum-preserving property. For example,

$$\begin{aligned}\theta_1\{\omega(AB_iC_jD) : i,j = 1,2\} &= \omega(AB_1C_1D) = 6 \\ &\neq 8 = \omega(AB_1C_2D) = \theta_1\{\omega(AB_1C_jD) : j = 1,2\}\end{aligned}$$

(iv) For a sequential decision process, the system $(W, G; \theta_1, \theta_2)$ can be defined as follows:

$$\begin{aligned}W &= \text{set of all possible weights} \\ \theta_1 &= \text{optimum choice rule} \\ \theta_2 &= \text{operation between weights} \\ G &= \text{the set of all states of the process}\end{aligned}$$

Define a network $g \in W \uparrow G^2$ such that, for any $x, y \in G$, $g(x,y) = w$ iff the cost (or profit) from the state x to the state y is w. If a state y is not accessible from a state x, then define $g(x,y) =$ the zero element in W. If the process has n stages, a_0 is the initial state and a_n is the terminal state. Then $g^{\langle n\rangle}(a_0, a_n)$ is simply the cost (or profit) of a global optimum decision. If $g^{\langle n\rangle}(a_0, a_n) = \omega(a_0a_1a_2\cdots a_n)$, then $a_0, a_1, \ldots, a_n$ is the optimum sequential decision required. For a subpolicy $a_i, a_{i+1}, \ldots, a_j$, where $0 \leq i < j \leq n$, the Generalized Principle of Optimality says that if θ_1 satisfies the optimum-preserving law and θ_2 satisfies the associative, cancellation, and distributive laws over θ_1, then $a_i, a_{i+1}, \ldots, a_j$ must be optimum with regard to the initial state a_i and the terminal state a_j.

In the shortest-path, inventory, and traveling-salesman problems (Dreyfus and Law, 1977), the system considered is $(\mathbb{R}, \mathbb{R}; \min, +)$. In the equipment-replacement models and the resource allocation problems (Dreyfus and Law, 1977), the system of interest is $(\mathbb{R}, \mathbb{R}; \max, +)$. It can be shown that $(\mathbb{R}, \mathbb{R}; \min, +)$ and $(\mathbb{R}, \mathbb{R}; \max, +)$ satisfy the Generalized Principle of Optimality (Theorem 7.3.1). Thus, Bellman's Principle of Optimality holds in these optimization problems, and can be solved by dynamic programming. At the same time, it can be shown that "min" and "max" satisfy the optimum-preserving law and that "+" satisfies the cancellation, associative, and distributive laws (over "min" or "max"). Thus, Theorem 7.5.1 says that in $(\mathbb{R}, \mathbb{R}; \min, +)$ and $(\mathbb{R}, \mathbb{R}; \max, +)$ the desired optimum solutions can be obtained by applying fundamental equations.

The following propositions and theorem reveal the relation between the fundamental equation approach and the generalized principle of optimality.

For convenience, a system $(W, G; \theta_1, \theta_2)$, as defined in Section 7.3, is recursive if for any $g \in W \uparrow G^2$, $n \in \mathbb{N}$, and $a, b \in G$, $g^{\langle n \rangle}(a, b) = f^{\langle n \rangle}(a, b)$. That is, being recursive is equivalent to having the fundamental equation approach hold. The system $(W, G; \theta_1, \theta_2)$ satisfies the optimum-preserving law if for any $g \in W \uparrow G^2$, $n \in \mathbb{N}$, and $a_0, a_1, \ldots, a_n \in G$, the following is true: if $g^{\langle n \rangle}(a_0, a_n) = \omega(a_0 a_1 \cdots a_n) \neq$ the zero element of W, then for any $0 \leq i \leq j \leq n$, $g^{\langle j-i \rangle}(a_i, a_j) = \omega(a_i a_{i+1} \cdots a_j)$. Intuitively speaking, $(W, G; \theta_1, \theta_2)$ satisfies the optimum-preserving law means that the generalized principle of optimality holds.

Proposition 7.8.1 [Wu and Wu (1986)]. *The recursiveness of the system* $(W, G; \theta_1, \theta_2)$ *does not imply that it satisfies the optimum-preserving law.*

Proof: It suffices to construct an example to show that a recursive system $(W, G; \theta_1, \theta_2)$ does not satisfy the optimum-preserving law. Let W is the set of all nonnegative real numbers, G be a subset of real numbers, θ_1 is the maximum, and θ_2 is the minimum. It is evident that θ_1 satisfies the optimum-preserving law and θ_2 satisfies the associative and distributive laws over θ_1. So, according to Theorem 7.5.1, $(W, G; \theta_1, \theta_2)$ is recursive, but it does not satisfy the optimum-preserving law. See the network in Fig. 7.2. ■

Proposition 7.8.2 [Wu and Wu (1986)]. *The optimum-preserving property of the system* $(W, G; \theta_1, \theta_2)$ *does not imply the recursiveness of the system.*

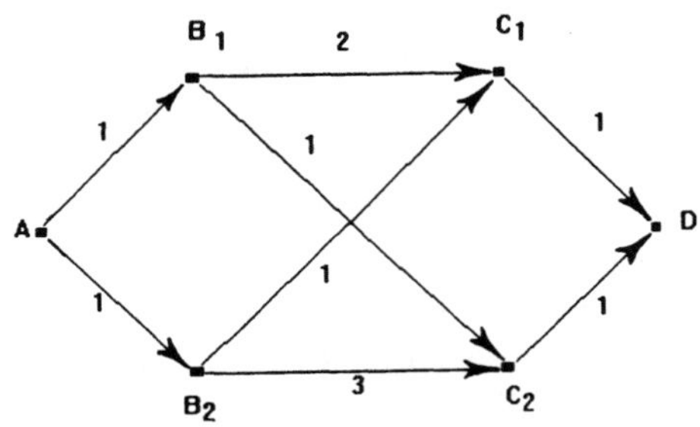

Figure 7.3. Optimum-preserving law does not imply recursiveness.

Proof: Again, it suffices to show the proposition by constructing a counterexample. Consider the system $(W, G; \theta_1, \theta_2)$, (Fig. 7.3), where

$$W = \text{the set of all nonnegative real numbers}$$

$$\theta_1(X) = \begin{cases} \text{subminimum of } X & \text{if } |X| > 1 \\ w & \text{if } X = \{w\} \end{cases}$$

$$\theta_2 = +(\text{regular addition})$$

Then it can be seen that the element $g \in W \uparrow G^2$ (Fig. 7.3) satisfies $g^{\langle 3 \rangle}(A,D) = 4$ and $A \to B_1 \to C_1 \to D$ and is the optimum policy, of which all subpolicies are optimum. That is, the system possesses a $(g; 3, A, D)$ optimum-preserving property. However, from the fundamental equation [Eq. (7.9)], it can be computed that

$$f^{\langle 3 \rangle}(A,D) = 5$$

with $A \to B_2 \to C_2 \to D$ its corresponding policy. Since

$$5 = f^{\langle 3 \rangle}(A,D) \neq g^{\langle 3 \rangle}(A,D) = 4$$

the system does not possess $(g; 3, A, D)$-recursiveness. ∎

Theorem 7.8.1 [Wu and Wu (1986)]. *In the system $(W, G; \theta_1, \theta_2)$ with θ_1 satisfying the optimum-preserving law and θ_2 satisfying the cancellation, associative, and distributive laws over θ_1 both the optimum-preserving law and the recursive law hold. That is, in this special system the fundamental equation will give a global optimum solution, and the generalized principle of optimality holds.*

The result is a consequence of Theorems 7.5.1 and 7.3.1.

7.9. Conclusion

This chapter generalizes the fundamental idea of dynamic programming to the widest sense available today in the research of systems science and pansystems analysis. For example, the first approach generalized the ideas of Bellman's Principle of Optimality and of the fundamental equations in dynamic programming to generalized weighted networks, where for any nonempty sets W and G each element $g \in W \uparrow G^2$ is called a generalized weighted network with weights in W and nodes in G. The second approach will surely open up a new era of research to study the idea of optimization in the theory of general systems.

In addition to the set of real numbers being used as the criteria set, generalized dynamic programming assume all other possible sets as criteria sets, such as vectors, matrices, tensors, and so on. Actually, there have appeared some studies and applications of vector-valued dynamic programming (Brown and Strauch, 1965; Furukawa, 1980; Denardo, 1967; Henig, 1983). In these works, Bellman's Principle of Optimality in the dynamic vector-valued models has also been discussed and similar results obtained. These facts are helpful in showing the universal significance of the approaches presented in this chapter.

Bellman's Principle of Optimality has been generalized to the Generalized Principle of Optimality and the Operation Epitome Principle. It may be expected that in dealing with optimization problems of the most general sense, especially with those out of the ordinary, the Generalized Principle of Optimality and the Operation Epitome Principle will play important roles.

Practically, the main point of dynamic programming is to use the fundamental equations to compute a desired optimal solution. However, in many practical optimization problems, especially those with nonscalar-valued performance criteria, their underlying structure of systems is much more complicated than $(\mathbb{R}, \mathbb{R}; \min, +)$ and $(\mathbb{R}, \mathbb{R}; \max, +)$. In those cases, the method of classical dynamic programming may not apply. In this chapter, a criterion has been established to decide whether the approach of fundamental equations can still be employed.

CHAPTER 8

Unreasonable Effectiveness of Mathematics: A New Tour

It is well known that the spectacular success of modern science and technology is due in large measure to the use of mathematics in the establishment and analysis of models for the various phenomena of interest. The application of mathematics brings life to all scientific theories, dealing with real and practical problems. It is mathematics that makes us better understand the phenomena under consideration. Mickens (1990) asked why this is the case. In fact, many great thinkers had thought of this question, when studying natural systems (Graniner, 1988). In this chapter, some problems related to the structure of mathematics, applications of mathematics and the origin of the universe are discussed. More specifically, in the first section, mathematics is compared with music, painting, and poetry. The structure of mathematics is studied in the second section. Then the structure of mathematics is written as a formal system in the third section. This new understanding is used to model and to analyze the states of materials and some epistemological problems in the fourth and fifth sections. In the sixth section a vase puzzle is given to show that mathematical modeling can lead to contradictory conclusions. Based on this fact, some impacts of the puzzle are mentioned. In the concluding section, problems concerning the material structure of human thought and the origin of the universe are posed.

8.1. What Is Mathematics?

Mathematics is not only a rigorously developed scientific theory, but it is also the fashion center of modern science. The reason is that a theory is not considered a real scientific theory if it has not been written in the language and symbols of mathematics, since such a theory does not have the capacity to predict the future. Indeed, no matter where we direct our vision — whether to the depths of outer space, where we are looking for our "brothers," to slides beneath microscopes, where we are trying to isolate the fundamental bricks of the world, or to the realm of imagination where much human ignorance is comforted, etc., we can always find the track and language of mathematics.

All through history, many diligent plowers of the great scientific garden have cultivated today's magnificent mathematical foliage. Mathematics is not like any other theories in history, which have faded as time goes on and for which new foundations and new fashions have had to be designed by coming generations in order to study new problems. History and reality have repeatedly showed that mathematical foliage is always young and flourishing.

Although mathematical abstraction has frightened many people, many unrecognized truths in mathematics have astonished truth pursuers by its sagacity. In fact, mathematics is a splendid abstractionism with the capacity to describe, investigate and predict the world. First, let us see how some of the greatest thinkers in history have enjoyed the beauty of mathematics.

Philosopher Bertrand Russell (Moritz, 1942):

> Mathematics, rightly viewed, possesses not only truth, but supreme beauty — a beauty cold and austere, like that of sculpture, without appeal to any part of our weaker nature, without the gorgeous trappings of painting and music, yet sublimely pure, and capable of a true perfection such as only the greatest art can show. The true spirit of delight, the exaltation, the sense of being more than man, which is the touchstone of the highest excellence, is to be found in mathematics as surely as in poetry. What is best in mathematics deserves not merely to be learned as a task, but to be assimilated as a part of daily thought, and brought again and again before the mind with ever-renewed encouragement. Real life is, to most men, a long second-best, a perpetual compromise between the real and the possible; but the world of pure reason knows no compromise, no practical limitations, no barrier to the creative activity embodying in splendid edifices the passionate aspiration after the perfect from which all great work springs. Remote from human passions, remote even from the pitiful facts of nature, the generations have gradually created an ordered cosmos, where pure thought can dwell as in its natural home, and where one, at least, of our nobler impulses can escape from the dreary exile of the natural world.

Mathematician Henri Poincare (Moritz, 1942):

> it [mathematics] ought to incite the philosopher to search into the notions of number, space, and time; and, above all, adepts find in mathematics delights analogous to those that painting and music give. They admire the delicate harmony of number and forms; they are amazed when a new discovery discloses for them an unlooked for perspective; and the joy they thus experience, has it not the esthetic character although the senses take no part in it? Only the privileged few are called to enjoy it fully, it is true; but is it not the same with all the noblest arts?

Organically arranged colors show us the beautiful landscape of nature so that we more ardently love the land on which we were brought up. Animated combinations of musical notes sometimes bring us to somewhere between the sky with rolling black clouds and the ocean with roaring tides, and sometimes comfort us in a fragrant bouquet of flowers with singing birds. How does mathematics depict the world for us? Mathematics has neither color nor sound, but with its special methods — numbers and forms — mathematics has been showing us the structure of the surrounding world. It is because mathematics is the combination of the marrow of human thought and aesthetics that it can be developed generation after generation.

Let us see, for example, the concept of numbers. What is "2"? Consider two books and two apples. The objects, book and apple, are completely different. Nevertheless, the meaning of the concept "2" is the same. Notice that the symbol "2" manifests the same inherent law in different things — quantity. We know that there are the same amount of books and apples, because by pairing a book and an apple together, we get two pairs. Otherwise, for example, there are two apples and three books. After forming two pairs, each of which contains exactly one apple and one book, there must be a book left. From this fact an ordering relation between the concepts of "2" and "3" can be expected.

Colors can describe figures of nature, and musical notes can express fantasies from the depths of the human mind. In the previous example, every one will agree that the symbol "2" displays a figure of nature and insinuates the marrow of human thought. By using intuitive pictures and logical deduction, it can be shown that the totality of even natural numbers and the totality of odd natural numbers have the same number of elements. That is because any even natural number can be denoted by $2n$, and any odd natural number can be described by $2n + 1$, where n is a natural number. Thus, for each natural number n, $2n$ and $2n + 1$ can be put together to form a pair. With the help of this kind of reasoning and abstraction, the domain of human thought is enlarged from "finitude" into "infinitude," just as human investigation about nature goes from daily life into the microcosm and outer space.

How can colors be used to display atomic structures? How can musical notes be used to imitate the association of stars in the universe? With mathematical symbols and abstract logic, the relationship between the stars and the attraction between an atomic nucleus and its electrons are written in detail in many books.

You might ask: Because mathematics, according to the previous description, is so universally powerful, can it be used to study human society, human relations, and communication and transportation between cities? The answer to this question is "yes." Even though the motivation for studying these kinds of problems appeared long before Aristotle's time, research along this line began not too long ago. Systems theory, one of the theories concerning these problems, was formally named in the second decade of this century, and the theoretical foundation began to be set up in the sixth decade. It is still not known whether the theory can eventually depict and answer the aforementioned problems. However, it has been used to explore some problems man has been studying since the beginning of history, such as: (1) Does there exist eternal truth in the universe? (2) Does the universe consist of fundamental particles? Systems theorists have eliminated the traditional research method — dividing the object under consideration into basic parts and processes, and studying each of them — and have begun to utilize systems methods to study problems with many cause–effect chains. For instance, the famous three-body problem of the sun, the moon, and the earth, the combination of vital basic particles, and social systems of human beings are examples of such problems.

Mathematics is analogous neither to painting, with its awesome blazing scenes of color, nor to music, which possesses intoxicatingly melodious sounds. It resembles poetry, which is infiltrated with exclamations and admirations about nature, a thirst for knowledge of the future, and the pursuit of ideals. If you have time to taste the flavor of mathematics meticulously, you will be attracted by the delicately arranged symbols to somewhere into mathematical depths. Even though gentle breezes have never carried any soft mathematical songs, man has firmly believed that mathematics shows truth since the first day it appeared. The following two poems by Shakespeare (Moritz, 1942) will give mathematics a better shape in your mind.

1

Music and poetry used to quicken you
The mathematics, and the metaphysics
Fall to them as you find your stomach serves you
No profit grows, where is no pleasure ta'en: —
In brief, sir, study what you most affect

2

I do present you with a man of mine
Cunning in music and in mathematics
To instruct her fully in those sciences
Whereof, I know, she is not ignorant.

8.2. The Construction of Mathematics

Analogous to paintings, which are based on a few colors, and to music, in which the most beautiful symphonies in the world can be written with a few basic notes, the mathematical world has been constructed on a few basic names, axioms, and the logical language.

The logical language K used in mathematics can be defined as follows. K is a class of formulas, where a formula is an expression (or sentence) which may contain (free) variables. Then K is defined by induction in the following manner, for details see (Kuratowski and Mostowski, 1976; Kunen, 1980).

(a) Expressions of the forms given below belong to K :

$$x \text{ is a set}, \quad x \in y, \quad x = y$$

as well as all expressions differing from them by a choice of variables, where the concept of "set" and the relation $\in$ of membership are primitive notations, which are the only undefined primitives in the mathematics developed in this section.

(b) If ϕ and ψ belong to the class K, then so do the expressions (ϕ or ψ), (ϕ and ψ), (if ϕ, then ψ), (ϕ and ψ are equivalent), and (not ϕ).

(c) If ϕ belongs to K and v is any variable, then the expressions (for any v, ϕ) and (there exists v such that ϕ) belong to K.

(d) Every element of K arises by a finite application of rules (a), (b), and (c).

The formulas in (a) are called atomic formulas. A well-known set of axioms, on which the whole classical mathematics can be built and which is called the ZFC axiom system, contains the following axioms.

Axiom 1. (Extensionality) If the sets A and B have the same elements, then they are equal or identical.

Axiom 2. (Empty set) There exists a set $\emptyset$ such that no x is an element of $\emptyset$.

Axiom 3. (Unions) Let A be a set of sets. There exists a set S such that x is an element of S if and only if x is an element of some set belonging to A. The set S is denoted by $\bigcup A$, $X \cup Y$ if $A = \{X, Y\}$, or $\bigcup_{i=1}^{n} X_i$ if $A = \{X_1, X_2, \ldots, X_n\}$.

Axiom 4. (Power sets) For every set A there exists a set of sets, denoted by $p(A)$, which contains exactly all the subsets of A.

Axiom 5. (Infinity) There exists a set of sets A satisfying the conditions: $\emptyset \in A$; if $X \in A$, then there exists an element $Y \in A$ such that Y consists exactly of all the elements of X and the set X itself.

Axiom 6. (Choice) For every set A of disjoint nonempty sets, there exists a set B which has exactly one element in common with each set belonging to A.

Axiom 7. (Replacement for the formula ϕ) If for every x there exists exactly one y such that $\phi(x,y)$ holds, then for every set A there exists a set B which contains those, and only those, elements y for which the condition $\phi(x,y)$ holds for some $x \in A$.

Axiom 8. (Regularity) If A is a nonempty set of sets, then there exists a set X such that $X \in A$ and $X \cap A = \emptyset$, where $X \cap A$ means the common part of the sets X and A.

The following axioms can be deduced from Axioms 1–8.

Axiom 2. (Pairs) For arbitrary a and b, there exists a set X which contains only a and b; the set X is denoted by $X = \{a, b\}$.

Axiom 6. (Subsets for the formula ϕ) For any set A there exists a set which contains the elements of A satisfying the formula ϕ and which contains no other elements.

From Axioms 2 and 4, we can construct a sequence $\{a_n\}_{n=0}^{\infty}$ of sets as follows:

$$\begin{aligned}
a_0 &= \emptyset \\
a_1 &= p(\emptyset) = \{\emptyset\} = \{a_0\} \\
a_2 &= p(a_1) = \{\emptyset, \{\emptyset\}\} = \{a_0, a_1\} \\
&\vdots \\
a_n &= p(a_{n-1}) = \{a_0, a_1, \ldots, a_{n-1}\} \\
&\vdots
\end{aligned}$$

Now it can be readily checked that the sequence $\{a_n\}_{n=0}^{\infty}$ satisfies Peano's axioms:

P1. 0 is a number.

P2. The successor of any number is a number.

P3. No two numbers have the same successor.

P4. 0 is not the successor of any number.

P5. If P is a property such that (a) 0 has the property P, and (b) whenever a number n has the property P, then the successor of n also has the property P, then every number has the property P.

In the construction of the sequence $\{a_n\}_{n=0}^{\infty}$, we understand the symbol 0 in P1 as $a_0 = \emptyset$, the notations of "number" in P2 as any element in the sequence $\{a_n\}_{n=0}^{\infty}$, and of "successor" in P2 as the power set of the number.

The entire arithmetic of natural numbers can be derived from the Peano axiom system. Axiom P5 embodies the principle of mathematical induction and illustrates in a very obvious manner the enforcement of a mathematical "truth" by stipulation. The construction of elementary arithmetic on Peano's axiom system begins with the definition of the various natural numbers — that is, the construction of the sequence $\{a_n\}_{n=0}^{\infty}$. Because of P3 (in combination with P5), none of the elements in $\{a_n\}_{n=0}^{\infty}$ will be led back to one of the numbers previously defined, and by P4, it does not lead back to 0 either.

As the next step, a definition of addition can be established, which expresses in a precise form the idea that the addition of any natural number to some given number may be considered as a repeated addition of 1; the latter operation is readily expressible by means of the successor relation (that is, the power set relation). This definition of addition goes as follows:

$$\text{(a)} \quad n+0=n; \qquad \text{(b)} \quad n+p(k)=p(n+k) \tag{8.1}$$

The two stipulations of this inductive definition completely determine the sum of any two natural numbers.

Multiplication of natural numbers may be defined by the following recursive definition, which expresses in rigorous form the idea that a product nk of two natural numbers may be considered as the sum of k terms, each of which equals n

$$\text{(a)} \quad n\cdot 0=0; \qquad \text{(b)} \quad n\cdot p(k)=nk+n \tag{8.2}$$

Within the Peano system of arithmetic, its true propositions flow not merely from the definition of the concepts involved but also from the axioms that govern these various concepts. If we call the axioms and definitions of an axiomatized theory the "stipulations" concerning the concepts of that theory, then we may now say that the propositions of the arithmetic of the natural numbers are true by virtue of the stipulations which have been laid down initially for the arithmetic concepts.

In terms of addition and multiplication, the inverse operations (i.e., subtraction and division) can then be defined. However they cannot always be performed; i.e., in contradistinction to addition and multiplication, subtraction and division are not defined for every pair of numbers; for example, $7-10$ and $7\div 10$ are undefined. This incompleteness of the definitions of subtraction and division suggests an enlargement of the number system by introducing negative and rational numbers.

Consider the Cartesian product $\mathbb{N}\times\mathbb{N}=\{(n,m): n,m\in\mathbb{N}\}$, where $\mathbb{N}$ is the collection of all natural numbers, defined as before, which is a set by Axiom 5, and (n,m) is the ordered pair of the natural numbers n and m, which is defined by $(n,m)=\{\{n\},\{n,m\}\}$. Consider a subset Z of the power set $p(\mathbb{N}\times\mathbb{N})$ such that for any two ordered pairs $(a,b),(c,d)\in\mathbb{N}\times\mathbb{N}$, (a,b) and (c,d) belong to the same set A in Z if and only if $a+d=b+c$. Then it can be seen that each ordered pair in $\mathbb{N}\times\mathbb{N}$ is contained in exactly one element belonging to Z and no two elements in Z have an element in common. Now, Z is the desired set on which the definition of subtraction of integers can be defined for each pair of elements from Z, where the set Z is called a set of all integers, and each element in Z is called an integer. In fact, for each ordered pair $(n,m)\in\mathbb{N}\times\mathbb{N}$, there exists exactly one element $A\in Z$ containing the pair (n,m). Thus, the element A can be denoted by $A = \overline{(n,m)}$. Then, addition, multiplication, and subtraction can be defined as follows:

$$\overline{(a,b)} + \overline{(c,d)} = \overline{(a+c,b+d)} \tag{8.3}$$
$$\overline{(a,b)} \cdot \overline{(c,d)} = \overline{(ac+bd,ad+bc)} \tag{8.4}$$
$$\overline{(a,b)} - \overline{(c,d)} = \overline{(a+d,b+c)} \tag{8.5}$$

for any elements $\overline{(a,b)}$ and $\overline{(c,d)}$ in Z. It is important to note that the subset $Z^+ = \{\overline{(a,b)} \in Z :$ there exists a number $x \in \mathbb{N}$ such that $b+x=a\}$ of Z now serves as the set of natural numbers, because Z^+ satisfies axioms P1–P5.

It is another remarkable fact that the rational numbers can be obtained from the ZFC primitives by the honest toil of constructing explicit definitions for them, without introducing any new postulates or assumptions. Similarly, rational numbers are defined as classes of ordered pairs of integers from Z. The various arithmetical operations can then be defined with reference to these new types of numbers, and the validity of all the arithmetical laws governing these operations can be proved from nothing more than Peano's axioms and the definitions of various arithmetical concepts involved.

The much broader system thus obtained is still incomplete in the sense that not every number in it has a square root, cube root, ..., and more generally, not every algebraic equation whose coefficients are all numbers of the system has a solution in the system. This suggests further expansions of the number system by introducing real and complex numbers. Again, this enormous extension can be affected by only definitions, without posing a single new axiom. On the basis thus obtained, the various arithmetical and algebraic operations can be defined for the numbers of the new system, the concepts of function, of limit, of derivatives and integral can be introduced, and the familiar theorems pertaining to these concepts can be proved; thus, finally the huge system of mathematics as here delimited rests on the narrow basis of the ZFC axioms: Every concept of mathematics can be defined by means of the two primitives in ZFC, and every proposition of mathematics can be deduced from the ZFC axioms enriched by the definitions of the nonprimitive terms. These deductions can be carried out, in most cases, by means of nothing more than the principles of formal logic, a remarkable achievement in systematizing the content of mathematics and clarifying the foundations of its validity.

Remark. A well-known result discovered by Gödel shows that the afore-described mathematics is an incomplete theory in the following sense. Even though all those propositions in classical mathematics can indeed be derived, in the sense just characterized, from ZFC, other propositions can be expressed in pure ZFC language which are true but which cannot be derived from the ZFC axioms. This fact does not, however, affect the results we have outlined because the most unreasonably effective part of mathematics in applications is a substructure of the mathematics we have constructed.

8.3. Mathematics from the Viewpoint of Systems

The concept of systems was introduced formally by von Bertalanffy in the 1920s according to the following understanding about the world: The world we live in is not a pile of uncountably many isolated "parts"; and any practical problem and natural phenomenon cannot be described perfectly by only one cause–effect chain. The basic character of the world is its organization and the connection between the interior and exterior of different things. Customary investigation of the isolated parts and processes cannot provide a complete explanation of the vital phenomena. This kind of investigation gives us no information about the coordination of parts and processes. Thus, the chief task of modern science should be a systematic study of the world. For details see (von Bertalanffy, 1934).

Mathematically speaking, S is said to be a system (Lin, 1987) if and only if S is an ordered pair (M, R) of sets, where R is a set of some relations defined on the set M. Each element in M is called an object of S, and M and R are called the object set and the relation set of S, respectively. In this definition, each relation $r \in R$ is defined as follows: There exists an ordinal number $n = n(r)$, called the length of the relation r, such that r is a subset of M^n, where M^n indicates the Cartesian product of n copies of M. Assume that the length of the empty relation $\emptyset$ is 0; i.e., $n = n(\emptyset) = 0$. In ordinary language, a system consists of a set of objects and a collection of relations between the objects.

Claim. If by a formal language we mean a language which does not contain sentences with grammatical mistakes, then any formal language can be described as a system.

Proof: Assume that L is a formal language; for convenience, suppose that L is English which does not contain sentences with grammatical mistakes, where an English grammar book B is chosen and rules in B are used as the measure to see if an English sentence is in L or not.

Each word in L consists of a finite combination of letters, that is, a finite sequence of letters. Let X be the collection of all 26 letters in L. From the finiteness of the collection X and Axiom 7, it follows that X is a set; otherwise the natural number 26 is not a set (from the discussion in the previous section), so instead of the collection X we can use the set 26. Let M be the totality of all finite sequences of elements from X. Then M is the union

$$M = \bigcup_{i=1}^{\infty} X^i$$

Hence, M is a set which contains all words in L.

Each sentence in L consists of a finite combination of some elements in M, i.e., a finite sequence of elements from M. If Q be the totality of finite sequences of elements from M, then

$$Q = \bigcup_{i=1}^{\infty} M^i$$

Thus, Q is a set. Each element in Q is called a sentence of L.

Write M as a union of finitely many subsets: $M = \bigcup\{M(i) : i \in J\}$, where $J = \{0, 1, 2, \ldots, n\}$ is a finite index set. Elements in $M(0)$ are called nouns; elements in $M(1)$ verbs; elements in $M(2)$ adjectives, etc., Let $K \subset Q$ be the collection of all sentences in the grammar book B. Then K must be a finite collection, and it follows from Axiom 7 that K is a set.

If each statement in K is assumed to be true, then the ordered pair (M, K) constitutes the English L. Here, (M, K) can be seen as a systems description of the formal language L. ■

Let (M, K) be the systems representation of the formal language L in the preceding proof, and T the collection of Axioms 1–6' in the previous section. Then $(M, T \cup K)$ is a systems description of mathematics.

From Klir's (1985) definition of systems — that a system is what is distinguished as a system — and Claim 1, it follows that if a system $S = (M, R)$ is given, then each relation $r \in R$ can be understood as an S-truth; i.e., the relation r is true among the objects in the set M. Therefore, any mathematical truth is a $(M, T \cup K)$-truth; i.e., it is derivable from the ZFC axioms, the principles of formal logic, and definitions of some nonprimitive terms. From this discussion about mathematical truths the following question is natural.

Question 8.3.1. If there existed two mathematical statements derivable from ZFC axioms with contradictory meanings, would they still be $(M, T \cup K)$-truths? Generally, what is the meaning of a system with contradictory relations, e.g., $\{x < y,\ x \geq y\}$ as the relation set?

The following natural epistemological problem was asked in (Lin, 1989c).

Question 8.3.2. How can we know whether there exist contradictory relations in a given system?

From Klir's definition of systems, it follows that generally, Questions 8.3.1 and 8.3.2 cannot be answered. Concerning this, a theorem of Gödel shows that it is impossible to show whether the systems description $(M, T \cup K)$ of mathematics is consistent (that is, any two statements derivable from ZFC will not have contradictory meanings) or not in ZFC. On the other hand, we still have not found any method outside ZFC which can be used to show that the system $(M, T \cup K)$ is inconsistent. This means that there are systems (for example, mathematics based on ZFC) in which we do not know whether there exist propositions with contradictory meanings. Therefore, this fact implies that perhaps not every system is consistent or that not every system has no contradictory relations.

The application of mathematics has showed us that mathematics is extremely effective in describing, solving, and predicting practical phenomena, problems, and future events, respectively. That means that in practice a subsystem of $(M, T \cup K)$ can always be found to match a situation under consideration. For example, a hibiscus flower has five petals. The mathematical word "five" provides a certain description of the flower. This description serves to distinguish it from flowers with three, four, six, ..., petals. When a watch chain is suspended from its ends, it assumes very nearly the shape of the mathematical curve known as the catenary. The equation of the catenary is $y = \frac{a}{2}(e^{x/a} + e^{-x/a})$, where a is a constant. At the same time, this equation can be used to predict (or, say, answer) the following problem: Suppose someone holds his 10-inch watch chain by its ends with his fingers at the same height and are 4 inches apart. How far below his fingers will the chain dip? Now, the following amazing question can be asked (this question was not originally posed by me).

Question 8.3.3. Why is mathematics so "unreasonably effective" when applied to the analysis of natural systems?

In order to discuss this question, we must go back to Axioms 1–6'. As a consequence of our discussion, the whole structure of mathematics might be said to be true by virtue of mere definitions (namely, of the nonprimitive mathematical terms) provided that the ZFC axioms are true. However, strictly speaking, we cannot, at this juncture, refer to the ZFC axioms as propositions which are true or false, for they contain free primitive terms, "set" and the relation of membership "$\in$," which have not been assigned any specific meanings. All we can assert so far is that any specific interpretation of the primitives which satisfies the axioms—i.e., turns them into true statements—will also satisfy all theorems deduced from them. For detailed discussion, see (Kuratowski and Mostowski, 1976; Kunen, 1980). But the partial structure of mathematics developed on the basis of Peano's axioms has several — indeed infinitely many — interpretations which will turn Peano's axioms, which contain the primitives "0," "number," and "successor," into true statements, and therefore, satisfy all the theorems deduced from them. (Note that this partial structure of mathematics constitutes the theoretical foundation for almost all successful applications of mathematics.) For example, let us understand by 0 the origin of a half-line, by the successor of a point on that half-line the point 1 inch behind it, counting from the origin, and by a number any point which is either the origin or can be reached from it by a finite succession of steps each of which leads from one point to its successor. It can then be readily seen that all Peano's axioms as well as the ensuing theorems turn into true propositions, although the interpretation given to the primitives is certainly not any of those given in the previous section. More generally, it can be shown that every progression of elements of any kind provides a true interpretation of Peano's axiom system. This example illustrates that mathematics permits many different interpretations, in everyday life as well as in the investigation of laws of the nature and from each different interpretation, we understand something more about the nature; and at the same time, because of this fact, we feel that mathematics is so unreasonably effective as applied to the study of natural problems.

We will give two examples to show how abstract mathematical structures developed on the basis of the ZFC axioms but outside Peano's axiom system can be used to describe and study problems in materials science and epistemology.

8.4. A Description of the State of Materials

An arbitrarily chosen experimental material X can be described as a system $S_x = (M_x, R_x)$ as follows, where M_x is the set of all molecules in the material and R_x contains the relations between the molecules in M_x, which describe the spatial structure in which the molecules in the material are arranged. If X is crystalline, then all cells in X have the same molecular spatial structure. That means that the molecules in X are aligned spatially in a periodic order, so we have the following.

Definition 8.4.1. The experimental material X is crystalline if for the system $S_x = (M_x, R_x)$, there exists a relation $f \in R_x$ such that $f|M^* \neq f$, for any proper subset M^* of M_x, and there exists a cover $\mathcal{B}$ for the set M_x such that for any A and B in $\mathcal{B}$, there exists a one-to-one correspondence $h : A \to B$ satisfying $f|A = f|B \circ h$, where $f|Y$ indicates the restriction of the relation f on the subset Y and $h(x_0, x_1, \ldots, x_\alpha, \ldots) = (h(x_0), h(x_1), \ldots, h(x_\alpha), \ldots), \alpha < \beta$ for any ordinal number β and any $(x_0, x_1, \ldots, x_\alpha, \ldots) \in A^\beta$.

For example, if X is crystalline, for any unit cell Y in X define M_Y as the set of all molecules contained in Y. Then $\mathcal{B}_x = \{M_Y : Y \text{ is a unit cell in } X\}$ is a cover of M_x, because each molecule in X must belong to one of the unit cells, and all unit cells look the same. It therefore follows that the main characteristics of crystalline materials are described by Definition 8.4.1.

Given a linearly ordered set W, a time system S over W is a function from W into a family of systems; say $S(w) = S_w = (M_w, R_w)$, for each $w \in W$, where each system S_w is called the state of the time system S at the moment w. If, in addition, a family $\{\ell_w^{w'} : w, w' \in W,\ w' \geq w\}$ of mappings is given such that for any r, s, and $t \in W$ satisfying $s \geq r \geq t$, $\ell_r^s : M_s \to M_r$ and

$$\ell_t^s = \ell_t^r \circ \ell_r^s \text{ and } \ell_t^t = \mathrm{id}_{M_t}$$

where id_{M_t} is the identity mapping on M_t, then S is called a linked time system and denoted by $\{S_w, \ell_t^s, W\}$. Each ℓ_t^s is called a linkage mapping of the linked time system. For details; see (Ma and Lin, 1987).

Let $\mathbb{R}$ be the set of all real numbers, $R_T (= \mathbb{R})$ indicate temperature and $R_t = [0, +\infty)$ and $R_p = [0, +\infty)$ (the interval of real numbers from 0 to positive infinity) indicate time and pressure, respectively. Define an "order" relation "$<$" on the Cartesian product $W = R_t \times R_T \times R_p$ as follows: For any $(x_1, y_1, z_1), (x_2, y_2, z_2) \in W$, $(x_1, y_1, z_1) < (x_2, y_2, z_2)$ if and only if either (i) $z_2 > z_1$, (ii) $z_1 = z_2$ and $y_1 > y_2$, or (ii) $z_1 = z_2, y_1 = y_2$ and $x_1 < x_2$. Then it can be shown that $<$ is a linear order.

We establish a systems theory model of crystallization process of the the material X. It is assumed that the size of molecules in X will not change as temperature and pressure change; i.e., any chemical reaction between the molecules during the process will be neglected.

In the state of time t, temperature T, and pressure P, the relation set, consisting of all relations describing spatial structure of the molecules in X, is F_w, where $w = (t, T, P)$. Then at the state w, X can be described as a system (M, F_w), where M is the set of all molecules in X so that a linked time system $S = \{(M, F_w), \mathrm{id}_M, W\}$, where $\mathrm{id}_M : M \to M$ is the identity mapping on M.

If X is going to crystallize as time goes on, pressure imposed on X goes up, and temperature goes down, there then is a state $w_0 = (t_0, T_0, P_0) \in W$ such that X is crystalline. That means that the state $S_{w_0} = (M, F_{w_0})$ of S satisfies the property in Definition 8.4.1.

The crystallization process of X is the process that X changes from an amorphous state to an ordered state or a crystalline state. The process is divided into two steps. The first step is called nucleation; i.e., when temperature is lower than the melting temperature of the material, the activity of the molecules decreases, which means that the attraction or interaction between molecules increases, so there would be a tendency for the molecules to get together to form a short-range order. This kind of locally ordered region is called a nucleus. The second step is called growth. Although at certain temperature and pressure values, many nuclei can be formed at the same time they will probably disintegrate because the ratio of the surface and the volume of each nucleus is still too great and the surface energy is too high. However, as soon as the nuclei grow large enough, which is caused by the fluctuation of local energy, their energy will be lower as more and more molecules join them. The molecules in the nearby region would prefer to join those nuclei so that they will grow larger and larger.

Question 8.4.1. In the previous mathematical model, does there exist a relation $f \in R_x$ satisfying the property in Definition 8.4.1 when the material X is actually in a melting state, where $S_x = (M_x, R_x)$ is the systems representation of X?

We assume that the answer to Question 8.4.1 is "yes." This assumption is based on

(i) A view of dialectical materialism — the internal movement of contradictions dominates the development of the matter under consideration

(ii) The beauty of the systems theory model described earlier

Under this assumption, it can be seen that under certain ideal conditions any material may crystallize, because the relation f will be observable under certain ideal conditions. Suppose Y is a cell in the crystalline material X. The spatial geometric relation (called the structure of Y) of the molecules in Y is termed the basic structure of X.

Generalizing the model in (Lin and Qiu, 1987), the following is our model for the arbitrarily fixed material X: no matter what state X is in (solid, liquid, or gas), there exist many pairwise disjoint groups of molecules in X such that each molecule in X is contained in exactly one of the groups, and every group is "topologically" isomorphic to the basic structure of X. The basic structure of the material determines all the physical properties that X has. The meaning of topologically isomorphic was explained in (Lin and Qiu, 1987) with examples. With this model many phenomena about materials can be explained naturally. For details, see (Lin and Qiu, 1987).

8.5. Some Epistemological Problems

In the ZFC axiom system, the following results can be shown:

(1) There is no system whose object set consists of all systems.

(2) A system $S_n = (M_n, R_n)$ is an nth-level object system of a system $S_0 = (M_0, R_0)$, if there exist systems $S_i = (M_i, R_i)$, $i = 1, 2, \ldots, n-1$, such that $S_i \in M_{i-1}$, $i = 1, 2, \ldots, n$. Then for any system S, there is no nth-level object system S_n of S such that $S_n = S$ for any natural number $n > 0$.

(3) A sequence $\{S_i\}_{i \in n}$ of systems, where n is an ordinal number, is said to be a chain of object systems of a given system S if S_0 is an i_0th-level object system of S, and for any $i, j \in n$, if $i < j$, there exists a natural number $i_j > 0$ such that S_j is an i_jth-level object system of S_i. Then for any fixed system S, each chain of object systems of S must be finite.

In (Lin, 1989c) the following epistemological problems were discussed: (i) the feasibility of the definition of the theory "science of science," (ii) the existence of fundamental particles in the world, and (iii) the existence of absolute truths.

The theory of science of science appeared in the 1930s. Since then more and more scientists have been involved in its research. The background in which the theory appeared is that, since human society entered the twentieth century, the development of different aspects of science and technology have sped up. The development of science shows a tendency to divide science into more and more disciplines while synthesizing results and methods in different fields to get new results and understanding about the world. Hence, a natural problem arose: Can science be studied as a social phenomenon so that man can develop scientific research with purposes and improve its recognition of the natural world? This problem is very important to administrators of scientific research, because so far each scientific achievement is obtained by either individual scientists or small groups of scientists, Thus, there are many scientific achievements obtained independently by many different scientists. If the answer to the aforementioned question is "yes," then it partially means that we can save at least some valuable scientific labor from doing the same things by arranging them to do some pretargeted research. In this way, the development of science will be sped up greatly. Some scientists think that the research object of the theory of science of science is the following: The theory of science of science is such a theory that instead of any real matter it studies science as a whole, the history and the present situation of each discipline, the relationship between disciplines, and the whole developing tendency of science.

Combining Claim 1 and result (2), we can see that the theory of science of science cannot exist. Because, roughly speaking, the theory has to contain a study of itself. For detailed discussion, see (Lin, 1989c).

From Klir's (1985) definition of systems and result (1), it follows that we cannot consider a relation which is true in a system that contains all systems as its objects. Does this imply that there is no universal truth? I believe so. I think that for any given truth, an environmental system is given in which the truth is a true relation.

Result (3) says that any system is built upon objects which are no longer systems. Does this imply that the world consists of fundamental particles, where a fundamental particle is a particle which cannot be divided into smaller particles? See (Lin, 1989c) for details.

8.6. The Vase Puzzle and Its Contradictory Mathematical Modelings

A conceptual problem, named the vase puzzle, is used to show that mathematical modelings can lead to contradictory inferences. Around this puzzle, some questions, concerning proof methodology in mathematics and understanding mathematical induction arise, including in general, methodology, epistemology, philosophy, and modeling time and space. Some unsettled problems are left open.

8.6.1. The Vase Puzzle and Its Modelings

The concept of infinity has been bothering mankind for centuries (Moore, 1991). In mathematics, different meanings are given to the concept, (Davis, 1977; Halmos, 1960; Jech, 1973). Many of the greatest thinkers in history (Moore, 1991), have been convinced that infinity is something we will never be able to understand because there are so many seemingly unsolvable puzzles about it.

If a person were concerned only with problems in real life, he or she would say that the concept of infinity had no interest. Unfortunately, it is no longer the case. For example, the study of quality control of the products of an automatic assembly line needs the concept. Specifically to study quality, we draw a random sample of the products and use sample statistics to make inferences on the continuously expanding population, which is theoretically the collection of all the products that have been and will be produced from the assembly line. To make the inferences more convincing, we often treat the ever-expanding population as an infinite population.

In this section a conceptual problem is constructed to show that mathematical modeling does not always work. In other words, different models can give contradictory results. This astonishing phenomenon forces us to rethink some principles and long-lasting questions in mathematical induction, mathematical proofs, methodology, epistemology, philosophy, etc. Even though the phenomenon has been studied for some time, we still hope that this presentation will cast the first brick in order to attract many beautiful jades.

8.6.1.1. The vase puzzle. Suppose a vase and an infinite number of pieces of paper are available. The pieces of paper are labeled by natural numbers $1,2,3,\ldots$, so that each piece has at most one label on it. The following recursive procedure is performed:

Step 1: Put the pieces of paper, labeled from 1 through 10, into the vase; then remove the piece labeled 1.

Step n: Put the pieces of paper, labeled from $10n-9$ through $10n$ into the vase; then remove the piece labeled n, where n is any natural number $1,2,3,\ldots$

Question 8.6.1. After the recursive procedure is finished, how many pieces of paper are left in the vase ?

Some comments are necessary. First, the vase need not be infinitely large — actually any size will do. Second, the total area of the infinite number of pieces of paper can also be any chosen size. For example, Fig. 8.1 shows how an infinite number of pieces of paper can be obtained. Third, number-labeling can be done according to the steps in the puzzle. Finally, the recursive procedure can be finished within any chosen period of time. For convenience, we write the mathematical induction procedure, which guarantees that the vase puzzle is well defined.

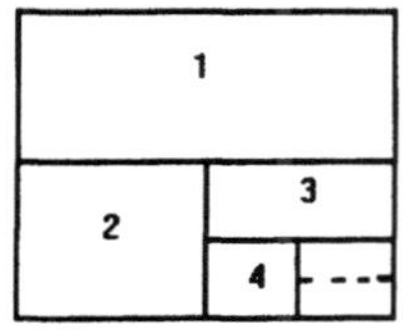

Figure 8.1. Obtain as many pieces of paper as needed out of a chosen total area of pieces of paper.

8.6.1.2. Mathematical induction. See (Smith et al., 1990). Let S be a subset of the set N of all natural numbers. If it can be shown that

- Initial step: $1 \in S$.
- Inductive step: If $n \in S$, then $n+1 \in S$.

then the subset S equals the set N.

To fit the recursive procedure in the vase puzzle into the format of mathematical induction, let us define a subset S of natural numbers as follows:

$$S = \{n : \text{step } n \text{ can be done and all pieces of paper with labels } 1 \text{ through } 10n \text{ have been in the vase at least once}\} \tag{8.6}$$

It is left to the reader to verify that S is the same as N.

8.6.1.3. An elementary modeling. To answer the question of the vase puzzle, let us define a function, based upon mathematical induction, by

$$f(n) = 9n \tag{8.7}$$

Equation (8.7) really tells how many pieces of paper are left in the vase after step n, where $n = 1, 2, 3, \ldots$. Therefore, if the recursive procedure can be finished, the number of pieces of paper left in the vase should equal the limit of $f(n)$ as n approaches ∞. The answer is that infinitely many pieces of paper are left in the vase.

8.6.1.4. A set-theoretical modeling. Based upon this modeling, the answer to the question of the vase puzzle is "no paper is left in the vase." This contradicts the conclusion, derived in the elementary modeling.

For each natural number n, define the set M_n of pieces of paper left in the vase as follows:

$$M_n = \{x : x \text{ has a label between } n \text{ and } 10n+1 \text{ exclusively}\} \tag{8.8}$$

Then after the recursive procedure is finished, the set of pieces of paper left in the vase equals the intersection

$$\bigcap_{n=1}^{\infty} M_n \tag{8.9}$$

That is, if x is a piece of paper left in the vase, x then has a label greater than all natural numbers. This contradicts the assumption that each piece of paper put into the vase has a natural number label. Thus, $\bigcap_{n=1}^{\infty} M_n$ is empty.

Lin, Ma, and Port (1990) pointed out theoretically that there must be an impassable chasm between pure mathematics and applied mathematics. The reason is that pure mathematics is established on a set of axioms, say, the ZFC axioms. The theory possesses a beauty analogous to painting, music, and poetry, and the harmony between numbers and figures [for details, see (Lin, 1990d)], while in applied mathematics each object of interest is always first given some mathematical meaning, and then conclusions are drawn based on the relations of those to the objects involved. It is the assignment of a mathematical meaning to each object that causes problems, because different interpretations can be given to the same object. As described in (Lin et al., 1990), some interpretations can result in contradictory mathematical models. In this section we chose the vase puzzle to emphasize the existence of the impassable chasm, for the following reasons: (1) The example is easy to describe. (2) The two contradictory modelings are readily comprehensible by readers with little background in mathematics. (3) The contradiction has very fruitful implications.

We give two more puzzles which will be relevant to forthcoming discussions. One concerns the structure of space, and the other concerns the concept of time.

Paradox [Paradox of the Hotel, (Moore, 1991)]. Suppose a hotel has infinitely many rooms, each occupied at a particular time. Then a newcomer can be accommodated without anybody having to move out; for if the person in the first room moves to the second, and the person in the second room moves to the third, and so on ad infinitum, this will release the first room for the newcomer.

Hilbert used to present this paradox in his lectures. The hotel, by the way, need only occupy a finite amount of space. For if each successive floor is half the height of the one below it, then the entire hotel will be only twice the height of the ground floor. However, this raises the following question about the nature of the concept of space.

Question 8.6.2 [Benardete (Moore, 1991)]. If the roof of the hotel were removed, what would a person looking at the hotel from above see?

Paradox [Paradox of a Moving Particle]. Suppose a particle moves from point 0 to point 1 on the real number line. If at the half-minute the particle is at point $1/2$, at $3/4$ minute it is at point $3/4$, and in general, at $(2^n - 1)/2^n$ minute, the particle is at point $(2^n - 1)/2^n$, where will the particle be at 1 minute?

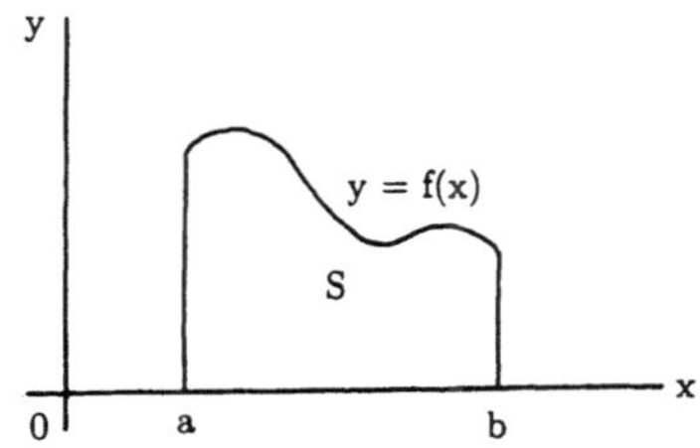

Figure 8.2. S is the area under $f(x)$ between a and b.

Question 8.6.3. If another particle moves from point 2 to point 0 on the same real number line, which point of the sequence

$$\frac{1}{2}, \frac{3}{4}, \frac{7}{8}, \frac{15}{16}, \ldots, \frac{(2^n - 1)}{2^n}, \ldots \tag{8.10}$$

should the particles meet first ?

8.6.2. "Experts" Explanations

The vase puzzle has been presented to a number of scholars with expertise in different areas over the years. We list two of the "explanations" given for the contradictory mathematical modelings.

Explanation. The elementary modeling is incorrect because it does not use all known information.

Explanation. There is no way to finish the recursive procedure in the puzzle. Therefore, the question in the puzzle is invalid.

8.6.3. Mathematical Proofs, Theory of Cardinal Numbers, and the Vase Puzzle

Some well-known theories and proofs of theorems in mathematics are used to argue that if explanation 2 were logically correct, then a large portion of mathematics would be incorrect.

Problem [The Area Problem, (Stewart, 1987)]. Find the area of the region S under the curve $y = f(x)$ from a to b. This means that S (see Fig. 8.2), is bounded by the graph of a function f (where $f(x) \geq 0$), the vertical lines $x = a$ and $x = b$, and the x-axis.

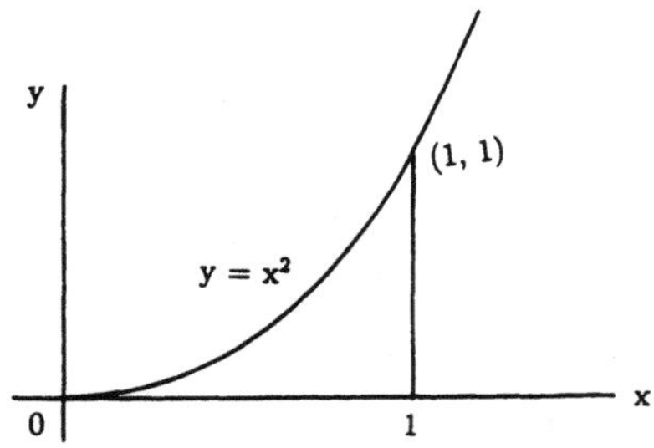

Figure 8.3. Region S under graph of $y = x^2$ from 0 to 1.

To solve this problem, we have to answer the question: What is the meaning of the word "area"? It is easy to answer this question for regions with straight sides. For a rectangle, area is defined as the product of length and width. The area of a triangle is half the base times the height. The area of a polygon is found by dividing it into triangles. However, it is not so easy to find the area of a region with curved sides. Theoretically, to find the area of a region S, we first approximate it by areas of polygons and then take the limit of the areas of these polygons. To make the method more specific, let us briefly go through an example.

Example 8.6.1. Find the area under the parabola $y = x^2$ from $x = 0$ to $x = 1$ (see Fig. 8.3).

Solution. Divide the interval $[0, 1]$ into subintervals of equal length and consider the rectangles whose bases are those subintervals and whose heights are the values of the function $y = x^2$ at the right-hand endpoints of these subintervals. Figsures 8.4–8.6 show the approximation of the parabolic segment by 4, 8, and n rectangles.

Let S_n be the sum of the areas of the n rectangles in Fig. 8.6. Therefore,

$$S_n = \frac{1}{n}\left(\frac{1}{n}\right)^2 + \frac{1}{n}\left(\frac{2}{n}\right)^2 + \cdots + \frac{1}{n}\left(\frac{n}{n}\right)^2$$

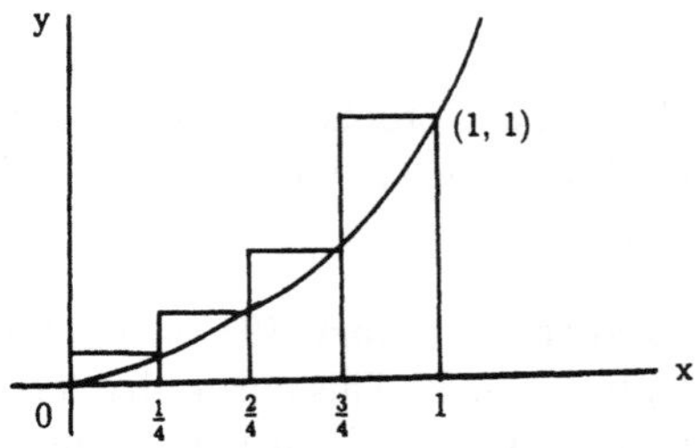

Figure 8.4. Four-rectangle approximation of the area of the region S.

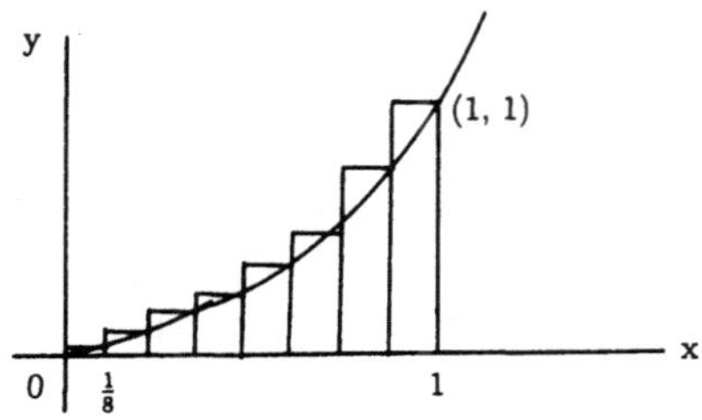

Figure 8.5. Eight-rectangle approximation of the area of the region S.

$$= \frac{1}{n^3}\sum_{i=1}^{n} i^2$$

$$= \frac{(n+1)(2n+1)}{6n^2} \tag{8.11}$$

From Figs. 8.4 to 8.6 it appears that as n increases S_n becomes a better and better approximation to the area of the parabolic segment. In calculus, the area A of the region S can be found by taking the limit:

$$A = \lim_{n\to\infty} S_n = \frac{1}{3} \tag{8.12}$$

Now compare the idea applied in the area problem with that in the vase puzzle do we have similar recursive procedures? In solving the area problem, we compute the area S_n for $n = 1, 2, 3, \ldots$; after all S_n's are computed (based on mathematical induction we can do this), by taking the limit of the S_n's we get the value for the area A of S. If Explanation 2 were correct, how could we finish computing the S_n's? If we are not able to, we could not take the limit of the S_n's, either. Some readers may get confused here and do not see the similarity because the concept of limit is involved.

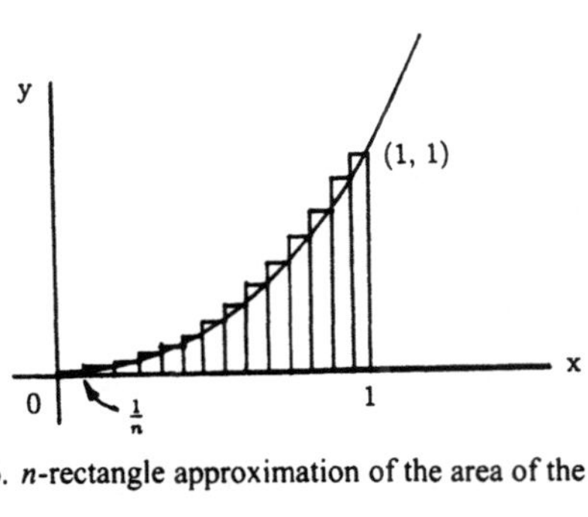

Figure 8.6. n-rectangle approximation of the area of the region S.

Example 8.6.2 [Department of Mathematics (1978)]. When studying the concept of convergence of a sequence, we have a well-known theorem: Each bounded sequence of numbers must have a convergent subsequence. Let us look at the proof of the theorem.

Case 1: Suppose the bounded sequence $\{x_n\}_{n=1}^{\infty}$ is made up of a finite set of numbers. Then one of the numbers, say A, must appear repeatedly infinitely many times. If the locations that the number A appears are the n_1st term, n_2nd term, ..., then

$$x_{n_1}, x_{n_2}, \ldots, x_{n_k}, \ldots \tag{8.13}$$

is a subsequence of $\{x_n\}_{n=1}^{\infty}$ satisfying $\lim_{k\to\infty} x_{n_k} = A$. Therefore, we have proved that $\{x_n\}_{n=1}^{\infty}$ has a convergent subsequence.

Case 2: Suppose $\{x_n\}_{n=1}^{\infty}$ is made up of infinitely many numbers. Then the set $E = \{x : \exists n(x = x_n)\}$ is a bounded infinite set. By a theorem of Weierstrass (1815–1897), the set has at least one cluster point c. We now show that there must exist a subsequence $\{x_{n_k}\}_{k=1}^{\infty}$ of $\{x_n\}_{n=1}^{\infty}$ which converges to c.

Since c is a cluster point of E, there must be a point in the set $(c-1, c+1) \cap (E - \{c\})$. Pick such a point y_1 in the set. This point must appear in the sequence $\{x_n\}_{n=1}^{\infty}$ at least once. Hence, there is an index n_1 such that

$$y_1 = x_{n_1} \tag{8.14}$$

In general, suppose we have found $x_{n_1}, x_{n_2}, \ldots, x_{n_k}$ such that $x_{n_i} \in (c - 1/i, c + 1/i) \cap (E - \{c\}), i = 1, 2, \ldots, k$. Since the set $(c - 1/(k+1), c + 1/(k+1)) \cap (E - \{c\})$ contains infinitely many points and the sequence $x_1, x_2, \ldots, x_{n_k}$ has only n_k terms, we can pick a point $y_{k+1} = x_{k+1} \in (c - 1/(k+1), c + 1/(k+1)) \cap (E - \{c\})$ such that $n_{k+1} > n_k$. According to mathematical induction, a subsequence $\{x_{n_k}\}_{k=1}^{\infty}$ of $\{x_n\}_{k=1}^{\infty}$ is obtained such that $x_{n_k} \in (c - 1/k, c + 1/k), k = 1, 2, 3, \ldots$. Thus,

$$|x_{n_k} - c| < 1/k \tag{8.15}$$

When $k \to \infty$, $1/k \to 0$. Thus, $\lim_{k\to\infty} x_{n_k} = c$.

Look at the proof here. If the subsequence $\{x_{n_k}\}_{k=1}^{\infty}$ can be obtained by mathematical induction, then the recursive procedure in the vase puzzle can analogously be finished. That is, if "there is no way to finish the procedure," as posited in Explanation 2, the proof given must be incorrect. Some may argue that the proof above is not the only one for the theorem. If so, let us look at the following well-known argument of Cantor.

Example 8.6.3 [Halmos (1960)]. One of the greatest proofs in set theory shows the following amazing result: The set of all real numbers that are greater than zero and smaller than 1 has greater cardinality than the set $\mathbb{N}$ of all natural numbers.

Proof: Obviously, the set $\mathbb{N}$ and the set $\mathbb{I}$ of all real numbers between 0 and 1 are infinite sets. By contradiction, suppose $|\mathbb{N}| = |\mathbb{I}|$. That is, there exists a one-to-one correspondence $f : \mathbb{N} \to \mathbb{I}$. For any $q \in \mathbb{I}$, let $a_n = q$ if $f(n) = q$ for some $n \in \mathbb{N}$. Thus, the elements in $\mathbb{I}$ can be written as a sequence

$$a_1, a_2, \ldots, a_n, \ldots \tag{8.16}$$

Since each $q \in \mathbb{I}$ satisfies $0 < q < 1$, it is possible to write the elements in $\mathbb{I}$ as

$$\begin{aligned} a_1 &= 0.a_{11}a_{12}a_{13}\cdots a_{1n}\cdots \\ a_2 &= 0.a_{21}a_{22}a_{23}\cdots a_{2n}\cdots \\ a_3 &= 0.a_{31}a_{32}a_{33}\cdots a_{3n}\cdots \\ &\vdots \\ a_m &= 0.a_{m1}a_{m2}a_{m3}\cdots a_{mn}\cdots \\ &\vdots \end{aligned} \tag{8.17}$$

where each a_{ij} is a digit in the set $W = \{0, 1, 2, 3, 4, 5, 6, 7, 8, 9\}$. Now let b be a number with expression

$$b = 0.b_1 b_2 b_3 \cdots b_m \cdots \tag{8.18}$$

such that each $b_k \in W$ and that for each $m \in \mathbb{N}$, $b_m \neq a_{mm}$. That is, $b \in \mathbb{I}$ and b is not in the list (8.17). This contradicts the assumption that each element of $\mathbb{I}$ is in the list (8.17). Hence, $|\mathbb{N}| \neq |\mathbb{I}|$. ■

Even though this proof does not mention mathematical induction, the existence of the number b does depend on the principle. To be more specific, the number b needs to be constructed as follows:

Initial step: Define b_1 being a number $\in A - \{a_{11}\}$.

Inductive step: Suppose for $n \in \mathbb{N}$, the b_k's have been defined for all $k < n$. Then let b_n be a number $\in A - \{a_{nn}\}$.

By mathematical induction, for each natural number n a number b_n can be chosen. If this inductive process "can be finished," the desired number b can be constructed in the form of Eq. (8.18). Once again, if "there is no way to finish the procedure," there is no way to finish choosing the b_n's. Consequently, there is no way to construct b in Eq. (8.18). That is, Cantor's diagonal method, which we have just described, is invalid.

A seasoned mathematician can see that the principle of mathematical induction has been and will be "used" or "abused" in ways as we have described. Thus, one or both of the following is true: (1) Explanation 2 is incorrect. (2) In the cases described, the principle of mathematical induction has been abused.

Conjecture 8.6.1. *Statement (1) is true while statement (2) is not.*

Table 8.1.

Age	Age	Age	Sum	Product
1	1	72	74	72
1	2	36	39	72
1	3	24	28	72
1	4	18	23	72
1	6	12	19	72
2	2	18	22	72
2	3	12	17	72
2	4	9	15	72
2	6	6	14	72
1	8	9	18	72
3	4	6	13	72
3	3	8	14	72

8.6.4. Some Threatening Impacts of the Vase Puzzle

The first explanation is a great and fruitful resource of ideas that connect the vase puzzle with problems in philosophy, methodology and epistemology. In this subsection, we discuss these connections.

8.6.4.1. Connections with methodology. In general, if known information about a phenomenon has not all been used, the phenomenon can either not be understood because of lack of knowledge or be understood at a more global level and more information is needed to be more specific. For example [see (Billstein et al., 1990) for details], when his students asked Mr. Factor what his children's ages were, he said, "I have three children. The product of their ages is 72, and the sum of their ages is the number of this room." The children then asked for the door to be opened to verify the room number, which is 14. Then, Sonja, the class math whiz, told the teacher that she needed more information to solve the problem. Mr. Factor said, "My oldest child is good at chess." Then, Sonja announced the correct ages of Mr. Factor's children as: $3,3,8$. Based on the given information, Sonja constructed Table 8.1. She knew the sum of the ages was 14, but could not determine the correct answer from the two choices $(2,6,6)$ and $(3,3,8)$. When Sonja was told that the oldest child was good at chess, she knew that $(2,6,6)$ could not be a possible combination because if the children were 2, 6, and 6 years old, there would not be an oldest child.

Notice that the vase puzzle does not satisfy the general methodology of recognition. The information about ordering really made understanding more controversial rather than more specific.

Our vase puzzle does not fit into categories of problems considered by Polya (1973). The problem is neither a "perfectly stated" nor "practical." It is conceptual and thought-provoking. It challenges traditional mathematical reasoning. It especially raises questions about the idea of mathematical modeling.

8.6.4.2. Connections with epistemology. Through history, the development of scientific theories and the understanding of realities have been hand in hand. To know more about nature, scientists establish hypotheses and develop theories. These theories further our understanding of nature and, in turn, push the development of the theories to a higher level.

Among the most common practices of modern scientific activities is the science and technology of data analysis. It is known in analysis that the more data that are collected, the more misleading the conclusion that will be obtained, measuring and collecting each datum are subject to errors or noise, and an extraordinary amount of noise can easily conceal the actual state of a phenomenon. The vase puzzle is another example showing that the more facts we know, the more confused we become. At the same time, the vase puzzle also questions the accuracy of scientific predictions, because the ignorance of a fact will lead to a completely different prediction. Therefore, one could question the value and the meaning of scientific research and scientific predictions.

8.6.4.3. Connections with philosophy. Studying nature and the structure of time and space is an ancient interest. The vase puzzle, the paradox of the hotel, and the paradox of a moving particle, once again, raise the interest of exploring the answer to the question, because by reassigning some numbers, an infinite procedure can be finished within a randomly chosen time frame, an infinitely large piece of paper can be shrunk to any desired size, and an infinitely high building can be rebuilt in order to fit into a limited space. Then we ask: Are the structures of space and time continuous? What is the meaning of volume?

Maybe the answer to the second question is that the concept of volume is meaningless mathematically because one can prove that it is possible to chop a solid ball into small pieces and then reassemble these pieces into as many balls as one desires of the same size as the original ball (Jech, 1973).

As for the first question, about the continuity of space and time, we think the structure of the set of all hyperreal numbers (Davis, 1977) describes and explains the hotel and moving particle paradoxes better than the commonly accepted one. For example, the commonly accepted model for time is the set of all real numbers, which is continuous and complete. The property of completeness of real numbers was applied to model the process of crystallization of polymers (Lin and Qiu, 1987). If the structure of hyperreal numbers is applied, the modeling for the crystallization in (Lin and Qiu, 1987) would no longer be correct. To make things more specific, let us describe the structure of the hyperreal numbers in some detail.

The concept of hyperreal numbers is a product of nonstandard analysis. This theory deals with ideal elements and is a time-honored and significant mathematical theory. These ideal elements include "numbers" that are infinitely close to the numbers we are interested in and those that are infinitely far away from all real numbers.

Briefly speaking, the set $^*\mathbb{R}$ of all hyperreal numbers has operations $+, -, \times, \div$ and a linear ordering relation $<$. Let $|x|$ be the absolute value of a hyperreal number $x \in {}^*\mathbb{R}$. Then $|x| = \max\{x, -x\}$, and $^*\mathbb{R}$ is made up of three parts:

$$F = \{x \in {}^*\mathbb{R} : |x| < n \text{ for some } n \in \mathbb{N}\} \tag{8.19}$$

$$I = \{x \in {}^*\mathbb{R} : x = 0 \text{ or } 1/x \in {}^*\mathbb{R} - F\} \tag{8.20}$$

and $^*\mathbb{R} - F$, where F contains all finite hyperreal numbers, I is the set of all infinitesimals, and $^*\mathbb{R} - F$ includes all infinite hyperreal numbers.

The set $\mathbb{R}$ of all real numbers is a dense subset of F. Geometrically, F can be obtained as follows: For each real number $r \in \mathbb{R}$, the number r in $^*\mathbb{R}$ has a neighborhood

$$r \oplus I = \{r + x : x \in I\} \tag{8.21}$$

such that for any real numbers $r, s \in \mathbb{R}$ and infinitesimals $x, y \in I$, $r < s$, if and only if $r + x < s + y$. This neighborhood $r \oplus I$ is called the monad of r. It can be seen that all monads of real numbers are densely ordered.

The set $^*\mathbb{R} - F$ of all infinite hyperreal numbers consists of F as the center "block," and, in either the positive or the negative directions, an ordered set of "blocks." These blocks are densely ordered with neither a first nor last element. Each block has the same geometric structure as F and is termed a galaxy.

If $^*\mathbb{R}$ is used to model time and the Cartesian product $^*\mathbb{R}^3$ to model space, Questions 8.6.2 and 8.6.3 can be answered as follows:

A. For (Benardete) Question 8.6.2: Suppose the hotel roof is r feet high. When the roof is removed, nothing can be seen if one looks at the hotel from above because what a person sees should be in the layers $r - x$ for each $x \in I$ and $x > 0$, and no guests live on these levels (no floor reaches these heights).

B. For Question 8.6.3: Since a seemly smooth moving particle really jumps forward from a monad to another, it could meet any point in the sequence (8.10) first.

8.7. Final Comments and Further Questions

From our discussion it can be seen that there are roughly two methods to introduce new mathematical concepts. One is that in the study of practical problems, new mathematical concepts are abstracted, and the other is that in the search for relations between mathematical concepts, new concepts are discovered. The application of mathematics, as shown by the examples, brings life to the development of mathematics so that when dealing with real problems, mathematics often makes us better understand the phenomena under consideration. If all well-developed mathematical theories do not fit a specific situation under consideration, a modified mathematical theory will be developed to study the situation. That is to say, mathematics, just as all other scientific branches, is developed in the process of examining, verifying and modifying itself.

When facing practical problems, new abstract mathematical theories are developed to investigate them. At the same time, predictions based on the newly developed theories are so accurate that the following question appears:

Question 8.7.1. Does the human way of thinking have the same structure as that of the material world?

On the other hand, the mathematical palace is built upon the "empty set." History gradually shows that almost all natural phenomena can be described and studied with mathematics. From these facts the following question arises.

Question 8.7.2. If all laws of nature can be written in the language of mathematics, can we conclude that the universe we live in rests on the empty set, just as *Genesis 1:1* states that "In the beginning, God created the heaven and the earth. And the earth was without form, and void"?

CHAPTER 9

General Systems: A Multirelation Approach

Based on the previous chapter, it is time to study the concept of general systems mathematically. The concept of systems is a generalization of that of structures. Generally speaking, a system consists of a set of objects and a set of relations between the objects, and has a structure of "layers." In this chapter, a model of general systems is introduced, based on the methods of set theory, and some basic global properties of systems are studied in detail. The chapter contains eight sections. In the first section, layer structures, the existence of fundamental objects, the finite chain condition, and centralized systems are studied. The second section consists of properties of relations between systems, the structure of quotient systems, and a characterization of centralizable systems. Methods of constructing new systems, based on a set of given systems, are introduced in the third section. These methods serve as a mathematical foundation for the analysis of "whole–parts" relations.

In the fourth section, the structure of connected systems is studied and used to check the connectedness of newly constructed systems. This structure shows whether a system can be seen as a set of systems or a single system only. In the fifth and the seventh sections, hierarchies of systems (or, say, large-scale systems) are studied. The concepts of controllabilities of general systems are introduced and studied in the sixth section. Some references are listed in the last section.

9.1. The Concept of General Systems

Let A be a set and n an ordinal number. An n-ary relation r on A is a subset of the Cartesian product A^n. In the sequel, A^n means either the Cartesian product of n copies of A or the set of all mappings from n into A. From Chapters 2 and 3 it is easy to see that the two definitions of A^n are equivalent.

If r is a relation on A, there exists an ordinal $n = n(r)$, a function of r, such that

$$r \subseteq A^n$$

The ordinal number $n = n(r)$ is called the length of the relation r. When r is the empty relation $\emptyset$, assume that $n(\emptyset) = 0$; i.e., the length of the relation $\emptyset$ is 0.

S is a (general) system if S is an ordered pair (M,R) of sets, where R is a set of some relations on the set M. Each element in M is called an object of S, and M and R are called the object set and the relation set of S, respectively. The system S is discrete if $R = \emptyset$ or $R = \{\emptyset\}$ and $M \neq \emptyset$. It is trivial if $M = \emptyset$.

Two systems $S_i = (M_i, R_i)$, $i = 1,2$, are identical, denoted $S_1 = S_2$, if

$$M_1 = M_2 \quad \text{and} \quad R_1 = R_2$$

The system S_1 is a partial system of the system S_2 if $M_1 \subseteq M_2$ and for each relation $r_1 \in R_1$ there exists a relation $r_2 \in R_2$ such that $r_1 = r_2|M_1$, where $r_2|M_1$ is the restriction of r_2 on M_1, and is defined by

$$r_2|M_1 = r_2 \cap M_1^{n(r_2)}$$

The system S_1 is a subsystem of S_2 if $M_1 \subseteq M_2$ and for each relation $r_1 \in R_1$ there exists a relation $r_2 \in R_2$, satisfying $r_1 \subseteq r_2|M_1$.

Theorem 9.1.1. *There is no system such that the object set of the system consists of all systems.*

This is a restatement of the Zermelo–Russel paradox.

Proof: Consider the class V of all sets. For each $x \in V$ define a system by $(x, \emptyset)$. Then a 1–1 correspondence h is defined such that $h(x) = (x, \emptyset)$. If there is a system $S = (M,R)$ whose object set contains all systems, then M is a set. Therefore, the Axiom of Comprehension shows that V is a set, contradiction. ■

A system $S_n = (M_n, R_n)$ is an nth-level object system of $S_0 = (M_0, R_0)$ if there exist systems $S_i = (M_i, R_i)$, for $i = 1, 2, \ldots, n-1$, such that S_i is an object in M_{i-1}, $1 \leq i \leq n$. Each element in M_n is called an nth-level object of S_0.

Theorem 9.1.2 [ZFC]. *Let $S_0 = (M_0, R_0)$ be a system and $S_n = (M_n, R_n)$ an nth-level object system of S_0. Then S_0 cannot be a subsystem of S_n for each natural number $n \in \mathbb{N}$.*

Proof: The theorem will be proved by contradiction. Suppose that for a certain natural number n, the system S_0 is a subsystem of the nth-level object system S_n.

Let $S_i = (M_i, R_i)$ be the systems, for $i = 1,2,\ldots,n-1$, such that $S_i \in M_{i-1}$, $i = 1,2,\ldots,n$. Define a set X by

$$X = \{M_i : 0 \leq i \leq n\} \cup \{S_i : 0 \leq i \leq n\} \cup \{\{M_i\} : 0 \leq i \leq n\}$$

From the Axiom of Regularity, it follows that there exists a set $Y \in X$ such that $Y \cap X = \emptyset$. There now exist three possibilities:

(i) $Y = M_i$ for some i.

(ii) $Y = S_i$ for some i.

(iii) $Y = \{M_i\}$ for some i.

If possibility (i) holds, $S_{i+1} \in Y \cap X$ for $i \leq n-1$; and $S_1 \in Y \cap X$ if $Y = M_n$. Therefore, $Y \cap X \neq \emptyset$, contradiction. If (ii) holds, $Y = S_i = (M_i, R_i) = \{\{M_i\}, \{M_i, R_i\}\}$ and $\{M_i\} \in Y \cap X \neq \emptyset$, contradiction. If (iii) holds, $M_i \in Y \cap X \neq \emptyset$, contradiction. These contradictions show that S_0 cannot be a subsystem of S_n. ■

A chain of object systems of a system S is a sequence $\{S_i : i < \alpha\}$, for some ordinal number α, of different-level object systems of S, such that for each pair $i,j < \alpha$ with $i < j$, there exists an integer $n = n(i,j)$, a function of i and j, such that S_j is an nth-level object system of S_i. A similar proof to that of Theorem 9.1.2 gives the following.

Theorem 9.1.3 [ZFC]. *Each chain of object systems of a S must be finite.*

Theorem 9.1.4 [ZFC]. *For any system S, there exists exactly one set $M(S)$ consisting of all fundamental objects in S, where a fundamental object in S is a level object of the system S which is no longer a system.*

Proof: Suppose that $S = (M_0, R_0)$. Define

$$^*M_0 = \{x \in M_0 : x \text{ is not a system}\}$$

and

$$\widetilde{M}_0 = \bigcup \{M_x : x = (M_x, R_x) \in M_0 - {}^*M_0\}$$

Then $^*M_0$ and $\widetilde{M}_0$ are sets.

Suppose that for a natural number n, two sequences $\{^*M_i : i = 0,1,2,\ldots,n\}$ and $\{\widetilde{M}_i : i = 0,1,2,\ldots,n\}$ have been defined such that

$$^*M_i = \{x \in \widetilde{M}_{i-1} : x \text{ is not a system}\}$$

and

$$\widetilde{M}_i = \bigcup\{M_x : x = (M_x, R_x) \in \widetilde{M}_{i-1} - {}^*M_i\}$$

We then define ${}^*M_{i+1} = \{x \in \widetilde{M}_i : x \text{ is not a system}\}$ and $\widetilde{M}_{i+1} = \bigcup\{M_x : x = (M_x, R_x) \in \widetilde{M}_i - {}^*M_{i+1}\}$.

By mathematical induction a sequence $\{{}^*M_i : i \in \omega\}$ can be defined. Let $M(S) = \bigcup\{{}^*M_i : i \in \omega\}$. Then Theorem 9.1.3 guarantees that $M(S)$ consists of all fundamental objects in S; and from the uniqueness of each ${}^*M_i$ it follows that $M(S)$ is uniquely determined by S. ■

Theorem 9.1.5. *Let $S = (M, R)$ be a system such that, for any $r \in R$, the length $n(r) = 2$. Then*

$$S \subseteq p^2(M) \cup p(p(M) \cup p^4(M)) \tag{9.1}$$

where $p(X)$ is the power set of X and $p^{i+1}(X) = p(p^i(X))$ for each $i = 1, 2, \ldots$

Proof: For each element $(x,y) \in r$, $(x,y) = \{\{x\}, \{x,y\}\} \in p^2(M)$ and $r \in p^3(M)$. Therefore, $R \in p^4(M)$. Since $\{M\} \in p(M)$ and $(M,R) \in p(p(M) \cup p^4(M))$, it follows that the inclusion in Eq. (9.1) holds. ■

Question 9.1.1. Give a structural representation for general systems similar to that in Theorem 9.1.5.

A system $S = (M, R)$ is called centralized if each object in S is a system and there exists a nontrivial system $C = (M_C, R_C)$ such that for any distinct elements $x, y \in M$, say $x = (M_x, R_x)$ and $y = (M_y, R_y)$, then

$$M_C = M_x \cap M_y \quad \text{and} \quad R_C \subseteq R_x|M_C \cap R_y|M_C$$

where $R_x|M_C = \{r|M_C : r \in R_x\}$ and $R_y|M_C = \{r|M_C : r \in R_y\}$. The system C is called a center of the centralized system S.

Theorem 9.1.6 [ZFC]. *Let κ be any infinite cardinality and $\theta > \kappa$ a regular cardinality such that, for any $\alpha < \theta$, $|\alpha^{<\kappa}| < \theta$. Assume that $S = (M, R)$ is a system satisfying $|M| \geq \theta$ and each object $m \in M$ is a system with $m = (M_m, R_m)$ and $|M_m| < \kappa$. If there exists an object contained in at least θ objects in M, then there exists a partial system $S' = (M', R')$ of S such that S' forms a centralized system and $|M'| \geq \theta$.*

Proof: Without loss of generality, we assume that $|M| = \theta$ and that there is a common element in all the object systems in M. Then $|\bigcup\{M_x : x = (M_x, R_x) \in M\}| \leq \theta$. Since the specific objects in each M_x, for each object $x = (M_x, R_x) \in M$, are irrelevant, we may assume that

$$\bigcup\{M_x : x = (M_x, R_x) \in M\} \subseteq \theta$$

Then, for each $x = (M_x, R_x) \in M$, the object set M_x has some order type $< \kappa$ as a subset of θ. Since θ is regular and $\theta > \kappa$, there exists a $\rho < \kappa$ such that $M_1 = \{x \in M : M_x \text{ has order type } \rho\}$ has cardinality θ. We now fix such a ρ and deal only with the partial system $S_1 = (M_1, R_1)$ of S, where R_1 is the restriction of the relation set R on M_1.

For each $\alpha < \theta$, $|\alpha^{<\kappa}| < \theta$ implies that less than θ objects of the partial system S_1 have object sets as subsets of α. Thus, $\bigcup\{M_x : x = (M_x, R_x) \in M_1\}$ is cofinal in θ. If $x \in M_1$ and $\xi < \rho$, let $M_x(\xi)$ be the ξth element of M_x. Since θ is regular, there is some ξ such that $\{M_x(\xi) : x \in M_1\}$ is cofinal in θ. Now fix ξ_0 to be the least such ξ. Then the condition that there exists a common element in each system in M_1 implies that we can guarantee that $\xi_0 > 0$. Let

$$\alpha_0 = \bigcup\{M_x(\eta) + 1 : x \in M_1 \text{ and } \eta < \xi_0\}$$

Then $\alpha_0 < \theta$ and $M_x(\eta) < \alpha_0$ for all $x \in M_1$ and all $\eta < \xi_0$.

By transfinite induction on $\mu < \theta$, pick $x_\mu \in M_1$ so that $M_{x_\mu}(\xi_0) > \alpha_0$ and $M_{x_\mu}(\xi_0)$ is above all elements of earlier x_ν; i.e.,

$$M_{x_\mu}(\xi_0) > \max\{\alpha_0, \bigcup\{M_{x_\nu}(\eta) : \eta < \rho \text{ and } \nu < \mu\}\}$$

Let $M_2 = \{x_\mu : \mu < \theta\}$. Then $|M_2| = \theta$ and $M_x \cap M_y \subseteq \alpha_0$ whenever $x = (M_x, R_x)$ and $y = (M_y, R_y)$ are distinct objects in M_2. Since for each $\alpha < \theta$,

$$\left|\alpha^{<\kappa}\right| < \theta$$

there exists an $r \subset \alpha_0$ and a $B \subset M_2$ with $|B| = \theta$ and for each $x \in B$, $M_x \cap \alpha_0 = r$, $S_2 = (B, R_B)$ forms a centralized system, where R_B is the restriction of the relation set R on B. ■

Corollary 9.1.1. *If S is a system with uncountable (or nondenumerable) object set, if each object in S is a system with a finite object set, and if there exists an object contained in at least $\aleph_1$ objects in S, there exists a partial system S^* of S with an uncountable object set and S^* forms a centralized system.*

Proof: It suffices from Theorem 9.1.6 to show that $\forall\alpha < \aleph_1(\left|\alpha^{<\aleph_0}\right| < \aleph_1)$. This is clear because for each $\alpha < \aleph_1$, $|\alpha| \leq \aleph_0$, so

$$\left|\alpha^{<\aleph_0}\right| = \aleph_0 < \aleph_1 \qquad (9.2)$$

■

A system $S = (M,R)$ is strongly centralized if each object in S is a system and there is a nondiscrete system $C = (M_C,R_C)$ such that for any distinct elements x and $y \in M$, say $x = (M_x,R_x)$ and $y = (M_y,R_y)$, $M_C = M_x \cap M_y$ and $R_C = R_x|M_C \cap R_y|M_C$. The system C is called an S-center of S.

Question 9.1.2. Give conditions under which a given system has a partial system which is strongly centralized and has an object set of the same cardinality as that of the given system.

9.1.1. A Brief Historical Remark

Roughly speaking, the idea of systems appeared as long ago as Aristotle. For example, Aristotle's statement "the whole is greater than the sum of its parts" could be the first definition of a basic systems problem. Later, many great thinkers used the languages of their times to study certain systems problems. Nicholas of Cusa, for example, a profound thinker of the fifteenth century, linking medieval mysticism with the first beginnings of modern science, introduced the notion of the *coincidentia oppositorum*. Leibniz's hierarchy of monads looks quite like that of modern systems. Gustav Fechner, known as the author of the psychophysical law, elaborated in the way of the native philosophers of the nineteenth century supraindividual organizations of higher order than the usual objects of observation, thus romantically anticipating the ecosystems of modern parlance; for details, see (von Bertalanffy, 1972).

The concept of systems was not introduced formally until the 1920s. Von Bertalanffy began to study the concept of systems formally in biology. Since then, more and more scholars have studied the concept of systems and related topics. For example, Tarski (1954–1955) defined the concept of relational systems. Hall and Fagan (1956) described a system as a set of objects and some relations between the objects and their attributes. They did not define mathematical meanings for objects nor for attributes of the objects. In 1964 Mesarovic began to study the model of general systems in the language of set theory. His final model (Mesarovic and Takahara, 1975) reads: A system S is a relation on nonempty sets:

$$S \subseteq \prod\{V_i : i \in I\} \tag{9.3}$$

After considering the interrelationship between the systems under concern and some environments of systems, Bunge (1979) gave a model of systems as follows: Let T be a nonempty set. Then the ordered triple $W = (C,E,S)$ is a system over T if and only if C and E are mutually disjoint subsets of T and S is a nonempty set of relations on $C \cup E$; the sets C and E are called the composition and an environment of the system W, respectively. Klir (1985) introduced a philosophical concept of general systems. The concept of general systems discussed intensively in this chapter was first introduced by Lin and Ma (1987; 1987).

We conclude this section with three mathematical structures related to the concept of general systems: structure, L-fuzzy system, and G-system.

9.1.2. Structures

Let A be a set and n a nonnegative integer. An n-ary operation on A is a mapping f from A^n into A. An n-ary relation r on the set A is a subset of A^n. A type τ of structures is an ordered pair

$$((n_0, \ldots, n_\nu, \ldots)_{\nu < 0_{0(\tau)}}, (m_0, \ldots, m_\nu, \ldots)_{\nu < 0_{1(\tau)}}) \tag{9.4}$$

where $0_0(\tau)$ and $0_1(\tau)$ are fixed ordinals and n_ν and m_ν are nonnegative integers. For every $\nu < 0_0(\tau)$ there exists a symbol f_ν of an n_ν-ary operation, and for every $\nu < 0_1(\tau)$ there exists a symbol r_ν of an m_ν-ary relation.

A structure U (Gratzer, 1978) is a triplet (A, F, R), where A is a nonempty set. For every $\nu < 0_0(\tau)$ we realize f_ν as an n_ν-ary operation $(f_\nu)_U$ on A, for every $\nu < 0_1(\tau)$ we realize r_ν as an m_ν-ary relation $(r_\nu)_U$ on A, and

$$F = \{(f_0)_U, \ldots, (f_\nu)_U, \ldots\}, \quad \nu < 0_0(\tau) \tag{9.5}$$

$$R = \{(r_0)_U, \ldots, (r_\nu)_U, \ldots\}, \quad \nu < 0_1(\tau) \tag{9.6}$$

If $0_1(\tau) = 0$, U is called an algebra and if $0_0(\tau) = 0$, U is called a relational system.

It can be seen that the concept of algebras is a generalization of the concepts of rings, groups, and the like. Any n-ary operation is an $(n+1)$-ary relation. Therefore, the concept of relational systems generalizes the idea of algebras, and the concept of structures combines those of algebras and relational systems. The fact that any topological space is a relational system shows that the research of relational systems will lead to the discovery of properties that topologies and algebras have in common.

Example 9.1.1. We show that any topological space is a relational system, thus a general system. Let (X, T) be a topological space [for details see (Engelking, 1975)]. Then for each open set $U \in T$, there exists an ordinal number $n = n(U) = 1$ such that $U \in X^n$. Therefore, (X, T) is also a system with object set X, and relation set T, and the length of each relation in T is 1. Now, applying the Axiom of Choice implies that the topology T can be well-ordered as

$$T = \{U_0, U_1, \ldots, U_\nu, \ldots\}, \quad \nu < |T| \tag{9.7}$$

where each open set is a 1-ary relation on X. So $(X, \emptyset, T)$ becomes a structure, which is a relational system.

A systematic introduction to the study of structures can be found in (Gratzer, 1978). Many great mathematicians worked in the field, including Whitehead, Birkhoff, Chang, Henkin, Jonsson, Keisler, Tarski, etc.

9.1.3. L-Fuzzy Systems

The concept of fuzzy systems (Lin, 1990a) is based on the concept of fuzzy sets introduced in 1965 by Zadeh. An ordered set $(L, \leq)$ is called a lattice if for any elements $a, b \in L$, there exists exactly one greatest lower bound of a and b, denoted by $a \wedge b$, and there exists exactly one least upper bound of a and b, denoted by $a \vee b$, such that

$$a \vee a = a, \quad a \wedge a = a, \quad a \vee b = b \vee a, \quad a \wedge b = b \wedge a$$
$$a \vee (b \vee c) = (a \vee b) \vee c, \quad a \wedge (b \wedge c) = (a \wedge b) \wedge c$$
$$a \wedge (a \vee b) = a, \quad a \vee (a \wedge b) = a$$

We call a lattice distributive if the lattice, in addition, satisfies

$$a \wedge (b \vee c) = (a \wedge b) \vee (a \wedge c)$$
$$a \vee (b \wedge c) = (a \vee b) \wedge (a \vee c)$$

We call a lattice complete if for any nonempty subset $A \subseteq L$, there exists exactly one least upper bound of the elements in A, denoted by $\vee A$ or $\sup A$, and there exists exactly one greatest lower bound of the elements in A, denoted by $\wedge A$ or $\inf A$.

We call a lattice completely distributive if for any $a \in L$ and any subset $\{b_i : i \in I\} \subseteq L$, where I is an index set, $\vee\{b_i : i \in I\}$ and $\wedge\{b_i : i \in I\}$ exist and satisfy

$$a \vee (\wedge\{b_i : i \in I\}) = \wedge\{a \vee b_i : i \in I\} \tag{9.8}$$

and

$$a \wedge (\vee\{b_i : i \in I\}) = \vee\{a \wedge b_i : i \in I\} \tag{9.9}$$

We say that L is a lattice with order-reversing involution if there exists an operator $'$ defined on L such that for any $a, b \in L$, if $a \leq b$, then $a' \geq b'$.

In the following, a completely distributive lattice L with order-reversing involution is fixed; assume that 0 and 1 are the minimum and maximum elements in L.

Let X be a set. A is an L-fuzzy set on X iff there exists a mapping $\mu_A \in L^X$ such that $A = \{(x, \mu_A(x)) : x \in X\}$. Without confusion, we consider that A is the mapping μ_A in L^X. If μ_A, or say A, has at most two values, 0 and 1, then the L-fuzzy set A can be seen as a subset of X.

Suppose that A and B are two L-fuzzy sets on X. Then $A = B$ iff $\mu_A(x) = \mu_B(x)$ for all $x \in X$. $A \subset B$ iff $\mu_A(x) \leq \mu_B(x)$, for all $x \in X$. If $Y \subset X$, then $A|Y = \{(x, \mu_A(x)) : x \in Y\}$.

Let $\{A_i : i \in I\}$ be a set of L-fuzzy sets on X. Then $C = \bigcup_{i \in I} A_i$ and $D = \bigcap_{i \in I} A_i$ are defined by

$$\mu_C(x) = \sup\{\mu_{A_i}(x) : i \in I\} \tag{9.10}$$

and

$$\mu_D(x) = \inf\{\mu_{A_i}(x) : i \in I\} \tag{9.11}$$

for all $x \in X$.

S is an L-fuzzy system iff S is an ordered pair (M,R) of sets, where R is a set of some L-fuzzy relations defined on the set M; i.e., for any L-fuzzy relation $r \in R$, there exists an ordinal number $n = n(r)$ such that r is an L-fuzzy set on the Cartesian product M^n. The ordinal n is called the length of the L-fuzzy relation r.

Theorem 9.1.7. *Suppose that an L-fuzzy system $S = (M,R)$ is such that, for any $r \in R$, $n(r) = 2$. Then*

$$S \subseteq p(M \cup p^3(p^2(M) \cup L)) \tag{9.12}$$

Proof: $S = (M,R) = \{\{M\},\{M,R\}\} \subseteq p(M \cup R)$, and for any $(x,y,i) \in r \subseteq M^2 \times L$,

$$\begin{gathered}
(x,y,i) = ((x,y),i) = \{\{(x,y)\}, \\
\{(x,y),i\}\} = \{\{\{\{x\},\{x,y\}\}\},\{i,\{\{x\},\{x,y\}\}\}\} \\
\{\{\{\{x\},\{x,y\}\}\},\{i,\{\{x\},\{x,y\}\}\}\} \in p(p^3(M) \cup p^2(M \cup L)) \\
p(p^3(M) \cup p^2(M \cup L)) = p(p(p^2(M \cup L))) = p^2(p^2(M) \cup L)
\end{gathered}$$

so $r \in p^3(p^2(M) \cup L)$ and $R \subseteq p^3(p^2(M) \cup L)$. Thus, the inclusion in Eq. (9.12) holds. ■

Suppose that $S_i = (M_i, R_i)$, $i = 1,2$, are L-fuzzy systems. The L-fuzzy system S_1 is a subsystem of the L-fuzzy system S_2 if $M_1 \subseteq M_2$ and for any L-fuzzy relation $r_1 \in R_1$, there exists an L-fuzzy relation $r_2 \in R_2$ such that $r_1 \subseteq r_2|M_1$.

Example 9.1.2. An example is constructed to show that the equation $S_1 = S_2$ will not follow from the condition that the L-fuzzy systems S_1 and S_2 are subsystems of each other.

Suppose that the completely distributive lattice with order-reversing involution lattice L is the closed interval with the usual order relation. Let $\mathbb{N} = \{1,2,\dots\}$ be the set of all natural numbers, and $\{a_i : i \in \mathbb{N}\}$ a sequence defined by $a_i = 1 - 1/i$. We now define two L-fuzzy systems, $S_1 = (\mathbb{N}, R_1)$ and $S_2 = (\mathbb{N}, R_2)$ as follows: $R_1 = \{r_{11}, r_{12}, \ldots, r_{1i}, \ldots\}$ and $R_2 = \{r_{20}, r_{21}, \ldots, r_{2i}, \ldots\}$, where

$$r_{1i}(x) = a_i \quad \text{for each } x \in \mathbb{N} \tag{9.13}$$

and for any $i, x \in \mathbb{N}$

$$r_{2i}(x) = \begin{cases} a_{i+1} & \text{if } x = 1 \\ a_i & \text{if } x \neq 1 \end{cases} \tag{9.14}$$

and

$$r_{20}(x) = 0 \quad \text{for all } x \in \mathbb{N} \tag{9.15}$$

Then it is easy to show that S_1 and S_2 are subsystems of each other, but $S_1 \neq S_2$.

9.1.4. G-Systems

In this part any set A will be seen as a partially ordered set $(A, \leq_A)$, where the ordering is defined by $\leq_A = \{(x,x) : x \in A\}$.

S is a G-system (Lin and Ma, 1989) of type $K = \{K_i\}_{i \in I}$, where I is an index set and, for each $i \in I$, K_i is a set partially ordered by $\leq_i$ iff S is an ordered pair of sets $(M, \{r_i\}_{i \in I})$ such that for any $i \in I$, $r_i \subseteq M^{K_i}$. The set K_i is called the type of the relation r_i, and the set K is called the type of the system S.

The concept of G-systems can be used to study more applied problems because it contains a more general meaning of relations. The interested reader is encouraged to consult (Lin and Ma, 1989).

9.2. Mappings from Systems into Systems

Let $S_i = (M_i, R_i)$, $i = 1,2$, be two systems and $h : M_1 \to M_2$ a mapping. By transfinite induction, two classes $\widehat{M}_i$, $i = 1,2$, and a class mapping $\widehat{h} : \widehat{M}_1 \to \widehat{M}_2$ can be defined with the following properties:

$$\widehat{M}_i = \bigcup_{n \in \mathbf{ON}} M_i^n \quad i = 1,2 \tag{9.16}$$

and for each $x = (x_0, x_1, \ldots, x_\alpha, \ldots) \in \widehat{M}_1$,

$$\widehat{h}(x) = (h(x_0), h(x_1), \ldots, h(x_\alpha), \ldots) \tag{9.17}$$

For each relation $r \in R_1, \widehat{h}(r) = \{\widehat{h}(x) : x \in r\}$ is a relation on M_2 with length $n(r)$. Without confusion, h will also be used to indicate the class mapping $\widehat{h}$, and h is a mapping from S_1 into S_2, denoted $h : S_1 \to S_2$. When $h : M_1 \to M_2$ is surjective, injective, or bijective, $h : S_1 \to S_2$ is also called surjective, injective, or bijective, respectively. Let X be a subsystem of S_1. Then $h|X$ indicates the restriction of the mapping on X. Without confusion, we use h for $h|X$.

The systems S_i are similar, if there exists a bijection $h : S_1 \to S_2$ such that $h(R_1) = \{h(r) : r \in R_1\} = R_2$. The mapping h is a similarity mapping from S_1 onto S_2. Clearly, the inverse mapping h^{-1} is a similarity mapping from S_2 onto S_1.A mapping $h : S_1 \to S_2$ is a homomorphism from S_1 into S_2 if $h(R_1) \subseteq R_2$.

Let E be an equivalence relation on the object set M of a system $S = (M, R)$. The quotient system S/E is defined

$$S/E = (M/E, R/E) \tag{9.18}$$

where $\mu : M \to M/E$ is the canonical mapping defined by $\mu(m) = [m]$. Then $R/E = \{\mu(r) : r \in R\}$.

Theorem 9.2.1. *Suppose that $h : S_1 \to S_2$ is a homomorphism from S_1 into S_2. Then $\overline{h} : S_1/E_h \to h(S_1) = (h(M_1), h(R_1))$ is a similarity mapping, where $E_h = \{(x,y) : h(x) = h(y)\}$ and $\overline{h}([m]) = h(m)$, for each $[m] \in M_1/E_h$.*

Proof: It is easy to check that E_h is an equivalence relation on the object set M_1. We next show that $\overline{h} : M_1/E_h \to h(M_1)$ is well defined. Let $[m_1], [m_2] \in M_1/E_h$ be such that $[m_1] = [m_2]$, then $m_2 \in [m_1]$, and so $h(m_2) = h(m_1)$. This implies that $\overline{h}([m_1]) = \overline{h}([m_2])$.

Finally, we show that $\overline{h}$ is a similarity mapping; $\overline{h} : M_1/E_h \to h(M_1)$ is a bijection. In fact, let $[m_1]$ and $[m_2] \in M_1/E_h$ be such that $[m_1] \neq [m_2]$ and $\overline{h}([m_1]) = \overline{h}([m_2])$. Therefore, $h(m_1) = h(m_2)$. This implies that $(m_1, m_2) \in E_h$, and so $[m_1] = [m_2]$, contradiction.

For any relation $r \in R_1/E_h$, there exists a relation $s \in R_1$ such that $r = \mu(s)$, where μ is the canonical mapping from M_1 into M_1/E_h. Then

$$\overline{h}(r) = \overline{h}(\mu(s)) = h(s) \tag{9.19}$$

Thus, $\overline{h}(r) \in h(R_1)$.

For any relation $r \in h(R_1)$, there exists a relation $s \in R_1$ such that $r = h(s)$. Therefore, $\overline{h}^{-1}(r) = \mu(s) \in R_1/E_h$. This completes the proof that $\overline{h}$ is a similarity mapping from S_1/E_h onto $h(S_1)$. ■

Let S_0 and A be two systems with no nonsystem kth-level objects for each $k < n$, where n is a fixed natural number. A mapping $h_{S_0} : S_0 \to A$ is an n-level homomorphism from S_0 into A if the following inductive conditions hold, and the system S_0 is n-level homomorphic to A.

(1) For each object S_1 in S_0, there exists a homomorphism h_{S_1} from the object system S_1 into the object system $h_{S_0}(S_1)$.

(2) For each $i < n$ and each ith-level object S_i of S_0, there exist level object systems S_k, $k = 0, 1, \ldots, i-1$, and homomorphisms $h_{S_k}, k = 1, 2, \ldots, i$, such that S_k is an object of the object system S_{k-1} and h_{S_k} is a homomorphism from S_k into $h_{S_{k-1}}(S_k)$, for $k = 1, 2, \ldots, i$.

When all homomorphisms in this definition are similarity mappings, the mapping h_{S_0} is termed as an n-level similarity, and the system S_0 is n-level similar to A.

Proposition 9.2.1. *Suppose that A, B, and C are systems.*

(a) If A is n-level homomorphic to B and B is m-level homomorphic to C, then A is $\min\{n,m\}$-level homomorphic to C.

(b) If A is n-level similar to B and B is m-level similar to C, then A is $\min\{n,m\}$-level similar to C.

The proof is straightforward and is omitted.

A system S is centralizable if it is 1-level homomorphic to a centralized system S_C under a homomorphism $h: S \to S_C$ such that for each object m in S, the object system m and $h(m)$ are similar. Each center of S_C is also called a center of S.

Theorem 9.2.2. *Let $S = (M, R)$ be a system such that each object in M is a system. Then S is centralizable, iff there exists a nontrivial system $C = (M_C, R_C)$ such that C is embeddable in each object of S. (C embeddable in an object S_1 in S means that C is similar to a partial system of S_1, and each similarity mapping here is called an embedding mapping from C into S_1.)*

Proof: Necessity. If the system $S = (M, R)$ is centralizable, there exists a centralized system $S^* = (M^*, R^*)$ such that S is 1-level homomorphic to S^*. Let C be a center of S^*. Then C is embeddable in each object of S.

Sufficiency. Without loss of generality, we can assume that the collection of all object sets of the objects in S consists of disjoint sets. For each $x = (M_x, R_x) \in M$, the system C is embeddable in x, so there exists a one-to-one mapping $h_x: M_C \to M_x$ such that $h_x(C) = (h_x(M_C), h_x(R_C))$ is a partial system of x. Define a new system $x^* = ({}^*M_x, {}^*R_x)$ as follows:

$$ {}^*M_x = M_C \cup (M_x - h_x(M_C)) \tag{9.20}$$

and let $g_x: M_x \to {}^*M_x$ be the bijection defined by

$$ g_x(m) = \begin{cases} m & \text{if } m \in M_x - h_x(M_C) \\ h_x^{-1}(m) & \text{if } m \in h_x(M_C) \end{cases} \tag{9.21}$$

Then define

$$^{*}R_x = \{g_x(r) : r \in R_x\} \tag{9.22}$$

So C is a partial system of x^*. Let h^* be a mapping defined by letting $h^*(x) = x^*$ for each $x \in M$. Then the system $h^*(S)$ is centralized with the system C as a center. Therefore, S is a centralizable system. ■

Example 9.2.1. An example is constructed to show that the hypothesis in Theorem 9.1.6 that there exists an object belonging to at least θ objects in S is essential for the result to follow and that not every centralizable system is a centralized system. Consequently, the concepts of centralized systems and centralizable systems are different.

Let ω_1 be the first uncountable ordinal number, and M a collection of n-element subsets of ω_1, for some natural number $n > 0$, such that $|M| = \omega_1$ and for any $x, y \in M$, $x \cap y = \emptyset$ if $x \neq y$.

We now consider a system S with object set M^* such that for any $x^* \in M^*$, x^* is a system with some element in M as its object set, and for any $x \in M$ there exists exactly one $x^* \in M^*$ such that x is the object set of the system x^*. It is clear that S has no partial system which forms a centralized system and that S is a centralizable system.

Theorem 9.2.3. *Let κ and θ be cardinalities satisfying the conditions in Theorem 9.1.6. Assume that S is a system with an object set of cardinality $\geq \theta$ and each object in S is a system with an object set of cardinality $< \kappa$. There then exists a partial system S' of S such that the object set of S' is of cardinality $\geq \theta$ and S' forms a centralizable system.*

The proof follows from Theorems 9.1.6 and 9.2.2.

A mapping h from a system $S_1 = (M_1, R_1)$ into a system $S_2 = (M_2, R_2)$ is S-continuous if for any relation $r \in R_2$,

$$h^{-1}(r) = \{h^{-1}(x) : x \in r\} \in R_1 \tag{9.23}$$

Proposition 9.2.2. *Suppose that $h : S_1 \to S_2$ is a mapping. Then*

(i) If the systems S_i, $i = 1, 2$, are topological spaces, h is S-continuous, iff it is continuous from S_1 into S_2.

(ii) If the systems S_i, $i = 1, 2$, are groups, the S-continuity of h implies that h is a homomorphism from S_1 into S_2.

(iii) If the systems S_i, $i = 1, 2$, are rings, the S-continuity of h implies that h is a homomorphism from S_1 into S_2.

Proof: (i) follows from the definition of S-continuity of mappings and Example 9.1.1.

For (ii), let $(X,+)$ be a group. A system (X,R) is defined as follows: $R=\{r_+\}$ and $r_+=\{(x,y,z)\in X^3 : x+y=z\}$. The system (X,R) will be identified with the group $(X,+)$, and the relation r_+ with $+$.

Let $S_i=(M_i,\{r_{+_i}\})$, $i=1,2$, be two groups and $h: S_1\to S_2$ an S-continuous mapping. Then $h^{-1}(r_{+_2})=r_{+_1}$. Thus, for any x, y, and $z\in M_1$, if $x+_1 y=z$, then $h(x)+_2 h(y)=h(z)$. This implies that h is a homomorphism from S_1 into S_2.

The proof of (iii) is a modification of that of (ii) and is omitted. ∎

Theorem 9.2.4. *A bijection $h: S_1\to S_2$ is a similarity mapping iff $h: S_1\to S_2$ and $h^{-1}: S_2\to S_1$ are S-continuous.*

The proof is straightforward and is left to the reader.

Theorem 9.2.5. *Suppose that $f: S_1\to S_2$ and $g: S_2\to S_3$ are S-continuous mappings. Then the composition $g\circ f: S_1\to S_3$ is also S-continuous.*

Proof: For each relation r in the system S_3, from the hypothesis that the mapping g is S-continuous we have

$$g^{-1}(r)\in R_2 \tag{9.24}$$

From the hypothesis that the mapping f is S-continuous, it follows that

$$f^{-1}(g^{-1}(r))\in R_1 \tag{9.25}$$

Therefore, from Eq. (9.25),

$$(g\circ f)^{-1}(r)=f^{-1}\circ g^{-1}(r)=f^{-1}(g^{-1}(r))\in R_1$$

Thus, $g\circ f$ is S-continuous. ∎

9.3. Constructions of Systems

Let $\{S_i=(M_i,R_i): i\in I\}$ be a set of systems such that $M_i\cap M_j=\emptyset$ for any distinct i and $j\in I$. The free sum of the systems S_i, denoted $\oplus\{S_i: i\in I\}$ or $S_1\oplus S_2\oplus\cdots\oplus S_n$, if $I=\{1,2,\ldots,n\}$ is finite, is defined to be the system

$$\oplus\{S_i: i\in I\}=(\bigcup\{M_i: i\in I\},\bigcup\{R_i: i\in I\}) \tag{9.26}$$

The concept of free sums of systems can be defined for any set $\{S_i : i \in I\}$ of systems such that the collection of all object sets of the systems in the set consists of not necessarily pairwise disjoint sets. To show this, we simply take a set $\{{}^*S_i : i \in I\}$ of systems with pairwise disjoint object sets such that ${}^*S_i$ is similar to S_i, for each $i \in I$, and define

$$\oplus\{S_i : i \in I\} = \oplus\{{}^*S_i : i \in I\} \tag{9.27}$$

Up to a similarity, the free sum of the systems S_i is unique. Therefore, it can be assumed that any set of systems has a free sum (uniquely determined up to a similarity), but in the proofs of some theorems in the future it will be tacitly assumed that the set under consideration consists of systems with pairwise disjoint object sets.

A system $S = (M, R)$ is an input–output system if two sets X and Y are given such that $X \cup Y = M$ and for each relation $r \in R$ there are nonzero ordinals $n = n(r)$ and $m = m(r)$ such that $r \subseteq X^n \times Y^m$. The sets X and Y are called the input space and the output space of the system S, respectively.

Proposition 9.3.1. *Let S_1 and S_2 be two similar systems. If S_1 is a free sum of some systems, then so is S_2.*

Proof: Suppose that $S_1 = \oplus\{S^i : i \in I\}$, where $\{S^i : i \in I\}$ is a set of systems $S^i = (M^i, R^i)$, and $h : S_1 \to S_2$ is a similarity mapping. Then

$$S_2 = \oplus\{h(S^i) : i \in I\} \tag{9.28}$$

because $h(R^i) \subseteq R_2$, for each $i \in I$, and for each $r \in R_2$ there exists a relation $s \in R^i$, for some $i \in I$, such that $h(s) = r$. ■

Theorem 9.3.1. *Suppose that $\{S_i = (M_i, R_i) : i \in I\}$ is a set of systems, where I is an index set. Then the free sum $\oplus\{S_i : i \in I\}$ is an input–output system iff each system $S_i = (M_i, R_i)$ is an input–output system, $i \in I$.*

Proof: Necessity. Suppose that $\oplus\{S_i : i \in I\}$ is an input–output system with input space X and output space Y. Then for any fixed $i \in I$ and any relation $r \in R_i$, there are nonzero ordinals $n = n(r)$ and $m = m(r)$ such that $r \subseteq X^n \times Y^m$. Hence, S_i is an input–output system with input space $M_i \cap X$ and output space $M_i \cap Y$.

Sufficiency. Suppose that $S_i = (M_i, R_i)$ is an input–output system with input space X_i and output space Y_i. Then for each relation r in $\oplus\{S_i : i \in I\}$, there exists an $i \in I$ such that $r \in R_i$; therefore,

$$\begin{aligned} r &\subseteq X_i^{n(r)} \times Y_i^{m(r)} \\ &\subseteq (\bigcup_{i \in I} X_i)^{n(r)} \times (\bigcup_{i \in I} Y_i)^{m(r)} \end{aligned} \tag{9.29}$$

Hence, the free sum $\oplus\{S_i : i \in I\}$ is an input–output system with input space $\bigcup_{i\in I} X_i$ and the output space $\bigcup_{i\in I} Y_i$. ■

Theorem 9.3.2. *If a system S is embeddable in the free sum $\oplus\{S_i : i \in I\}$ of some systems S_i, then S is also a free sum of some systems.*

The proof is straightforward and is omitted.

Theorem 9.3.3. *If the free sum $\oplus\{S_i : i \in I\}$ of nontrivial systems S_i is a centralized system, then each S_i is centralized, for $i \in I$.*

Proof: The theorem follows from the fact that each center of the free sum $\oplus\{S_i : i \in I\}$ is a center of each system S_i, $i \in I$. ■

Theorem 9.3.4. *Suppose that $\{S_i : i \in I\}$ is a set of centralizable systems such that each system S_i has a center $C_i = (M_{C_i}, R_{C_i})$, $i \in I$, and that $|M_{C_i}| = |M_{C_j}|$, for all i and $j \in I$. Then the free sum $\oplus\{S_i : i \in I\}$ is a centralizable system.*

Proof: Let $\aleph = |M_{C_i}|$ for some $i \in I$. Then the system $(\aleph, \emptyset)$ is embeddable in each object in the free sum $\oplus\{S_i : i \in I\}$. Theorem 9.2.2 implies that the free sum is centralizable. ■

It is easy to construct an example to show that the free sum of two centralized systems may not be a centralizable system.

Theorem 9.3.5. *Let $h : S \to \oplus\{S_i : i \in I\}$ be a homomorphism. Then the system S is also a free sum of some systems.*

Proof: Suppose that $S = (M, R)$ and $S_i = (M_i, R_i)$ for each $i \in I$. Then

$$M = \bigcup\{h^{-1}(M_i) : i \in I\} \tag{9.30}$$

Define

$$^*R_i = \{r \in R : h(r) \in R_i\} \tag{9.31}$$

From the hypothesis that h is a homomorphism it follows that

$$R = \bigcup\{^*R_i : i \in I\} \tag{9.32}$$

Therefore, $S = \oplus\{(h^{-1}(M_i), {}^*R_i) : i \in I\}$. ■

Theorem 9.3.6. *Let $f : \oplus\{S_i : i \in I\} \to S$ be an onto S-continuous mapping. Then there are partial systems $\{{}^*S_i : i \in I\}$ of S such that*

$$S = (\bigcup\{{}^*M_i : i \in I\}, \bigcup\{{}^*R_i : i \in I\}) \tag{9.33}$$

where ${}^*S_i = ({}^*M_i, {}^*R_i)$, for each $i \in I$, and $f : S_i \to {}^*S_i$ is S-continuous.

Proof: For each $i \in I$ we define a system ${}^*S_i = ({}^*M_i, {}^*R_i)$ as follows:

$${}^*M_i = f(M_i) \tag{9.34}$$

and

$${}^*R_i = \{r \in R : f^{-1}(r) \in R_i\} \tag{9.35}$$

From the hypothesis that f is surjective, it follows that $M = \bigcup\{{}^*M_i : i \in I\}$. Since f is S-continuous, $R = \bigcup\{{}^*R_i : i \in I\}$. Then Eq. (9.33) follows and the S-continuity of the mapping $f : S_i \to {}^*S_i$ follows from Eq. (9.35). ∎

Question 9.3.1. Under what conditions will the system S in Theorem 9.3.6 be a free sum of some systems?

Theorem 9.3.7. *Suppose that $h : S_1 \to S_2$ is a homomorphism from a system S_1 into an input–output system S_2. Then S_1 is also an input–output system.*

Proof: Suppose that S_2 has input space X and output space Y. Define

$$X^* = h^{-1}(X) \quad \text{and} \quad Y^* = h^{-1}(Y) \tag{9.36}$$

Then S_1 is an input–output system with input space X^* and output space Y^*. In fact, let r be a relation in S_1. Then $h(r) \in R_2$. Hence, there are nonzero ordinals n and m such that $h(r) \subseteq X^n \times Y^m$. This implies that $r \subseteq (X^*)^n \times (Y^*)^m$. ∎

Theorem 9.3.8. *Let $f : S_1 \to S_2$ be an S-continuous mapping from an input–output system S_1 into a system S_2. Then S_2 is also an input–output system.*

Proof: Suppose that S_1 has input space X_1 and output space Y_1. Define

$$X_2 = f(X_1) \quad \text{and} \quad Y_2 = f(Y_1) \tag{9.37}$$

Then S_2 is an input–output system with input space X_2 and output space Y_2. In fact, for each relation $r \in R_2, f^{-1}(r) \in R_1$. Hence, there are nonzero ordinals n and m such that $f^{-1}(r) \subseteq X_1^n \times Y_1^m$. Therefore, $r \subseteq X_2^n \times Y_2^m$. ∎

Let S_1 and S_2 be two systems. We define

$$\mathrm{Hom}(S_1,S_2)=\{h:S_1\to S_2: h \text{ is a homomorphism}\} \tag{9.38}$$

Suppose that $\{S_i : i\in I\}$ is a set of systems. A system $S=(M,R)$ is a product of the systems S_i if there exists a family of homomorphisms $\{\Phi_i\in \mathrm{Hom}(S,S_i): i\in I\}$ such that for each system S' and a family $\{\Psi_i\in\mathrm{Hom}(S',S_i): i\in I\}$, there exists a unique homomorphism $\lambda\in\mathrm{Hom}(S',S)$ so that

$$\Psi_i=\Phi_i\circ\lambda \tag{9.39}$$

for all $i\in I$.

Theorem 9.3.9. *For each set $\{S_i : i\in I\}$ of systems, a product of the systems S_i always exists, which is unique up to a similarity and denoted by $\prod_{i\in I}S_i$ or $\prod\{S_i : i\in I\}$.*

Proof: For the set of systems $\{S_i : i\in I\}$, we define a system $S=(M,R)$ as follows:

$$M=\prod_{i\in I}M_i \tag{9.40}$$

$$\widehat{R}=\{\widehat{r}=(r_i)_{i\in I}\in\prod_{i\in I}R_i : n(r_i)=n(r_j),\quad i,j\in I\} \tag{9.41}$$

(9.42)

and

$$R=\{r:\widehat{r}\in\widehat{R}\} \tag{9.43}$$

where for any $\widehat{r}=(r_i)_{i\in I}\in\widehat{R}$, let $n(\widehat{r})=n(r_i)$, for $i\in I$, and

$$r=\{x\in M^{n(\widehat{r})}:\exists(x_i)_{i\in I}\in\prod_{i\in I}r_i(\forall\alpha<n(\widehat{r})(x(\alpha)=(x_i(\alpha))_{i\in I}))\} \tag{9.44}$$

Let $p_i:\prod_{i\in I}M_i\to M_i$ be the projection for each $i\in I$. Then $p_i\in\mathrm{Hom}(S,S_i)$.

We now suppose that $S'=(M',R')$ is a system and $\{\Psi_i\in\mathrm{Hom}(S',S_i): i\in I\}$. Define a mapping $\lambda : S'\to S$ as follows: For any object $m'\in M'$, $\lambda(m')=\{\Psi(m'): i\in I\}\in M$. It then can be shown that $\lambda\in\mathrm{Hom}(S',S)$.

If λ' is another such homomorphism, then $\Psi_i(m')=p_i\circ\lambda'(m')=m_i$ if $\lambda'(m')=(m_i)_{i\in I}\in M$. Thus, $\lambda'=\lambda$ as mappings from the object set M' into the object set M. Therefore, they induce the same homomorphism. Therefore, S is a product of the systems S_i, $i\in I$. ■

Let $\{S_i : i\in I\}$ be a set of systems. A system $S=(M,R)$ is a Cartesian product of the systems S_i if there exists a family of S-continuous mappings $\{p_i :$

$S \to S_i : i \in I\}$ such that for each system S' and a family $\{q_i : S' \to S_i : i \in I\}$ of S-continuous mappings, there exists a unique S-continuous mapping $\lambda : S' \to S$ so that $q_i = p_i \circ \lambda$ for each $i \in I$.

Theorem 9.3.10. *For each set $\{S_i : i \in I\}$ of systems, a Cartesian product of the systems S_i always exists, which is unique up to a similarity and denoted by $\prod_{\mathrm{cp}}\{S_i : i \in I\}$.*

Proof: Let $S_i = (M_i, R_i)$ for each $i \in I$. Define a system $S = (M, R)$ as follows:

$$M = \prod\{M_i : i \in I\} \tag{9.45}$$

and

$$R = \{r : \exists j \in I \exists r_j \in R_j (r = \{(x_\alpha) \in M^{n(r_j)} : (x_\alpha(j))_{\alpha < n(r_j)} \in r_j\})\} \tag{9.46}$$

and define mappings $p_i : S \to S_i$ by letting $p_i((m_j)_{j \in I}) = m_i$ for any $(m_j)_{j\in I} \in M$, for each $i \in I$. Then it can be readily checked that each mapping $p_i : S \to S_i$ is S-continuous.

Let $S' = (M', R')$ be another system, and $\{q_i : S' \to S_i : i \in I\}$ a family of S-continuous mappings. Define a mapping $\lambda : S' \to S$ by

$$\lambda(m') = (q_i(m'))_{i \in I} \tag{9.47}$$

for each object $m' \in M'$. Then λ is S-continuous. In fact, for an arbitrary relation $r \in R$, let $r_j \in R_j$ be a relation such that

$$r = \{((m_{ik})_{i\in I})_{k<n(r_j)} \in M^{n(r_j)} : (m_{jk})_{k<n(r_j)} \in r_j\}$$

Then

$$\begin{aligned}
\lambda^{-1}(r) &= \bigcup\{\lambda^{-1}(m) : m \in r\} \\
&= \bigcup\{\lambda^{-1}[((m_{ik})_{i\in I})_{k<n(r_j)}] : n = ((m_{ik})_{i\in I})_{k<n(r_j)} \in r\} \\
&= \bigcup\{\prod\{\lambda^{-1}[(m_{ik})_{i\in I}] : k < n(r_j)\} : ((m_{ik})_{i\in I})_{k<n(r_j)} \in r\} \\
&= \bigcup\{\prod\{q_i^{-1}(m_{ik}) : i \in I\} : k < n(r_j)\} : ((m_{ik})_{i\in I})_{k<n(r_j)} \in r\} \\
&= \bigcup\{\prod\{q_j^{-1}(m_{jk}) : k < n(r_j)\} : ((m_{ik})_{i\in I})_{k<n(r_j)} \in r\} \\
&= \bigcup\{q_j^{-1}(m_j) : m_j = (m_{jk})_{k<n(r_j)} \in r_j\} \\
&= q_j^{-1}(r_j) \\
&\in R'
\end{aligned}$$

To complete the proof, we show that all Cartesian products of the systems S_i are similar to each other.

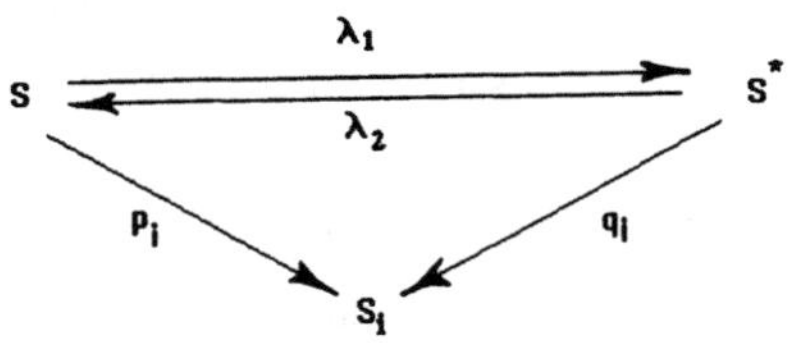

Figure 9.1. Uniqueness of the Cartesian product.

Let $S^* = (M^*, R^*)$ be another Cartesian product of S_i, $i \in I$. Then there are S-continuous mappings $\lambda_1 : S \to S^*$ and $\lambda_2 : S^* \to S$ such that the diagram (Fig. 9.1) commutes for each $i \in I$. That is, for each $i \in I$, we have

$$q_i = p_i \circ \lambda_2 \quad \text{and} \quad p_i = q_i \circ \lambda_1 \tag{9.48}$$

It suffices to show that

$$\lambda_2 \circ \lambda_1(m) = m \tag{9.49}$$

and

$$\lambda_1 \circ \lambda_2(m^*) = m^* \tag{9.50}$$

for each $m \in M$ and $m^* \in M^*$. In fact, Eq. (9.49) implies that the mapping λ_1 is 1–1, and Eq. (9.50) implies that λ_2 is 1–1; $\lambda_1 = \lambda_2^{-1}$ and $\lambda_2 = \lambda_1^{-1}$. Therefore, λ_1 and λ_2 are similarity mappings.

Now, we prove Eqs. (9.49) and (9.50). For Eq. (9.49), let $m = (m_i)_{i \in I} \in M$, then

$$\begin{aligned} p_i(\lambda_2(\lambda_1(m))) &= p_i \circ \lambda_2(\lambda_1(m)) \\ &= q_i \circ \lambda_1(m) \\ &= p_i(m) \\ &= m_i \end{aligned}$$

Therefore, $\lambda_2 \circ \lambda_1(m) = \lambda_2(\lambda_1(m)) = (m_i)_{i \in I}$. A similar argument shows Eq. (9.50). ■

The reason why the system S in Theorem 9.3.10 is called a Cartesian product of the systems S_i is that when S_i are topological spaces, S is the Cartesian product of the spaces S_i. The system $\prod\{S_i : i \in I\}$ in Theorem 9.3.9 is the box product of the spaces S_i. When the systems S_i are groups, $\prod\{S_i : i \in I\}$ in Theorem 9.3.9 is the direct sum of the groups S_i. The mappings p_i in the proofs of Theorems 9.3.9 and 9.3.10 are called projections from the product systems into the factor systems.

Let $\{S_i : i \in I\}$ be a set of systems such that each object in any one system S_i in the set is a system again. The 1-level product system, denoted by ${}^1\prod\{S_i : i \in I\}$, is defined by ${}^1\prod\{S_i : i \in I\} = (M,R)$, where the object set M and the relation set R are defined as follows: There exists a bijection $h : \prod\{M_i : i \in I\} \to M$ such that

$$h((m_i)_{i\in I}) = \prod\{m_i : i \in I\} \tag{9.51}$$

and

$$R = \{h(r) : r \text{ is a relation in } \prod\{S_i : i \in I\}\} \tag{9.52}$$

The 1-level Cartesian product, denoted by ${}^1\prod_{\text{cp}}\{S_i : i \in I\}$, is the system (M,R) defined by the following: There exists a bijection $h : \prod\{M_i : i \in I\} \to M$ such that

$$h((m_i)_{i\in I}) = \prod\nolimits_{\text{cp}}\{m_i : i \in I\} \tag{9.53}$$

and

$$R = \{h(r) : r \text{ is a relation in } \prod\nolimits_{\text{cp}}\{S_i : i \in I\}\} \tag{9.54}$$

Theorem 9.3.11. *Let $\{S_i : i \in I\}$ be a set of centralizable systems. Then the 1-level product system ${}^1\prod\{S_i : i \in I\}$ and the 1-level Cartesian product system ${}^1\prod_{\text{cp}}\{S_i : i \in I\}$ are also centralizable systems.*

Proof: For each $i \in I$ suppose that $C_i = (M_{C_i}, R_{C_i})$ is a center of the system S_i. For the 1-level product system ${}^1\prod\{S_i : i \in I\}$, it suffices by Theorem 9.2.2 to show that the product system $\prod\{C_i : i \in I\}$ is embeddable in each product $\prod\{m_i : i \in I\}$ for all $\{m_i\}_{i\in I} \in \prod\{M_i : i \in I\}$.

From Theorem 9.2.2 we will not lose generality by assuming that each system C_i is embeddable in each object in M_i for $i \in I$. For each $\{m_i\}_{i\in I} \in \prod\{M_i : i \in I\}$. Let $m_i = (M_{m_i}, R_{m_i})$ and $h_i : C_i \to m_i$ be an embedding mapping. Then the product mapping

$$\prod_{i\in I} h_i : \prod\{C_i : i \in I\} \to \prod\{m_i : i \in I\} \tag{9.55}$$

defined by

$$(\prod_{i\in I} h_i)((x_i)_{i\in I}) = (h_i(x_i))_{i\in I} \tag{9.56}$$

for each object $(x_i)_{i\in I}$ in $\prod\{C_i : i \in I\}$ is an embedding mapping. In fact, first, the product mapping $\prod_{i\in I} h_i$ is 1–1, because if $(x_i)_{i\in I}$ and $(y_i)_{i\in I}$ are two distinct objects in $\prod\{C_i : i \in I\}$, there exists an $i \in I$ such that $x_i \neq y_i$. Therefore, since h_i is an embedding mapping it follows that $h_i(x_i) \neq h_i(y_i)$. Therefore,

$$\begin{aligned}(\prod_{i\in I} h_i)((x_i)_{i\in I}) &= (h_i(x_i))_{i\in I} \\ &\neq (h_i(y_i))_{i\in I} \\ &= (\prod_{i\in I} h_i)((y_i)_{i\in I})\end{aligned} \tag{9.57}$$

Second, for each relation r in $\prod\{C_i : i \in I\}$, there exists a relation s in $\prod\{m_i : i \in I\}$ such that

$$(\prod_{i\in I} h_i)(r) = s|\prod\{h_i(M_{C_i}) : i \in I\} \tag{9.58}$$

In fact, for each relation r in $\prod\{C_i : i \in I\}$, there exist relations $r_i \in R_{C_i}$, for each $i \in I$, such that $n(r) = n(r_i)$ for all $i \in I$ and

$$\begin{aligned}r = \{x \in (\prod\{M_{C_i} : i \in I\})^{n(r)} :\\ \exists(x_i)_{i\in I} \in \prod_{i\in I} r_i(\forall\alpha < n(r)(x(\alpha) = (x_i(\alpha))_{i\in I}))\}\end{aligned} \tag{9.59}$$

From the hypothesis that each h_i is an embedding mapping, it follows that there exists a relation $s_i \in R_{m_i}$, for each $i \in I$, such that

$$h_i(r_i) = s_i|h_i(M_{C_i}) \tag{9.60}$$

Equation (9.60) implies that $n(s_i) = n(r)$ for each $i \in I$. Define a relation s in $\prod\{m_i : i \in I\}$ as follows:

$$\begin{aligned}s = \{x \in (\prod\{M_{m_i} : i \in I\})^{n(r)} :\\ \exists(x_i)_{i\in I} \in \prod_{i\in I} s_i(\forall\alpha < n(r)(x(\alpha) = (x_i(\alpha))_{i\in I}))\}\end{aligned} \tag{9.61}$$

We now show that s satisfies Eq. (9.58). In fact,

$$\begin{aligned}(\prod_{i\in I} h_i)(r) &= \{(\prod_{i\in I} h_i)(x) : x \in r\} \\ &= \{((\prod_{i\in I} h_i)(x_\alpha))_{\alpha<n(r)} : x = (x_\alpha)_{\alpha<n(r)} \in r\} \\ &= \{((h_i(x_{\alpha_i}))_{i\in I})_{\alpha<n(r)} : x = (x_\alpha)_{\alpha<n(r)} \in r \text{ and} \\ &\qquad \forall\alpha < n(r)(x_\alpha = (x_{\alpha_i})_{i\in I} \in \prod\{M_{C_i} : i \in I\})\} \\ &= \{((h_i(x_{\alpha_i}))_{i\in I})_{\alpha<n(r)} : \\ &\qquad \exists(x_i)_{i\in I} \in \prod_{i\in I} r_i(\forall\alpha < n(r)(x_{\alpha_i} = (x_i(\alpha))))\} \\ &= \{(y_\alpha)_{\alpha<n(r)} : \\ &\qquad \exists(y_i)_{i\in I} \in \prod_{i\in I} s_i|h_i(M_{C_i})(\forall\alpha < n(r)(y_\alpha = (y_i(\alpha))_{i\in I}))\}\end{aligned} \tag{9.62}$$

where $y_\alpha = (h_i(x_{\alpha_i}))_{i\in I}, y_i(\alpha) = h_i(x_{\alpha_i})$ for each $\alpha < n(r)$ and $i \in I$.

We complete the proof of the theorem by observing that

$$\{(y_\alpha)_{\alpha<n(r)} : \exists (y_i)_{i\in I} \in \prod_{i\in I} s_i|h_i(M_{C_i})(\forall \alpha < n(r)(y_\alpha = (y_i(\alpha))_{i\in I}))\} =$$
$$= s|\prod\{h_i(M_{C_i}) : i \in I\} \tag{9.63}$$

The centralizability of the 1-level Cartesian product system ${}^1\prod_{\text{cp}}\{S_i : i \in I\}$ can be proved in a similar way. ■

Theorem 9.3.12. *A mapping h from a system $S = (M,R)$ into a Cartesian product system $\prod_{\text{cp}}\{S_i : i \in I\}$ is S-continuous iff the composition $p_i \circ h$ is S-continuous for each $i \in I$.*

Proof: Necessity. Suppose that $h : S \to \prod_{\text{cp}}\{S_i : i \in I\}$ is S-continuous. From Theorem 9.2.5 and the proof of Theorem 9.3.10, it follows that each composition $p_i \circ h$ is S-continuous, $i \in I$.

Sufficiency. Suppose that each composition $p_i \circ h$ is S-continuous, $i \in I$. Let $S_i = (M_i, R_i)$ for each $i \in I$. For each relation r in $\prod_{\text{cp}}\{S_i : i \in I\}$ there exist an $i \in I$ and a relation $r_i \in R_i$ satisfying that

$$r = \{(x_\alpha)_{\alpha<n(r_i)} : x_\alpha \in \prod\{M_i : i \in I\} \text{ and } (x_\alpha(j))_{\alpha<n(r_i)} \in r_i\}$$

Therefore,

$$(p_i \circ h)^{-1}(r_i) = h^{-1}(r) \tag{9.64}$$

This implies that $h^{-1}(r) \in R$; i.e., h is S-continuous. ■

Theorem 9.3.13. *A mapping h from a system $S = (M,R)$ into a product system $\prod\{S_i : i \in I\}$ is a homomorphism iff the composition $p_i \circ h$ is a homomorphism for every $i \in I$.*

Proof: Necessity. Suppose that $h : S \to \prod\{S_i : i \in I\}$ is a homomorphism. From the fact that each projection $p_i : \prod\{S_i : i \in I\} \to S_i$ is a homomorphism, it follows that $p_i \circ h$ is a homomorphism for every $i \in I$.

Sufficiency. Suppose that $p_i \circ h$ is a homomorphism for every $i \in I$. Let $r \in R$. We need to show that $h(r)$ is a relation in $\prod\{S_i : i \in I\}$. For each $i \in I$ we have $p_i \circ h(r) \in R_i$ since $p_i \circ h$ is a homomorphism. Therefore,

$$h(r) = \bigcap_{i\in I} p_i^{-1}(p_i \circ h(r)) \tag{9.65}$$

and for each $i \in I$ and $s \in R_i$,

$$p_i^{-1}(s) = \{x \in (\prod\{M_i : i \in I\})^{n(s)} : p_i(x) \in s\}$$
$$= \{(x_\alpha)_{\alpha<n(s)} \in (\prod\{M_i : i \in I\})^{n(s)} : (x_\alpha(i))_{\alpha<n(s)} \in s\} \tag{9.66}$$

Equation (9.66) implies that for every $s_i \in R_i$ with $n(s) = n(s_i)$,

$$\bigcap_{i \in I} p_i^{-1}(s_i) = \{(x_\alpha)_{\alpha<n(s)} \in (\prod\{M_i : i \in I\})^{n(s)} :$$
$$\forall i \in I((x_\alpha(i))_{\alpha<n(s)} \in s_i)\} \tag{9.67}$$

So Eq. (9.65) implies that

$$h(r) = \{x \in (\prod\{M_i : i \in I\})^{n(r)} :$$
$$\forall i \in I((x_\alpha(i))_{\alpha<n(s)} \in p_i \circ h(r))\} \tag{9.68}$$

This implies that the relation $h(r)$ belongs to the relation set of $\prod\{S_i : i \in I\}$. ■

9.4. Structures of Systems

Let $S = (M,R)$ be a system and $r \in R$ a relation. The support of r, denoted $\mathrm{Supp}(r)$, is defined by

$$\mathrm{Supp}(r) = \{m \in M : \exists x \in r \exists \beta < n(r)(x(\beta) = m)\} \tag{9.69}$$

The system S is connected if it cannot be represented in the form $S_1 \oplus S_2$, where S_1 and S_2 are nontrivial subsystems of S.

Theorem 9.4.1. *For every system $S = (M,R)$ the following conditions are equivalent:*

(i) The system S is connected.

(ii) For any two objects x and $y \in M$, there exist a natural number $n > 0$ and n relations $r_i \in R$ such that

$$x \in \mathrm{Supp}(r_1) \quad \textit{and} \quad y \in \mathrm{Supp}(r_n) \tag{9.70}$$

and

$$\mathrm{Supp}(r_i) \cap \mathrm{Supp}(r_{i+1}) \neq \emptyset \tag{9.71}$$

for each $i = 1, 2, \ldots, n-1$.

Proof: (i) $\rightarrow$ (ii). We prove by contradiction. Suppose that the system S is connected and there exist two objects x and $y \in M$ such that there do not exist relations $r_i \in R$, $i = 1,2,\ldots,n$, for any natural number $n \geq 1$, such that

$$x \in \text{Supp}(r_1) \quad \text{and} \quad y \in \text{Supp}(r_n) \tag{9.72}$$

and

$$\text{Supp}(r_i) \cap \text{Supp}(r_{i+1}) \neq \emptyset \tag{9.73}$$

for each $i = 1,2,\ldots,n-1$. From the hypothesis that S is connected, it follows that there must be relations $r_1, s_1 \in R$ such that $x \in \text{Supp}(r_1)$ and $y \in \text{Supp}(s_1)$. Then our hypothesis implies that

$$\text{Supp}(r_1) \cap \text{Supp}(s_1) = \emptyset \tag{9.74}$$

Let $U_0 = \text{Supp}(r_1)$ and $V_0 = \text{Supp}(s_1)$, and for each natural number $n \in \mathbb{N}$ let

$$U_n = \bigcup\{\text{Supp}(r) : r \in R \text{ and } \text{Supp}(r) \cap U_{n-1} \neq \emptyset\} \tag{9.75}$$

and

$$V_n = \bigcup\{\text{Supp}(s) : s \in R \text{ and } \text{Supp}(s) \cap V_{n-1} \neq \emptyset\} \tag{9.76}$$

Then $U_0 \subseteq U_1 \subseteq \cdots \subseteq U_n \subseteq \cdots$, $V_0 \subseteq V_1 \subseteq \cdots \subseteq V_n \subseteq \cdots$, and $U_n \cap V_m = \emptyset$, for all natural numbers $n, m \in \mathbb{N}$.

We now define two subsystems $S_i = (M_i, R_i)$, $i = 1,2$, of S such that

$$M_1 = \bigcup_{n=0}^{\infty} U_n \tag{9.77}$$

$$M_2 = M - M_1 \tag{9.78}$$

$$R_1 = \{r \in R : \text{Supp}(r) \cap M_1 \neq \emptyset\} \tag{9.79}$$

and

$$R_2 = \{r \in R : \text{Supp}(r) \subseteq M_2\} \tag{9.80}$$

Then $R_1 \cup R_2 = R$ and $R_1 \cap R_2 = \emptyset$. In fact, for each relation $r \in R$, if $r \notin R_1$, $\text{Supp}(r) \cap M_1 = \emptyset$ and so $\text{Supp}(r) \subseteq M_2$; thus $r \in R_2$. Therefore, $S = (M,R) = (M_1 \cup M_2, R_1 \cup R_2) = S_1 \oplus S_2$, contradiction.

(ii) $\rightarrow$ (i). The proof is again by contradiction. Suppose condition (ii) holds and S is disconnected. Thus, there exist nontrivial subsystems S_1 and S_2 of S such that $S = S_1 \oplus S_2$. Suppose that $S_i = (M_i, R_i)$, $i = 1,2$. Pick an object $m_i \in M_i$,

$i = 1,2$. Then there are no relations $r_j \in R, j = 1,2,\ldots,n$, for any fixed $n \in \mathbb{N}$, such that

$$m_1 \in \text{Supp}(r_1) \quad \text{and} \quad m_2 \in \text{Supp}(r_n) \tag{9.81}$$

and

$$\text{Supp}(r_j) \cap \text{Supp}(r_{j+1}) \neq \emptyset \tag{9.82}$$

for each $j = 1,2,\ldots,n-1$, contradiction. ■

Let $M^* \subseteq M$, where $S = (M,R)$ is a system. Define the star neighborhood of the subset M^*, denoted Star(M^*), as follows:

$$\text{Star}(M^*) = \bigcup \{\text{Supp}(r) : r \in R \quad \text{and} \quad \text{Supp}(r) \cap M^* \neq \emptyset\} \tag{9.83}$$

By applying induction on $n \in \mathbb{N}$, we can define

$$\text{Star}^1(M^*) = \text{Star}(M^*) \tag{9.84}$$

and

$$\text{Star}^{n+1}(M^*) = \text{Star}(\text{Star}^n(M^*)) \tag{9.85}$$

Corollary 9.4.1. *For every system $S = (M,R)$ the following conditions are equivalent:*

(i) The system S is connected.

(ii) For each object $x \in M$, $\text{Star}^\infty(\{x\}) = \bigcup_{n=1}^{\infty} \text{Star}^n(\{x\}) = M$.

Proof: (i) $\rightarrow$ (ii) follows from Theorem 9.4.1 directly.

(ii) $\rightarrow$ (i). For any two objects $x, y \in M$, from the hypothesis that $\text{Star}^\infty(\{p\}) = M$, for every object $p \in M$, it follows that there exist $n, m \in \mathbb{N}$ and relations $r_1, r_2, \ldots, r_n$, and $s_1, s_2, \ldots, s_m \in R$ such that

$$p \in \text{Supp}(r_1) \quad \text{and} \quad p \in \text{Supp}(s_1) \tag{9.86}$$

$$x \in \text{Supp}(r_n) \quad \text{and} \quad y \in \text{Supp}(s_m) \tag{9.87}$$

$$\text{Supp}(r_i) \cap \text{Supp}(r_{i+1}) \neq \emptyset \quad \text{for } i = 1,2,\ldots,n-1$$

and

$$\text{Supp}(s_i) \cap \text{Supp}(s_{i+1}) \neq \emptyset \quad \text{for } i = 1,2,\ldots,m-1$$

Thus, it follows that Theorem 9.4.1(ii) holds and from Theorem 9.4.1 that S is connected. ■

An object m in a system $S = (M,R)$ is isolated if there is not relation $r \in R$ such that the cardinality $|\text{Supp}(r)| > 1$ and $m \in \text{Supp}(r)$.

Corollary 9.4.2. *There is no isolated object in a connected system.*

Theorem 9.4.2. *Suppose that $S = (M,R)$ is a connected system and $D = (M_D,R_D) = (M_1,R_1) \oplus (M_2,R_2)$ is a disconnected system. If there exists a mapping $f : S \to D$ with $f^{-1}(R_D) = \{f^{-1}(r) : r \in R_D\} \supseteq R$, then only one of the following holds:*

(i) $f(M) \subseteq M_1$.

(ii) $F(M) \subseteq M_2$.

Proof: We prove by contradiction. Suppose there exists a mapping $f : S \to D$ such that $f(M) \cap M_1 \neq \emptyset \neq f(M) \cap M_2$ and $f^{-1}(R_D) \supseteq R$. Pick elements $x \in f^{-1}(M_1)$ and $y \in f^{-1}(M_2)$. From Theorem 9.4.1 it follows that there exist $n \in \mathbb{N}$ and n relations $r_i \in R$, $i = 1,2,\ldots,n$, such that

$$x \in \text{Supp}(r_1) \quad \text{and} \quad y \in \text{Supp}(r_n) \tag{9.88}$$

and

$$\text{Supp}(r_i) \cap \text{Supp}(r_{i+1}) \neq \emptyset \tag{9.89}$$

for each $i = 1,2,\ldots,n-1$. Pick an i with $1 \leq i \leq n$ such that

$$\text{Supp}(r_i) \cap f^{-1}(M_1) \neq \emptyset \neq \text{Supp}(r_i) \cap f^{-1}(M_2) \tag{9.90}$$

Since $f^{-1}(R_D) \supseteq R$, it follows that there exists a relation $r \in R_D$ such that $f(r_i) = r$. Now there are two possibilities: (1) $r \in R_1$; (2) $r \in R_2$. If (1) holds, we must have $\text{Supp}(r_i) \cap f^{-1}(M_2) = \emptyset$, contradiction. If (2) holds, we will have $\text{Supp}(r_i) \cap f^{-1}(M_1) = \emptyset$, again a contradiction. The contradictions imply that every mapping $f : S \to D$ with $f^{-1}(R_D) \supseteq R$ must satisfy $f(M) \subseteq M_1$ or $f(M) \subseteq M_2$. ■

Corollary 9.4.3. *Let $h : S_1 \to S_2$ be a homomorphism from a connected system S_1 into a system S_2. Then $h(S_1) = (h(M_1), h(R_1))$ must be a connected partial system of S_2.*

Theorem 9.4.3. *Suppose that $h : S_1 \to S_2$ is an S-continuous mapping from a system S_1 into a system S_2. If S_2 is connected, the S_1 is also.*

Proof: Suppose that $S_i = (M_i, R_i)$, $i = 1,2$. Let us pick two arbitrary objects $x,y \in M_1$. There are relations $r_j \in R_2, j = 1,2,\ldots,n$, for some fixed $n \in \mathbb{N}$, such that $h(x) \in \text{Supp}(r_1), h(y) \in \text{Supp}(r_n)$, and $\text{Supp}(r_i) \cap \text{Supp}(r_{i+1}) \neq \emptyset$, for each

$i = 1,2,\dots,n-1$. From the hypothesis that h is S-continuous, it follows that $h^{-1}(r_i) \in R_1$ for $i = 1,2,\dots,n$. Thus,

$$x \in \text{Supp}(h^{-1}(r_1)) \quad \text{and} \quad y \in \text{Supp}(h^{-1}(r_n)) \tag{9.91}$$

and

$$\text{Supp}(h^{-1}(r_i)) \cap \text{Supp}(h^{-1}(r_{i+1})) \neq \emptyset \tag{9.92}$$

for $i = 1,2,\dots,n-1$. From Theorem 9.4.1 it follows that S_1 is connected. ■

Example 9.4.1. We construct an example to show that the connectedness of the system S_1 cannot guarantee the connectedness of the system S_2 in Theorem 9.4.3.

Let $S_2 = (\mathbb{Z}, R_2)$ be a system where $\mathbb{Z}$ is the set of all integers and $R_2 = \{r_i : i = \pm1,\pm2,\pm3,\dots\}$, where for each $i = \pm1,\pm2,\pm3,\dots$, $r_i = \{(i,i+1)\}$. Then the system $S_2 = (\mathbb{Z}^-, R_2^-) \oplus (\mathbb{Z}^+, R_2^+)$ is disconnected, where $\mathbb{Z}^- = \{0,-1,-2,\dots\}$, $\mathbb{Z}^+ = \{1,2,3,\dots\}$, $R_2^- = \{r_i : i = -1,-2,\dots\}$ and $R_2^+ = \{r_i : i = 1,2,\dots\}$.

Define a system $S_1 = (\mathbb{Z}, R_1)$ as follows: $R_1 = \{r_i : i = 0,\pm1,\pm2,\dots\}$, where $r_i = \{(i,i+1)\}$ for each $i \in \mathbb{Z}$. Then S_1 is connected. In fact, for any two objects $i,j \in \mathbb{Z}$, there exist relations $r_i, r_{i+1},\dots,r_j$, where we assumed that $i < j$, such that $i \in \text{Supp}(r_i) = \{i,i+1\}, j \in \text{Supp}(r_j) = \{j,j+1\}$, and $\text{Supp}(r_k) \cap \text{Supp}(r_{k+1}) = \{k+1\}$, for $k = i,i+1,\dots,j-1$. Therefore, it follows from Theorem 9.4.1 that S_1 is connected.

Let $h : S_1 \to S_2$ be the identity mapping defined by $h(i) = i$ for each $i \in \mathbb{Z}$. Then h is S-continuous, and S_1 is connected, and S_2 is disconnected.

Theorem 9.4.4. *If a subsystem C of a system S is connected, and for every pair S_1, S_2 of subsystems of S with disjoint object sets such that C is a subsystem of $S_1 \oplus S_2$, then C is a subsystem of S_1 or S_2.*

Proof: We show the result by contradiction. Suppose that C is not a subsystem of S_1 nor of S_2. Let $C = (M_C, R_C)$ and $S_i = (M_i, R_i)$, $i = 1,2$. Then

$$M_C \cap M_1 \neq \emptyset \neq M_C \cap M_2 \tag{9.93}$$

Pick elements $x \in M_C \cap M_1$ and $y \in M_C \cap M_2$. Since C is connected and from Theorem 9.4.1, it follows that there exist a natural number $n > 0$ and n relations $r_i \in R_C$, for $i = 1,2,\dots,n$, such that

$$x \in \text{Supp}(r_1) \quad \text{and} \quad y \in \text{Supp}(r_n) \tag{9.94}$$

and

$$\text{Supp}(r_i) \cap \text{Supp}(r_{i+1}) \neq \emptyset \quad \text{for each } i = 1,\dots,n-1 \tag{9.95}$$

Therefore, there must be an i satisfying $0 < i \leq n$ such that

$$\mathrm{Supp}(r_i) \cap M_1 \neq \emptyset \neq \mathrm{Supp}(r_i) \cap M_2 \tag{9.96}$$

From the definition of subsystems it follows that there exists a relation r in the free sum $S_1 \oplus S_2$ such that $r_i \subseteq r|M_C$. It follows from the definition of free sums that r is contained either in R_1 or in R_2, but not both. Therefore, if $r \in R_1$, $\mathrm{Supp}(r_i) \cap M_2 = \emptyset$; if $r \in R_2$, $\mathrm{Supp}(r_i) \cap M_1 = \emptyset$, contradiction. This implies that C is a subsystem of S_1 or S_2. ■

Theorem 9.4.5. *Let $\{C_i = (M_i, R_i) : i \in I\}$ be a set of connected subsystems of a system $S = (M, R)$. If there exists an index $i_0 \in I$ such that each system $(M_i \cup M_{i_0}, R|(M_i \cup M_{i_0}))$ is connected, for each $i \in I$, then the system $(\bigcup_{i \in I} M_i, R|\bigcup_{i \in I} M_i)$ is connected.*

Proof: Let us pick two arbitrary objects x and y from $\bigcup_{i \in I} M_i$. There are two indices i and j such that $x \in M_i$ and $y \in M_j$. Choose an object $z \in M_{i_0}$. From the hypothesis that the systems $(M_i \cup M_{i_0}, R|(M_i \cup M_{i_0}))$ and $(M_j \cup M_{i_0}, R|(M_j \cup M_{i_0}))$ are connected, it follows that there are natural numbers n and m and relations r_k and $q_p \in R$, for $k = 1, 2, \ldots, n$ and $p = 1, 2, \ldots, m$, such that

$$x \in \mathrm{Supp}(r_1) \quad \text{and} \quad z \in \mathrm{Supp}(r_n) \tag{9.97}$$

$$\mathrm{Supp}(r_k) \cap \mathrm{Supp}(r_{k+1}) \neq \emptyset \quad \text{for each } k = 1, 2, \ldots, n-1 \tag{9.98}$$

and

$$y \in \mathrm{Supp}(q_1) \quad \text{and} \quad z \in \mathrm{Supp}(q_m) \tag{9.99}$$

$$\mathrm{Supp}(q_p) \cap \mathrm{Supp}(q_{p+1}) \neq \emptyset \quad \text{for each } p = 1, 2, \ldots, m-1 \tag{9.100}$$

Therefore, Theorem 9.4.1 implies that the system $(\bigcup_{i \in I} M_i, R|\bigcup_{i \in I} M_i)$ is connected. ■

Corollary 9.4.4. *If the set $\{C_i = (M_i, R_i) : i \in I\}$ of connected subsystems of the system $S = (M, R)$ satisfies $\bigcap_{i \in I} M_i \neq \emptyset$, then the subsystem $(\bigcup_{i \in I} M_i, R|\bigcup_{i \in I} M_i)$ is connected.*

Let $S = (M, R)$ be a system and $D \subset M$ a subset. The subset D is dense in the object set M if the following conditions hold:

(i) For any relation $r \in R$ with $r \neq \emptyset$, $\mathrm{Supp}(r) \cap D \neq \emptyset$.

(ii) Each isolated object in M belongs to D.

A subsystem $S^* = (M^*, R^*)$ of S is a dense subsystem of S if the object set M^* is dense in M.

Theorem 9.4.6. *If a system $S = (M, R)$ contains a connected dense subsystem, then S is connected.*

Proof: Suppose that $S^* = (M^*, R^*)$ is a connected dense subsystem of S. To show that S is connected, we pick two arbitrary objects $x, y \in M$. There are three possibilities: (1) $x, y \in M - M^*$; (2) one of x and y is contained in M^* and the other is contained in $M - M^*$; (3) x, y are contained in M^*.

If possibility (1) holds, x and y are not isolated objects in M. Therefore, there are relations r_x and $r_y \in R$ such that

$$x \in \text{Supp}(r_x) \quad \text{and} \quad y \in \text{Supp}(r_y) \tag{9.101}$$

It follows from the definition of dense subsets that there must be an object $x^* \in \text{Supp}(r_x) \cap M^*$ and an object $y^* \in \text{Supp}(r_y) \cap M^*$. Theorem 9.4.1 implies that there are a natural number n and n relations $r_i \in R^*$, $i = 1, 2, \ldots, n$, such that

$$x^* \in \text{Supp}(r_1) \quad \text{and} \quad y^* \in \text{Supp}(r_n) \tag{9.102}$$

and

$$\text{Supp}(r_i) \cap \text{Supp}(r_{i+1}) \neq \emptyset \quad \text{for each } i = 1, 2, \ldots, n-1 \tag{9.103}$$

For each $i = 1, 2, \ldots, n$, let q_i be a relation in R such that $r_i \subset q_i | M^*$. Then it has to be that

$$x \in \text{Supp}(r_x) \quad \text{and} \quad y \in \text{Supp}(r_y) \tag{9.104}$$

$$\text{Supp}(r_x) \cap \text{Supp}(q_1) \neq \emptyset \neq \text{Supp}(r_y) \cap \text{Supp}(q_n) \tag{9.105}$$

and

$$\text{Supp}(q_i) \cap \text{Supp}(r_{i+1}) \neq \emptyset \quad \text{for each } i = 1, 2, \ldots, n-1 \tag{9.106}$$

Theorem 9.4.1 implies that S is connected.

If possibility (2) holds, without confusion, let $x \in M^*$ and $y \in M - M^*$. Then y is not an isolated object in M. Let $r_y \in R$ be a relation such that

$$y \in \text{Supp}(r_y) \tag{9.107}$$

Pick $y^* \in \text{Supp}(r_y) \cap M^*$. From the hypothesis that S^* is connected, it follows that there are a natural number $n > 0$ and n relations $r_i \in R^*$, $i = 1, 2, \ldots, n$, such that

$$x \in \text{Supp}(r_1) \quad \text{and} \quad y^* \in \text{Supp}(r_n) \tag{9.108}$$

and

$$\text{Supp}(r_i) \cap \text{Supp}(r_{i+1}) \neq \emptyset \quad \text{for each } i = 1,2,\ldots,n-1 \tag{9.109}$$

For each $i = 1,2,\ldots,n$, let q_i be a relation in R such that $r_i \subset q_i|M^*$. Then

$$x \in \text{Supp}(q_1) \quad \text{and} \quad y \in \text{Supp}(r_y) \tag{9.110}$$

$$\text{Supp}(q_i) \cap \text{Supp}(q_{i+1}) \neq \emptyset \quad \text{for each } i = 1,2,\ldots,n-1 \tag{9.111}$$

and

$$\text{Supp}(q_n) \cap \text{Supp}(r_y) \neq \emptyset \tag{9.112}$$

Applying Theorem 9.4.1 we have showed that S is connected. ■

A connected subsystem S^* of S is maximal in S if there is not connected subsystem S^+ of S such that S^* is a subsystem of S^+ and $S^* \neq S^+$. Each maximal connected subsystem of S is called a component. Then, for each object $m \in M$, the partial system $(\text{Star}^\infty(\{m\}), R|\text{Star}^\infty(\{m\}))$ of S is a component.

Theorem 9.4.7. *Suppose that $S = (M,R)$ is a system and E an equivalence relation on the object set M. Then the quotient system $S/E = (M/E, R/E)$ is connected iff for any two components $X = (M_X, R_X)$ and $Y = (M_Y, R_Y)$ of S, there exist a natural number n and n components $S_i = (M_i, R_i)$ of S, $i = 1,2,\ldots,n$, such that there exist $x \in M_X, y \in M_Y$ and $m_{x_i}, m_{y_i} \in M_i$, $i = 1,2,\ldots,n$, satisfying $(x, m_{x_1}) \in E$, $(m_{x_{i+1}}, m_{y_i}) \in E$, $i = 1,2,\ldots,n-1$, and $(m_{y_n}, y) \in E$.*

Proof: Necessity. Suppose that the quotient system $S/E = (M/E, R/E)$ is connected, and there exist two components $X = (M_X, R_X)$ and $Y = (M_Y, R_Y)$ of S such that there do not exist components $S_i = (M_i, R_i)$ of $S, i = 1,2,\ldots,n$, for each natural number n, such that there exist $x \in M_X$ and $y \in M_Y$ and $m_{x_i}, m_{y_i} \in M_i$, for each i, satisfying $(x, m_{x_1}) \in E$, $(m_{x_{i+1}}, m_{y_i}) \in E$, $i = 1,2,\ldots,n-1$, and $(m_{y_n}, y) \in E$.

Pick objects $x \in M_X$ and $y \in M_Y$. We then make a claim.

Claim. The subsystems

$$(\text{Star}^\infty(\{\mu(x)\}),\ R/E|\text{Star}^\infty(\{\mu(x)\}))$$

and

$$(\text{Star}^\infty(\{\mu(y)\}),\ R/E|\text{Star}^\infty(\{\mu(y)\}))$$

of the quotient system S/E are not identical, where $\mu : M \to M/E$ is the canonical mapping.

This claim implies that the quotient system S/E is not connected. In fact, from the hypothesis that the quotient system S/E is connected, it follows that for each object $\mu(m) \in M/E$, the connected component of S/E containing the object $\mu(m)$ must equal the whole system S/E.

We now see the proof of Claim 1. It suffices to show that $\text{Star}^\infty(\{\mu(x)\}) \neq \text{Star}^\infty(\{\mu(y)\})$. To this end, it suffices to show that

$$\text{Star}^\infty(\{\mu(x)\}) \cap \text{Star}^\infty(\{\mu(y)\}) = \emptyset \tag{9.113}$$

which we do by contradiction. Suppose that

$$\text{Star}^\infty(\{\mu(x)\}) \cap \text{Star}^\infty(\{\mu(y)\}) \neq \emptyset \tag{9.114}$$

for some $x \in M_X$ and some $y \in M_Y$. Choose $[z] \in \text{Star}^\infty(\{\mu(x)\}) \cap \text{Star}^\infty(\{\mu(y)\})$, where m is an object in M. Then there exist components $S_{x_i} = (M_{x_i}, R_{x_i})$ and $S_{y_j} = (M_{y_j}, R_{y_j})$, for $i = 1,2,\ldots,n$ and $j = 1,2,\ldots,m$, for some natural numbers n and m, such that there are objects $x_i, m_{x_i} \in M_{x_i}$ and $y_j, m_{y_j} \in M_{y_j}$, for each $i = 1,2,\ldots,n$ and $j = 1,2,\ldots,m$, such that (x,x_1), $(x_{i+1}, m_{x_i}) \in E$, for $i = 1,2,\ldots,n-1$, and (m_{x_n}, z), (m_{y_m}, z), (y,y_1), $(y_{j+1}, m_{y_j}) \in E$, for $j = 1,2,\ldots,m-1$. This contradicts the assumption in the first paragraph of the necessary part of the proof.

Sufficiency. Suppose that for any two components $X = (M_X, R_X)$ and $Y = (M_Y, R_Y)$ of S, there exist a natural number n and components $S_i = (M_i, R_i)$ of S, $i = 1,2,\ldots,n$, such that there exist $x \in M_X$, $y \in M_Y$, and $m_{x_i}, m_{y_i} \in M_i$, for each $i = 1,2,\ldots,n$, satisfying (x, m_{x_1}), (m_{y_n}, y), $(m_{x_{i+1}}, m_{y_i}) \in E$, $i = 1,2,\ldots,n-1$. To show that S/E is connected, it suffices to verify that, for each object $x \in M$,

$$\text{Star}^\infty(\{\mu(x)\}) = M/E \tag{9.115}$$

by Corollary 9.4.1.

It is clear that $\text{Star}^\infty(\{\mu(x)\}) \subset M/E$. Suppose $M/E \neq \text{Star}^\infty(\{\mu(x)\})$. We can then pick an object $[m] \in M/E - \text{Star}^\infty(\{\mu(x)\})$, where m is an object in M. This implies that there are two connected components $X = (M_X, R_X)$ and $Y = (M_Y, R_Y)$, such that $x \in M_X$ and $m \in M_Y$ and for any two objects $m_x \in M_X$ and $m_y \in M_Y$ there do not exist connected components $S_i = (M_i, R_i)$ of S, $i = 1,2,\ldots,n$, for any $n \in \mathbb{N}$, such that there exist $m_{x_i}, m_{y_i} \in M_i$, $i = 1,2,\ldots,n$, satisfying (m_x, m_{x_1}), $(m_{x_{i+1}}, m_{y_i})$, $(m_{y_n}, m_y) \in E$, $i = 1,2,\ldots,n-1$, contradiction which implies that $M/E = \text{Star}^\infty(\{\mu(x)\})$. ■

Theorem 9.4.8. *Let $\{S_i : i \in I\}$ be a set of systems such that there exists an $i_0 \in I$ such that the system S_{i_0} is connected. Then the Cartesian product system $\prod_{cp}\{S_i : i \in I\}$ is connected.*

Proof: Pick objects $x = (x_i)_{i \in I}$ and $y = (y_i)_{i \in I}$ from the Cartesian product system $\prod_{cp}\{S_i : i \in I\}$. Then $p_{i_0}(x) = x_{i_0}$ and $p_{i_0}(y) = y_{i_0} \in M_{i_0}$. From Theorem

9.4.1 it follows that there are a natural number $n > 0$ and n relations $r_j \in R_{i_0}$, $j = 1, 2, \ldots, n$, such that

$$x_{i_0} \in \mathrm{Supp}(r_1) \quad \text{and} \quad y_{i_0} \in \mathrm{Supp}(r_n) \tag{9.116}$$

and

$$\mathrm{Supp}(r_i) \cap \mathrm{Supp}(r_{i+1}) \neq \emptyset \quad \text{for each } i = 1, 2, \ldots, n-1 \tag{9.117}$$

Therefore, the relations $(p_{i_0})^{-1}(r_j)$ in $\prod_{\mathrm{cp}}\{S_i : i \in I\}, j = 1, 2, \ldots, n$, satisfy

$$x \in \mathrm{Supp}((p_{i_0})^{-1}(r_1)) \quad \text{and} \quad y \in \mathrm{Supp}((p_{i_0})^{-1}(r_n)) \tag{9.118}$$

and

$$\mathrm{Supp}((p_{i_0})^{-1}(r_i)) \cap \mathrm{Supp}((p_{i_0})^{-1}(r_{i+1})) \neq \emptyset \tag{9.119}$$

for each $i = 1, 2, \ldots, n-1$. Applying Theorem 9.4.1 we have proved the theorem.■

Theorem 9.4.9. *Let $\{S_i : i \in I\}$ be a set of systems such that the product $\prod\{S_i : i \in I\}$ is connected. Then each factor system S_i is connected.*

Proof: Pick two arbitrary objects x_i and y_i from a system S_i for any fixed $i \in I$. There then are two objects x and y in the product system such that $p_i(x) = x_i$ and $p_i(y) = y_i$. From the hypothesis that the product system is connected, it follows that there are a natural number $n > 0$ and n relations r_j from the product system, $j = 1, 2, \ldots, n$, such that

$$x \in \mathrm{Supp}(r_1) \quad \text{and} \quad y \in \mathrm{Supp}(r_n) \tag{9.120}$$

and

$$\mathrm{Supp}(r_j) \cap \mathrm{Supp}(r_{j+1}) \neq \emptyset \tag{9.121}$$

for each $j = 1, 2, \ldots, n-1$. Then the relations $p_i(r_j) \in R_i, j = 1, 2, \ldots, n$, satisfy

$$x_i \in \mathrm{Supp}(p_i(r_1)) \quad \text{and} \quad y_i \in \mathrm{Supp}(p_i(r_n)) \tag{9.122}$$

and

$$\mathrm{Supp}(p_i(r_j)) \cap \mathrm{Supp}(p_i(r_{j+1})) \neq \emptyset \tag{9.123}$$

for each $j = 1, 2, \ldots, n-1$. Theorem 9.4.1 implies that S_i is connected. ■

9.5. Hierarchies of Systems

Let $(T, \leq)$ be a partially ordered set with order type α. An α-type hierarchy S of systems over the partially ordered set $(T, \leq)$ is a function defined on T such that for each $t \in T$, $S(t) = S_t = (M_t, R_t)$ is a system, called the state of the α-type hierarchy S at the moment t. Without causing confusion, we omit the words "over the partially ordered set $(T, \leq)$."

For an α-type hierarchy S of systems, let $\ell_{tr} : S_r \to S_t$ be a mapping from the system S_r into the system S_t, for any $r, t \in T$, with $r \geq t$ such that

$$\ell_{ts} = \ell_{tr} \circ \ell_{rs} \quad \text{and} \quad \ell_{tt} = \mathrm{id}_{S_t} \tag{9.124}$$

where r, s, t are arbitrary elements in T satisfying $s \geq r \geq t$, and $\mathrm{id}_{S_t} = \mathrm{id}_{M_t}$ is the identity mapping on the set M_t. The family $\{\ell_{ts} : t, s \in T, s \geq t\}$ is a family of linkage mappings of the α-type hierarchy S, and each mapping ℓ_{ts} a linkage mapping from S_s into S_t.

An α-type hierarchy of systems S, denoted $\{S, \ell_{ts}, T\}$ or $\{S(t), \ell_{ts}, T\}$, is referred to as a linked α-type hierarchy (of systems) if a family $\{\ell_{ts} : t, s \in T, s \geq t\}$ of linkage mappings is given.

Theorem 9.5.1 [ZFC]. *Let S be a nontrivial α-type hierarchy of systems; i.e., each state S_t is a nontrivial system. Then there exists a family $\{\ell_{ts} : t, s \in T,\ s \geq t\}$ of linkage mappings of S.*

Proof: Suppose $S_t = (M_t, R_t)$, for each $t \in T$. From the Axiom of Choice it follows that there exists a choice function $C : T \to \bigcup\{M_t : t \in T\}$ such that $C(t) \in M_t$ for each $t \in T$. A family $\{\ell_{ts} : t, s \in T, s \geq t\}$ of linkage mappings of S can now be defined. For any $s, t \in T$ with $s > t$, let

$$\ell_{ts}(x) = C(t) \tag{9.125}$$

for all $x \in M_s$ and $\ell_{tt} = \mathrm{id}_{M_t}$. ■

Question 9.5.1. How many families of linkage mappings are there for a given α-type hierarchy of systems?

Suppose $\{S, \ell_{ts}, T\}$ is a linked α-type hierarchy of systems, where $S_t = (M_t, R_t)$ for all $t \in T$. An element $\{x_w\}_{w \in T} \in \prod\{M_t : t \in T\}$ is called a thread of the linked α-type hierarchy S if $\ell_{ts}(x_s) = x_t$ for all $s, t \in T$ with $s \geq t$. Let $\sum \subset \prod\{M_t : t \in T\}$ be the subset of all threads, which will be called the thread set of the hierarchy $\{S, \ell_{ts}, T\}$.

Theorem 9.5.2 [ZFC]. *Let κ be a cardinal number and S an α-type hierarchy of systems such that $|M_t| \geq \kappa$ for each $t \in T$. Then for any nonzero cardinal number $\kappa' \leq \kappa$, there exists a family $\{\ell_{ts} : t,s \in T,\ s \geq t\}$ of linkage mappings of S such that $|\Sigma| \geq \kappa'$.*

Proof: From the hypothesis that for any $t \in T$, $|M_t| \geq \kappa'$, and the Axiom of Choice, it follows that, for any $t \in T$, there exist $A_t \subset M_t$ satisfying $|A_t| = \kappa'$, and an equivalence relation E_t on the object set M_t such that $M_t/E_t = \{[a]_{E_t} : a \in A_t\}$ and that for any distinct $a, b \in A_t$,

$$[a]_{E_t} \neq [b]_{E_t} \tag{9.126}$$

Again, from the Axiom of Choice it follows that there exists a set of well-order relations $\{\leq_t : t \in T\}$ such that, for each $t \in T$, the relation $\leq_t$ well-orders A_t as follows:

$$a_{0t}, a_{1t}, \ldots, a_{\alpha t}, \ldots, \alpha < \kappa' \tag{9.127}$$

For any $s, t \in T$ with $s > t$, we now define a mapping $\ell_{ts} : S_s \to S_t$. For any $x \in M_s$, there exists exactly one $\alpha < \kappa$ such that $x \in [a_{\alpha s}]_{E_s}$. Then define

$$\ell_{ts}(x) = a_{\alpha t} \tag{9.128}$$

Then $\{\ell_{ts} : t,s \in T,\ s \geq t\}$ is a family of linkage mappings, where $\ell_{tt} = \mathrm{id}_{S_t}$, and $|\Sigma| \geq |A_t|$, for any $t \in T$. Thus $|\Sigma| \geq \kappa'$. ■

Question 9.5.2. Suppose that S is an α-type hierarchy of systems over a partially ordered set $(T, \leq)$ and $\{\ell_{ts} : s,t \in T^*,\ s \geq t\}$ is a family of linkage mappings of the hierarchy of systems $\{S_t : t \in T^*\}$, where T^* is a proper subset of T. Under what conditions does there exist a family $\{\overline{\ell}_{ts} : t,s \in T,\ s \geq t\}$ of linkage mappings for S such that

$$\overline{\ell}_{ts} = \ell_{ts} \tag{9.129}$$

for all s, $t \in T^*$ with $s \geq t$?

Lemma 9.5.1 [ZFC]. *Let $D \subset T$ be a cofinal and coinitial subset in a partially ordered set $(T, \leq)$. Then there exists a subset $T^* = D_-^* \cup D_+^*$ of D such that*

(a) $D_+^* = \bigcup\{D_\alpha^+ : \alpha < \eta\}$ *and* $D_-^* = \bigcup\{D_\alpha^- : \alpha < \xi\}$ *for some ordinals η and ξ such that*

$$D_0^+ = D_0^1 \tag{9.130}$$

(b) *Let* $E_\beta^+ = \{x \in T : \exists d \in \bigcup\{D_\alpha^+ : \alpha < \beta\}(x \leq d)\}$ *then* D_β^+ *is a maximal antichain in* $D - E_\beta^+$ *for any* $\beta < \eta$.

(c) *Let* $E_\beta^- = \{x \in T : \exists d \in \bigcup\{D_\alpha^- : \alpha < \beta\}(d \leq x)\}$. *Then* D_β^- *is a maximal antichain in* $D - E_\beta^-$ *for each* $\beta < \xi$.

(d) T^* *is cofinal and coinitial in* T.

Proof: First, note the construction of D_+^*. From the Axiom of Choice it follows that there exists an order relation $\leq *$ on T which well-orders T as

$$t_0, t_1, \ldots, t_\alpha, \ldots, \alpha < |T| \tag{9.131}$$

where $\leq$ and $\leq *$ are generally different. Let d_0 be the $< *$-initial element of D, and $D_0^1 \subset D$ such that

$$D_0^1 = \{d \in D : d_0 < {}^*d \text{ and } d_0 \text{ and } d \text{ are incomparable under } \leq\} \tag{9.132}$$

If $D_0^1 = \emptyset$, let $D_0^+ = \{d_0\}$; otherwise, let d_1 be the initial element of $(D_0^1, < *)$.

Suppose that sequences $\{d_\alpha : \alpha < \eta\}$ and $\{D_0^\alpha : 0 \neq \alpha < \eta\}$ have been defined, for some ordinal $\eta > 1$, such that

(i) $\{d_\alpha : \alpha < \eta\}$ is an antichain under the order $<$.

(ii) $\emptyset \neq D_0^\alpha \subset D_0^\beta$ for all $\alpha, \beta < \eta$ with $\alpha > \beta$.

(iii) For each $\alpha < \eta$, d_α is the initial element of the well-ordered set $(D_0^\alpha, \leq *)$.

(iv) For each $\alpha < \eta$, $d_\alpha < {}^*d$, for all $d \in D_0^\beta$, and d_α is incomparable with d under the relation $<$, for each $d \in D_0^\beta$, for each β with $\alpha < \beta < \eta$.

If η is a limit ordinal, let

$$D_0^\eta = \bigcap\{D_0^\alpha : \alpha < \eta\} \tag{9.133}$$

If $D_0^\eta = \emptyset$, let $D_0^+ = \{d_\alpha : \alpha < \eta\}$. Then D_0^+ is a maximal antichain in D under $\leq$. If $D_0^\eta \neq \emptyset$, let d_η be the initial element of $(D_0^\eta, < *)$. By (iv) $\{d_\alpha : \alpha < \eta + 1\}$ is an antichain under the relation $\leq$; and for each $\alpha < \eta$, $d_\alpha < {}^*d$ for each $d \in D_0^\eta$, d_α and d are incomparable under $\leq$ for each $d \in D_0^\eta$.

If $\eta = \xi + 1$ is isolated, let $D_0^\eta \subset D_0^\xi$ be defined by

$$D_0^\eta = \{d \in D_0^\xi : d_0^\xi < {}^*d, d_0^\eta \text{ and } d \text{ are incomparable under } \leq\} \tag{9.134}$$

If $D_0^\eta = \emptyset$, let $D_0^+ = \{d_\alpha : \alpha < \eta\}$; then D_0^+ is a maximal antichain in D under the order $\leq$. If $D_0^\eta \neq \emptyset$, let d_η be the initial element of $(D_0^\eta, \leq *)$. Then $\{d_\alpha : \alpha < \eta + 1\}$ is an antichain in D under $\leq$, and $D_0^\eta \subset D_0^\beta$ for all $\beta < \eta$. For each $\alpha < \eta + 1$,

$d_\alpha < {}^*d$, for all $d \in D_0^\beta$, and d_α is incomparable with $d \in D_0^\beta$ under $\leq$ for any β, with $\alpha < \beta < \eta + 1$.

By transfinite induction, sequences $\{d_\alpha : \alpha < \eta\}$ and $\{D_0^\alpha : \alpha < \eta\}$ with properties (i)–(iv) can be defined, such that if a subset D_0^η of $\bigcap\{D_0^\alpha : \alpha < \eta\}$ with property (iv) can be defined in the way discussed earlier, it must be that $D_0^\eta = \emptyset$.

Let $D_0^+ = \{d_\alpha : \alpha < \eta\}$. Then D_0^+ is a maximal antichain in $(D, \leq)$. Using transfinite induction to consider sets $U = \{x \in T : \exists d \in D_0^+ \text{ such that } x > d\}$ and $V = \{x \in T : \exists d \in D_0^+ (x < d)\}$ and applying the same technique we used before, we can show that ordinals η and ξ and subsets $D_\alpha^+ \subset U$ and $D_\beta^- \subset V$, for $0 \neq \alpha < \eta$ and $0 \neq \beta < \xi$, exist such that properties (a)–(d) in the lemma hold. ■

Theorem 9.5.3 [ZFC]. *Suppose that S is an α-type hierarchy of systems over a partially ordered set $(T, \leq)$ such that there exists a cofinal and coinitial subset $D \subset T$ such that $\{\ell_{ts} : t, s \in D, s \geq t\}$ is a family of linkage mappings for the hierarchy of systems $\{S_t : t \in D\}$ satisfying*

$$\ell_{ts}(M_s) = \ell_{tr}(M_r) \tag{9.135}$$

for any $s, r, t \in D$ with $s \geq r > t$. Then there exists a subset $T^* \subset D$ such that

(i) *T^* is cofinal and coinitial in T.*

(ii) *There is a family $\{\overline{\ell}_{ts} : t, s \in T, s \geq t\}$ of linkage mappings for S.*

(iii) *For any $s, t \in T^*$ with $s \geq t$, $\overline{\ell}_{ts} = \ell_{ts}$.*

Proof: Let $T^* \subset D$ be the subset defined in the proof of Lemma 9.5.1. Let us rewrite $T = T_L \cup T_R$, where

$$T_L = \{x \in T : \exists d \in D_0^- (x \leq d)\} \tag{9.136}$$

and

$$T_R = \{x \in T : \exists d \in D_0^+ (d \leq x)\} \tag{9.137}$$

and pick one $t \in T$. There are two possibilities: (1) $t \in T_L$ and (2) $t \in T_R$.

Case 1: $t \in T_L$. There then exists an ordinal $\alpha < \xi$ such that there exists $e_\alpha \in D_\alpha^-$, with $e_\alpha \leq t$, and assume that α is the initial one in ξ with this property. If $t = e_\alpha$, let $\overline{\ell}_{tt} = \ell_{tt}$. Now we assume that $e_\alpha < t$.

If $\alpha = \delta + 1$ is isolated, there exists $e_\delta \in D_\delta^-$ such that $e_\alpha < t < e_\delta$. From the Axiom of Choice it follows that for any $s \in T$ satisfying $e_\alpha < s < e_\delta$, there exist an $A_s \subset M_s$ such that $|A_s| = \left|\ell_{e_\alpha e_\delta}(M_{e_\delta})\right| \leq \kappa$; an equivalence relation E_s on M_s such that $M_s/E_s = \{[a]_{E_s} : a \in A_s\}$ and satisfies, for any distinct $a, b \in A_s$,

$$[a]_{E_s} \neq [b]_{E_s} \tag{9.138}$$

and a set $\{<_s: s \in T, e_\alpha \leq s < e_\delta\}$ of well-order relations such that, for any $s \in T$ satisfying $e_\alpha < s < e_\delta$, the relation $<_s$ well-orders M_s/E_s as follows:

$$[a_{s,0}]_{E_s}, \ldots, [a_{s,i}]_{E_s}, \ldots, i < |A_s| \tag{9.139}$$

where $a_{s,i} \in A_s$, for all $i < |A_s|$, and the relation $<_{e_\alpha}$ well-orders $\ell_{e_\alpha e_\delta}(M_{e_\delta})$ as follows:

$$m_0, m_1, \ldots, m_\alpha, \ldots, \alpha < \left|\ell_{e_\alpha e_\delta}(M_{e_\delta})\right| \tag{9.140}$$

We now define mappings $\overline{\ell}_{st} : M_t \to M_s$ for all $s, t \in T$, with $e_\alpha \leq s < t < e_\delta$ as follows: For any $x \in M_t$ there exists exactly one $a_{ti} \in A_t$ such that $x \in [a_{ti}]_{E_t}$. If $s > e_\alpha$, let

$$\overline{\ell}_{st}(x) = a_{si} \tag{9.141}$$

If $s = e_\alpha$, let

$$\overline{\ell}_{st}(x) = m_i \tag{9.142}$$

Define mappings $\overline{\ell}_{se_\delta} : M_{e_\delta} \to M_s$ for all $s \in T$ with $e_\alpha < s < e_\delta$ as follows:

$$\overline{\ell}_{se_\delta}(x) = a_{si} \quad \text{if } \ell_{e_\alpha e_\delta}(x) = m_i \tag{9.143}$$

If α is a limit ordinal, then by definition $\alpha > 0$. Let

$$T_\alpha = \{s \in T : e_\alpha < s < e_\delta, \forall \delta < \alpha\} \tag{9.144}$$

From the Axiom of Choice it follows that for any $s \in T_\alpha$, there exists an $A_s \subset M_s$ satisfying $|A_s| = |\ell_{e_\alpha q}(M_q)|$, where $q \in D_0^-$ satisfies $t < q$. An equivalence relation E_s on M_s exists such that $M_s/E_s = \{[a]_{E_s} : a \in A_s\}$ satisfies, for any distinct $a, b \in A_s$, $[a]_{E_s} \neq [b]_{E_s}$. Let $\{<_s: s \in \{e_\alpha\} \cup T_\alpha\}$ be a set of order relations such that, for any $s \in T_\alpha$, the relation $<_s$ well-orders M_s/E_s as follows:

$$[a_{s0}]_{E_s}, \ldots, [a_{si}]_{E_s}, \ldots, i < |A_s| \tag{9.145}$$

where $a_{si} \in A_s$, for all i, and the relation $<_{e_\alpha}$ well-orders $\ell_{e_\alpha q}(M_q)$ as follows:

$$m_0, m_1, \ldots, m_i, \ldots, i < |\ell_{e_\alpha q}(M_q)| \tag{9.146}$$

We now define mappings $\overline{\ell}_{ij} : M_j \to M_i$, for all $i, j \in \{e_\alpha\} \cup T_\alpha$, as follows: For any $x \in M_j$ there exists exactly one $a_{jk} \in A_j$ such that $x \in [a_{jk}]_{E_j}$. If $j > e_\alpha$, let

$$\overline{\ell}_{ij}(x) = a_{ik} \tag{9.147}$$

If $j = e_\alpha$, let

$$\overline{\ell}_{ij}(x) = m_k \tag{9.148}$$

Define mappings $\overline{\ell}_{se_\delta} : M_{e_\delta} \to M_s$ for all $s \in T_\alpha$ and all $\delta < \alpha$ by

$$\overline{\ell}_{se_\delta}(x) = a_{sk} \quad \text{if } \ell_{e_\alpha e_\delta}(x) = m_k \tag{9.149}$$

Then it is easy to check that $\{\overline{\ell}_{st} : s,t \in T_L,\ t \geq s\}$ is a family of linkage mappings, where $\overline{\ell}_{ss} = \mathrm{id}_{M_s}$. Furthermore, for any $s,t \in T_L$ with $s > t$, if $\overline{\ell}_{ts}$ is undefined above, then there exist $\alpha, \beta < \xi$ such that there exist $d_\alpha \in D_\alpha^-$ and $d_\beta \in D_\beta$, with $d_\beta < t < d_\alpha < s$. Let α and β be the least ordinals with this property. (1) If $\beta = \delta + 1$, then there exists exactly one $d_\delta \in D_\delta^-$ such that $d_\beta < t \leq d_\delta$. Define

$$\overline{\ell}_{ts} = \overline{\ell}_{sd_\delta} \circ \overline{\ell}_{d_\delta d_\alpha} \circ \overline{\ell}_{d_\alpha s} \tag{9.150}$$

(2) If β is a limit ordinal, define

$$\overline{\ell}_{ts} = \overline{\ell}_{td_\alpha} \circ \overline{\ell}_{d_\alpha s} \tag{9.151}$$

Case 2: $t \in T_R$. A similar method can be used to define a family $\{\overline{\ell}_{ts} : t,s \in T,\ t \geq s\}$ of linkage mappings such that $\overline{\ell}_{st} = \ell_{st}$ for any $s,t \in D_+^*$ with $s \leq t$.

Now let $s \in T_R$ and T_L satisfying $s > t$. There exists exactly one $q \in D_0^+$ such that $t \leq q \leq s$. Define $\overline{\ell}_{ts} = \overline{\ell}_{tq} \circ \overline{\ell}_{qs}$. Then $\{\overline{\ell}_{st} : s,t \in T,\ t \geq s\}$ is a family of linkage mappings for S such that $\overline{\ell}_{ts} = \ell_{ts}$ for all $s,t \in T^*$ with $s \geq t$. ■

Let $H = \{S, \ell_{ts}, T\}$ and $H' = \{S', \ell_{t's'}, T'\}$ be two hierarchies of systems. A mapping from the hierarchy H into the hierarchy H' is a set $\{\phi, f_{t'}\}$ of mappings, where ϕ is a nondecreasing mapping from T' into T with $\phi(T')$ cofinal in T, and for each $t' \in T', f_{t'}$ is a mapping from $S_{\phi(t')}$ into $S'_{t'}$ such that

$$\ell_{t's'} \circ f_{s'} = f_{t'} \circ \ell_{\phi(t')\phi(s')} \tag{9.152}$$

i.e., such that the diagram in Fig. 9.2 commutes for any $s', t' \in T'$ satisfying $s' \geq t'$.

A mapping $\{\phi, f_{t'}\}$ from H into H' is a similarity mapping if ϕ is a similarity mapping from T' onto T and each $f_{t'} : S_{\phi(t')} \to S'_{t'}$ is a similarity mapping. It is easy to show that $\{\phi^{-1}, f_{t'}^{-1}\}$ is a similarity mapping from H' into H.

When the partially ordered set $(T, \leq)$ is a pseudo-tree (i.e., for each $t \in T$ the set $\{y \in T : y \leq x\}$ is a chain), each hierarchy S of systems over T is called a treelike hierarchy of systems; each linked hierarchy $\{S, \ell_{ts}, T\}$ of systems is called a linked treelike hierarchy of systems. A treelike hierarchy B of systems over a tree T is a partial treelike hierarchy of a treelike hierarchy A of systems over the same tree T if for each $t \in T$, the state B_t is a partial system of the state A_t.

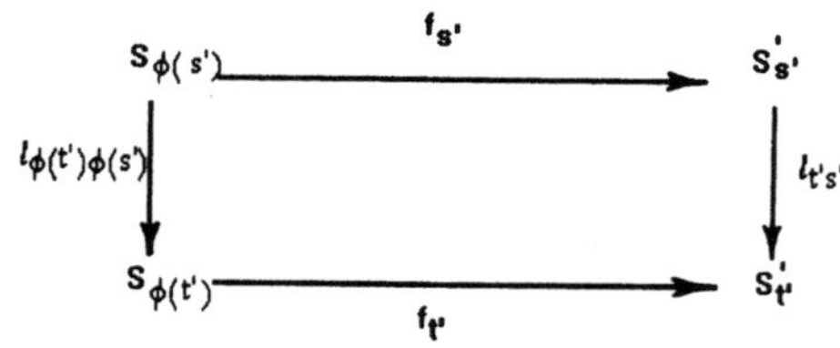

Figure 9.2. A hierarchy mapping guarantees the diagram is commutative.

Theorem 9.5.4 [ZFC]. *Let $\{S, \ell_{ts}, T\}$ be a linked treelike hierarchy of systems such that the partially ordered set $(T, \leq)$ satisfies*

(i) T has a subset T_0 consisting of all minimal elements in T such that for each $t \in T$, there exists a $t_0 \in T_0$ such that $t_0 \leq t$.

(ii) There exists a cofinal subset $D \subset T$ such that ℓ_{ts} is an onto mapping for all $s, t \in D \cup T_0$ with $s \geq t$.

Then there exists a nontrivial partial linked hierarchy $\{S', \widetilde{\ell}_{ts}, T\}$ of $\{S, \ell_{ts}, T\}$ such that the state $S'(t)$ is embeddable in the state $S'(s)$ for all $t, s \in T$ with $s \geq t$.

Proof: From Lemma 9.5.1 it follows that there exists $D^* \subset D$ such that

(a) $D^* = \bigcup\{D^*_\alpha : \alpha < \eta\}$ for some ordinal $\eta > 0$.

(b) D^*_β is a maximal antichain in $D - E_\beta$, for each $\beta < \eta$, where $E_\beta = \{x \in T : \exists d \in \bigcup\{D^*_\alpha : \alpha < \beta\}(x \leq d)\}$.

(c) D^* is cofinal in T.

Let $\leq^*$ indicate the order relation on T defined in the proof of Lemma 9.5.1. From the hypothesis that ℓ_{ts} is onto for any $s, t \in D \cup T_0$ with $s \geq t$, it follows that each linkage mapping $\ell_{t_0 r}$ is onto, for all $r \in T$ and $t_0 \in T_0$ satisfying $r \geq t_0$. Let d_0 be the initial element of the well-ordered set $(D^*, \leq^*)$. We define an equivalence relation E_{d_0} on M_{d_0} such that the equivalence class

$$[m]_{E_{d_0}} = \{x \in M_{d_0} : \forall t_0 \in T_0(t_0 \leq d_0 \rightarrow \ell_{t_0 d_0}(x) = \ell_{t_0 d_0}(m))\} \tag{9.153}$$

for each $m \in M_{d_0}$. By transfinite induction, for each element d in $(D^*, \leq^*)$ an equivalence relation E_d on the object set M_d of the state S_d can be defined such that the equivalence class

$$[m]_{E_d} = \{x \in M_d : \forall t_0 \in T_0(t_0 \leq d \rightarrow \ell_{t_0 d}(x) = \ell_{t_0 d}(m))\} \tag{9.154}$$

The linkage mapping ℓ_{tr}, for any $t, r \in D^*$ with $r \geq t$, can induce a mapping $\overline{\ell}_{tr}$ from S_r/E_r onto S_t/E_t such that

$$\overline{\ell}_{tr}([m_r]_{E_r}) = [m_t]_{E_t} \tag{9.155}$$

where $[m_t]_{E_t} \in M_t/E_t$ satisfies $\ell_{tr}(m_r) = m_t$. In fact, for each $[m_r]_{E_r} \in M_r/E_r$, there exists exactly one $[m_t]_{E_t} \in M_t/E_t$ such that

$$\forall t_0 \in T_0(t_0 \leq t \to \ell_{t_0t}(m_t) = \ell_{t_0r}(m_r)) \tag{9.156}$$

Therefore, $\ell_{tr}([m_r]_{E_r}) \subset [m_t]_{E_t}$, and $\widetilde{\ell}_{tr}$ is well defined.

For any maximal linearly ordered set $B \subset D^* \cup T_0$, called a branch, Lemma 9.5.1 implies that $(B, \leq)$ is well-ordered. Let b be the initial element in B. For any fixed element $x \in M_b$, from the fact that (B, $\leq$) is well-ordered and AC, it follows that there exists an element $\{[n_d]\}_{d \in B} \in \prod\{M_d/E_d : d \in B\}$ such that

$$\ell_{bd}([n_d]) = \{x\} \tag{9.157}$$

and that there exists an element $\{g_d\}_{d \in B} \in \prod\{[n_d] : d \in B\}$ such that

$$\ell_{ji}(g_i) = g_j \tag{9.158}$$

for all $i,j \in B$ with $i \geq j$.

From AC it follows that there exists a subset $M(B) \subset \bigcup\{\prod\{[n_d]_{d \in B} : \forall d \in B(\ell_{bd}(n_d) = x)\} : x \in M_b\}$ such that, for each $x \in M_b$,

$$M(B) \cap \prod\{[n_d]_{d \in B} : \forall d \in B(\ell_{bd}(n_d) = x)\} = \{(g_d)_{d \in B}\} \tag{9.159}$$

where $\ell_{ts}(g_s) = g_t$, for all $s, t \in B$ with $s \geq t$. Let $p_d : M(B) \to M_d$, for each $d \in B$, be the projection, and define

$$S'_d(B) = (p_d(M(B)), \{\emptyset\}) \tag{9.160}$$

Then $\{S'_d(B), \ell_{ts}|S'_s(B), B\}$ is a hierarchy of systems and a partial linked hierarchy of $\{S, \ell_{ts}, B\}$.

Let X be the set of all branches in the set $D^* \cup T_0$. By transfinite induction, hierarchies $\{S'_d(B), \ell_{ts}, B\}$ of systems, for each $B \in X$, can be defined such that for any branches $C, B \in X$ and $d \in C \cap B$,

$$S'_d(B) = S'_d(C) \tag{9.161}$$

Here we used the hypothesis that T is a pseudotree. Therefore, for each $d \in D^* \cup T_0$, we can define

$$S'_d = S'_d(B) \tag{9.162}$$

where B is a branch containing d. Then S'_d is a partial system of S_d and $\{S', \ell_{ts}|S'_s, D^* \cup T_0\}$ is a nontrivial linked hierarchy of systems such that the state $S'(t) = S'_t$ is embeddable in the state $S'(s) = S'_s$, for any $t, s \in D^* \cup T_0$ satisfying $t \leq s$.

For any $t \in T - (D^* \cup T_0)$, choose $s \in D^*$ such that $s \in D^*_\alpha$, where α is the least ordinal in η with this property, and $s > t$. We take a partial system S'_t of the system S_t defined by

$$S'_t = (\ell_{ts}(M'_s), \{\emptyset\}) \tag{9.163}$$

Then it is not hard to check that $\{S', \ell_{ts}|S'_s,\ T\}$ is a nontrivial partial linked hierarchy of $\{S, \ell_{ts}, T\}$ such that the state $S'(t)$ is embeddable in the state $S'(s)$ for any $s, t \in T$, with $s \geq t$. ■

When the partially ordered set $(T, \leq)$ is a total ordering, each hierarchy S of systems over T is called a time system; each linked hierarchy $\{S, \ell_{ts}, T\}$ of systems is called a linked time system. In this case, T is called the time set (or time axis) of the time system S.

An ordered triple $(T, +, \leq)$ is called an ordered Abelian group if $(T, +)$ is an Abelian group and $(T, \leq)$ is a linearly ordered set such that for any elements $t_i, s_i \in T$, $i = 1, 2$,

$$t_1 \leq t_2 \quad \text{and} \quad s_1 \leq s_2 \rightarrow s_1 + t_1 \leq s_2 + t_2 \tag{9.164}$$

We conclude this section with a discussion of time systems over ordered Abelian groups. This kind of time system is called a system over a group. If the time system is also a linked time system, we call the system a linked system over a group.

Suppose $(T, +, \leq)$ is an ordered Abelian group. We let 0 be the identity in T, and, for any $x \in T$, $|x|$ indicates the absolute value of x, defined by

$$|x| = \max\{-x, x\} \tag{9.165}$$

where $-x$ is the additive inverse of x. For any integer n,

$$nx = \begin{cases} 0 & \text{if } n = 0 \\ \underbrace{x + x + \cdots + x}_{n \text{ times}} & \text{if } n > 0 \\ \underbrace{-x - x - \cdots - x}_{n \text{ times}} & \text{if } n < 0 \end{cases} \tag{9.166}$$

The following notation will be used:

$$T_{[a,b)} = \{x \in T : a \leq x < b\}$$
$$T_{(a,b]} = \{x \in T : a < x \leq b\}$$
$$T_{(\leftarrow,a)} = \{x \in T : x < a\}$$
$$T_{(\leftarrow,a]} = \{x \in T : x \leq a\}$$
$$T_{(a,\rightarrow)} = \{x \in T : x > a\}$$
$$T_{[a,\rightarrow)} = \{x \in T : x \geq a\}$$

Let S be a system over a group and $t_0 \in T$ a fixed element. The right transformation of S, denoted S^{-t_0}, is defined by $S^{-t_0}(t) = S(t - t_0)$ for each $t \in T$; the left transformation of S, denoted S^{+t_0}, is defined by $S^{+t_0}(t) = S(t + t_0)$ for each $t \in T$.

Theorem 9.5.5. *Suppose that S is a system over a group. Then for any $a, b \in T$,*

$$(S^{+a})^{+b} = S^{+(a+b)} \quad \textit{and} \quad (S^{-a})^{+b} = S^{-(a-b)}$$

A system S over a group T is periodic if there exists $p \in T$ such that $p \neq 0$ and $S^{+p} = S$. The element p is called a period of S.

Theorem 9.5.6. *Let S be a system over a group T. If S is periodic with $p \in T$ as its period, then for each integer n, $S^{np} = S$, where $S^{np} = S^{+np}$ if $n \geq 0$, and $S^{np} = S^{-[(-n)p]}$ if $n < 0$.*

Proof: We show the theorem by induction on n.

Case 1: $n \geq 0$. If $n = 0$, the result is clear. Suppose $S^{+kp} = S$ for each integer k satisfying $0 \leq k < n$. Then it follows that

$$\begin{aligned} S^{np} &= S^{+[(n-1)p+p]} \\ &= [S^{+[(n-1)p]}]^{+p} \\ &= S^{+p} \\ &= S \end{aligned}$$

Thus, by induction, we have $S^{+np} = S$ for each $n \geq 0$.

Case 2: $n < 0$. If $n = -1$, in the condition that $S(t+p) = S(t)$, for each $t \in T$, let $t = x - p$. Then $S(x) = S(x - p)$ for each $x \in T$. This says that $S^{-p} = S$. Suppose $S^{-kp} = S$ for each integer k satisfying $0 \leq k < -n$. Then it follows that

$$\begin{aligned} S^{np} &= S^{-[(-n)p]} \\ &= S^{-[(-n-1)p+p]} \\ &= (S^{-[(-n-1)p]})^{-p} \\ &= S^{-p} \\ &= S \end{aligned}$$

Hence, by induction we have showed that $S^{np} = S$ for each integer $n < 0$.

Combining cases 1 and 2 completed the proof. ■

Corollary 9.5.1. *If a system S over a group is periodic, there then is a positive period for S.*

A linked system $\{S, \ell_{ts}, T\}$ over a group is periodic, if S is a periodic system over a group with a period τ such that $\ell_{ts} = \ell_{t-\tau,s-\tau}$ for any $t,s \in T$ satisfying $s \geq t$. This element τ in T is called a period of the linked system $\{S, \ell_{ts}, T\}$.

Theorem 9.5.7. *If τ is a period of a linked system $\{S, \ell_{ts}, T\}$ over a group, then so is $n\tau$ for each integer $n \neq 0$.*

The proof is straightforward and is omitted.

Theorem 9.5.8. *Suppose that a linked system $\{S, \ell_{ts}, T\}$ over a group is periodic with $\tau > 0$ as a period. Then the hierarchy $H = \{S, \ell_{ts}, T_{[0,\tau]}\}$ is similar to the hierarchy $K = \{S, \ell_{ts}, T_{[n\tau,(n+1)\tau]}\}$ for any integer n.*

Proof: First we show that $n\tau < (n+1)\tau$ for each integer n. In fact, if $n\tau \geq (n+1)\tau$ for some integer n, then

$$0 = n\tau - n\tau \geq (n+1)\tau - n\tau = \tau \tag{9.167}$$

This contradicts the hypothesis that $\tau > 0$.

Second, we define a mapping $\phi : T_{[n\tau,(n+1)\tau]} \to T_{[0,\tau]}$ by

$$\phi(t) = t - n\tau \tag{9.168}$$

Then ϕ is a similarity mapping from the ordered set $T_{[n\tau,(n+1)\tau]}$ onto the ordered set $T_{[0,\tau]}$. We next define mappings $f_x : S_{\phi(x)} \to S_x$, for each $x \in T_{[n\tau,(n+1)\tau]}$, as identity mappings. Then $\{\phi, f_x\}$ is a similarity mapping from the hierarchy H into the hierarchy K. ■

Corollary 9.5.2. *In Theorem 9.5.8, the ordered sets $T_{[0,\tau]}$ and $T_{[n\tau,(n+1)\tau]}$ can be replaced by each of the following pairs:*

(1) $T_{(0,\tau]}$ and $T_{(n\tau,(n+1)\tau]}$

(2) $T_{(0,\tau)}$ and $T_{(n\tau,(n+1)\tau)}$

(3) $T_{[0,\tau)}$ and $T_{[n\tau,(n+1)\tau)}$.

Generally, for a periodic system $\{S, \ell_{ts}, T\}$ over a group, the hierarchies $\{S, \ell_{ts}, T_{(\leftarrow,0)}\}$ and $\{S, \ell_{ts}, T_{(0,\rightarrow)}\}$ are not similar.

9.6. Controllabilities

The concept of controllability of systems was originally introduced as a property of input–output systems. The main idea behind the concept says that an input–output system S is controllable provided that, if we need to obtain some output of the system with certain predetermined properties, we then can find some input such that the corresponding output satisfies the required properties. In this section we study different concepts of controllability not only of input–output systems but also of general systems.

Let $S = (M, R)$ be an input–output system with input space X and output space Y. In the following, the symbols (x,y) and $(x_0,x_1,\ldots,x_\alpha,\ldots,y_0,y_1,\ldots,y_\beta,\ldots)$ will be used congruently, where $x = (x_0,x_1,\ldots,x_\alpha,\ldots)$ and $y = (y_0,y_1,\ldots,y_\alpha,\ldots)$. We now define

$$o(X) = \{n : \exists r \in R \exists m \in \mathbf{ON}(r \subseteq X^n \times Y^m)\} \tag{9.169}$$

$$o(Y) = \{m : \exists r \in R \exists n \in \mathbf{ON}(r \subseteq X^n \times Y^m)\} \tag{9.170}$$

for any $i \in o(X)$,

$$X_{i,D} = \{x \in X^i : \exists r \in R \exists j \in o(Y)(r(x) \subseteq Y^j)\} \tag{9.171}$$

for any $j \in o(Y)$,

$$Y_{j,R} = \{y \in Y^j : \exists r \in R \exists x \in X_{i,D}((x,y) \in r)\} \tag{9.172}$$

and

$$\text{Dom}(X) = \bigcup\{X_{i,D} : i \in o(X)\} \tag{9.173}$$

and

$$\text{Ran}(Y) = \bigcup\{Y_{j,R} : j \in o(Y)\} \tag{9.174}$$

Suppose V is a set and G a mapping such that

$$G : \text{Dom}(X) \times \text{Ran}(Y) \to V \tag{9.175}$$

For an element $v \in V$, the system S is v-controllable relative to G if and only if there exists a mapping $\text{cg} : \text{Dom}(X) \to \text{Dom}(X)$ such that, for any $x \in \text{Dom}(X)$, there exists a $y_x \in \text{Ran}(Y)$ such that

$$G(\text{cg}(x), y_x) = v \tag{9.176}$$

The mapping cg is called a control function of S.

For example, suppose we are processing a chemical reaction. To reach a collection of desired results (denoted by y_x) which satisfy an assigned target (denoted by v), we could choose various methods (denoted by x) for processing the wanted results. But, in order to control the whole reaction procedure, we may add relative materials [denoted by $\mathrm{cg}(x)$] together with those "choices" we have had. We would like to deem the "input" (the relative materials) as "control" for the whole reaction procedure.

Theorem 9.6.1. *For each element $v \in V$, the input–output system S is v-controllable relative to the mapping G iff $v \in G(\mathrm{Dom}(X) \times \mathrm{Ran}(Y))$.*

Proof: The necessity part of the proof follows from Eq. (9.176).

Sufficiency. Let $v \in G(\mathrm{Dom}(X) \times \mathrm{Ran}(Y))$. Then there exists $(x,y) \in \mathrm{Dom}(X) \times \mathrm{Ran}(Y)$ such that $G(x,y) = v$. Define a control function of S, $\mathrm{cg} : \mathrm{Dom}(X) \to \mathrm{Dom}(X)$, as follows: $\mathrm{cg}(w) = x$ for all $w \in \mathrm{Dom}(X)$. Then for each $w \in \mathrm{Dom}(X)$ there exists $y_w = y \in \mathrm{Ran}(Y)$ such that

$$G(\mathrm{cg}(w),y_w) = G(x,y) = v$$

■

Let $V' \subset V$ be a subset. The input–output system S is controllable on V' relative to $G : \mathrm{Dom}(X) \times \mathrm{Ran}(Y) \to V$ iff there exists a mapping $\mathrm{cg} : \mathrm{Dom}(X) \to \mathrm{Dom}(X)$ such that, for each $v \in V'$ and any $x \in \mathrm{Dom}(X)$, there exists $y_z = y(v,x) \in \mathrm{Ran}(Y)$ such that

$$G(\mathrm{cg}(x),y_z) = v \tag{9.177}$$

The mapping cg is called a control function of S.

Theorem 9.6.2. *For a subset $V' \subset V$, the input–output system S is controllable on V' relative to the mapping G iff there exists an element $x \in \mathrm{Dom}(X)$ such that $G(\{x\} \times \mathrm{Ran}(Y)) \supset V'$.*

Proof: Necessity. Let $\mathrm{cg} : \mathrm{Dom}(X) \to \mathrm{Dom}(X)$ be a control function of the system S. Then, for each $v \in V'$ and each $x \in \mathrm{cg}(\mathrm{Dom}(X))$, there exists a $y \in \mathrm{Ran}(Y)$ such that $G(x,y) = v$. Hence, $G(x, \mathrm{Ran}(Y)) \supset V'$.

Sufficiency. Suppose there exists an element $x \in \mathrm{Dom}(X)$ such that $G(\{x\} \times \mathrm{Ran}(Y)) \supset V'$. Define a control function $\mathrm{cg} : \mathrm{Dom}(X) \to \mathrm{Dom}(X)$ by letting $\mathrm{cg}(w) = x$ for all $w \in \mathrm{Dom}(X)$. Clearly, each $v \in V'$ and any $w \in \mathrm{Dom}(X)$, there exists a $y \in \mathrm{Ran}(Y)$ such that

$$G(\mathrm{cg}(w),y) = G(x,y) = v$$

Therefore, S is controllable on V' relative to G. ■

For $V' \subset V$, the input–output system S is collectively controllable on V' relative to the mapping $G : \mathrm{Dom}(X) \times \mathrm{Ran}(Y) \to V$ iff there exists a mapping $\mathrm{cg} : \mathrm{Dom}(X) \to p(\mathrm{Dom}(X))$, where $p(\mathrm{Dom}(X))$ is the power set of $\mathrm{Dom}(X)$, such that for any $v \in V'$ and any $x \in \mathrm{Dom}(X)$, there exists $y = y(v,x) \in \mathrm{Ran}(Y)$ such that

$$G(z,y) = v \quad \text{for some } z \in \mathrm{cg}(x) \tag{9.178}$$

Theorem 9.6.3. *For a subset $V' \subset V$, the input–output system S is collectively controllable on V' relative to the mapping $G : \mathrm{Dom}(X) \times \mathrm{Ran}(Y) \to V$ iff there exists a nonempty subset $A \subseteq \mathrm{Dom}(X)$ such that $G(A \times \mathrm{Ran}(Y)) \supseteq V'$.*

Proof: Necessity. Let $\mathrm{cg} : \mathrm{Dom}(X) \to p(\mathrm{Dom}(X))$ be a control function for S. Then for any $v \in V'$ and each $x \in \mathrm{Dom}(X)$, there exists a $y \in \mathrm{Ran}(Y)$ such that $G(z,y) = v$ for some $z \in \mathrm{cg}(x)$. Hence, there exists an $x \in \mathrm{Dom}(X)$ such that $G(\mathrm{cg}(x) \times \mathrm{Ran}(Y)) \supseteq V'$, where $\mathrm{cg}(x) \neq \emptyset$.

Sufficiency. Suppose there exists a nonempty subset $A \subseteq \mathrm{Dom}(X)$ such that $G(A \times \mathrm{Ran}(Y)) \supseteq V'$. Let us define a control function $\mathrm{cg} : \mathrm{Dom}(X) \to p(\mathrm{Dom}(X))$ as follows: $\mathrm{cg}(x) = A$ for all $x \in \mathrm{Dom}(X)$. Then for any $v \in V'$ there exists $(z,y) \in A \times \mathrm{Ran}(Y)$ such that $G(z,y) = v$. This implies that, for any $v \in V'$ and any $x \in \mathrm{Dom}(X)$, there exists $y = y(v,x) \in \mathrm{Ran}(Y)$ such that $G(z,y) = v$. This completes the proof of the theorem. ■

Examples can be given to show that not all systems are input–output systems. In the following, we study the concept of controllability of general systems.

Suppose $S = (M,R)$ is a general system. For any relation $r \in R$ and any ordinal number $\mu > 0$, we define the μ-initial section of r, denoted by $I(r_\mu)$, as follows: For any $x \in I(r_\mu) \subset M^\mu$, there exists $y \in M^\nu$ such that $(x,y) \in r$, where ν is a nonzero ordinal number such that $\mu + \nu = n(r)$ such that $r \subset M^{n(r)}$. We define the μ-final section of r, denoted by $E(r_\mu)$, as follows: for any $x \in E(r_\mu) \subset M^\nu$, there exists $y \in I(r_\mu)$ such that $(y,x) \in r$.

Let

$$^{I}r = \{x : \exists \mu \in n(r)(\mu \neq 0 \wedge x \in I(r_\mu))\} \tag{9.179}$$

$$^{E}r = \{x : \exists \mu \in n(r)(\mu \neq 0 \wedge x \in E(r_\mu))\} \tag{9.180}$$

$$\tag{9.181}$$

and

$$^{I}R = \bigcup\{^{I}r : r \in R\} \quad \text{and} \quad ^{E}r = \bigcup\{^{E}r : r \in R\} \tag{9.182}$$

If $M^* \subset M$,

$$^{I}R|M^* = {}^{I}R(M^*) \quad \text{and} \quad {}^{E}R|M^* = {}^{E}R(M^*) \tag{9.183}$$

are used to indicate all the elements in ^{I}R and all the elements in ^{E}R, respectively, in which only the elements of M^* appear as coordinates.

Let V be an another set and $V' \subset V$. The general system S is generally controllable on V' relative to a mapping

$$G : {}^{I}R \times {}^{E}R \to V \tag{9.184}$$

if, for each $v \in V'$, there exist a set W_v, an $M_v \subset M$, and a mapping

$$C_v : W_v \to p({}^{I}R(M_v)) \tag{9.185}$$

such that for each $u \in W_v$ there exists a $y_u \in {}^{E}R$ such that

$$G(z, y_u) = v \quad \text{for some } z \in C_v(u) \tag{9.186}$$

The mapping C_v is called a control function of S.

The main idea of general controllability of systems is the following. Suppose we are given a set of desired targets V. For each $v \in V$, there exists a set of "commands" given to a system S so that S can realize the desired target v. When S receives the commands, there will be some "well-organized" objects of S to accomplish the task. Therefore, there should be some "receivers" and "accomplishers" in S such that the whole process of executing the task is under a good control with a "fine" cooperation between the "receivers" and the "accomplishers."

Theorem 9.6.4. *Let S be a general system and $V' \subset V$ a subset. Then S is generally controllable on V' relative to a mapping $G : {}^{I}R \times {}^{E}R \to V$ iff there exists a nonempty subset $M_{V'} \subset V$ such that for some nonempty $A \in p({}^{I}R(M_{V'}))$, $G(A \times {}^{E}R) \supset V'$.*

Proof: Necessity. Suppose S is generally controllable on V' relative to G. Then for each $v \in V'$ there exist $M_v \subset M$ and a nonempty $A_v \in p({}^{I}R(M_v))$ such that

$$G(A_v \times {}^{E}R) \supset \{v\} \tag{9.187}$$

Let $M_{V'} = \bigcup\{M_v : v \in V'\}$ and $A = \bigcup\{A_v : v \in V'\}$. Then

$$A \in p(\bigcup\{{}^{I}R(M_v) : v \in V'\}) \subset p({}^{I}R(M_{V'})) \tag{9.188}$$

and from Eq. (9.187) we have $G(A \times {}^{E}R) \supset V'$.

Sufficiency. Suppose there exists a nonempty subset $M_{V'} \subset M$ such that for some nonempty $A \in p({}^IR(M_{V'}))$, $G(A \times {}^ER) \supset V'$. Thus, for each $v \in V'$ there are subsets $A_v \subset A$ and $B_v \subset {}^ER$ such that

$$G(A_v \times B_v) = \{v\} \tag{9.189}$$

Define $M_v \subset M$ by $M_v = \{m \in M : \exists x \in A_v \exists \mu \in \mathbf{ON}(m = x_\mu)\}$. Let W_v be any nonempty set and $C_v : W_v \to p({}^IR(M_v))$ a mapping defined by $C_v(u) = A_v$ for each $u \in W_v$. Then for each $u \in W_v$ there exists a $y_u \in B_v \subset {}^ER$ such that

$$G(z,y_u) = v \quad \text{for some } z \in C_v(u)$$

This proves the assertion. ■

Proposition 9.6.1. *Suppose that $S = (M,R)$ is an input–output system with input space X and output space Y, and that S is v-controllable relative to a mapping $G : \mathrm{Dom}(X) \times \mathrm{Ran}(Y) \to V$. Then S is generally controllable on $\{v\}$ relative to the mapping $G^* : {}^IR \times {}^ER \to V$, where G^* is an extension of G.*

Proof: Because $\mathrm{Dom}(X) \subset {}^IR$ and $\mathrm{Ran}(Y) \subset {}^ER$, we can define extensions $G^* : {}^IR \times {}^ER \to V$ for the mapping G.

We now show the general controllability of S. For $v \in \{v\}$ there exist a set $W_v = \mathrm{Dom}(X)$ and a mapping

$$C_v : W_v \to p({}^IR) \tag{9.190}$$

defined by $C_v(u) = \{\mathrm{cg}(u)\}$, where $\mathrm{cg} : \mathrm{Dom}(X) \to \mathrm{Dom}(X)$ is a control function for S. Hence, for each $u \in W_v$ there exists a $y_u \in \mathrm{Ran}(Y) \subset {}^ER$ such that

$$G^*(\mathrm{cg}(u),y_u) = G(\mathrm{cg}(u),y_u) = v \tag{9.191}$$

where $\mathrm{cg}(u) \in C_v(u)$. ■

Proposition 9.6.2. *Let $V' \subset V$ be a subset such that an input–output system $S = (M,R)$ is controllable on V' relative to a mapping $G : \mathrm{Dom}(X) \times \mathrm{Ran}(Y) \to V$, where X and Y are the input and the output spaces of S, respectively. Then S is generally controllable on V' relative to each extension G^* of G.*

Proof: For each $v \in V'$ there exist sets $W_v = \mathrm{Dom}(X)$ and $M_v = M$ and a mapping

$$C_v : W_v \to p({}^IR) \tag{9.192}$$

defined by $C_v(u) = \{\mathrm{cg}(u)\}$ for each $u \in W_v$. Then for each $u \in W_u$, there exists $y_u = y(v,u) \in \mathrm{Ran}(Y) \subset {}^E R$ such that

$$G^*(\mathrm{cg}(u), y_u) = G(\mathrm{cg}(u), y_u) = v \tag{9.193}$$

where $\mathrm{cg}(u) \in C_v(u)$. ■

Proposition 9.6.3. *Suppose that $S = (M,R)$ is an input–output system with input space X and output space Y such that S is collectively controllable on V' relative to a mapping $G : \mathrm{Dom}(X) \times \mathrm{Ran}(Y) \to V$. Let $G^* : {}^I R \times {}^E R \to V$ be an extension of G. Then S is generally controllable on V' relative to G^*.*

Proof: For each $v \in V'$ there exist sets $W_v = \mathrm{Dom}(X)$ and $M_v = M$ and a mapping

$$C_v : W_v \to p({}^I R) \tag{9.194}$$

defined by $C_v(u) = \mathrm{cg}(u)$ such that, for each $u \in W_v$, there exists $y_u = y(v,u) \in \mathrm{Ran}(Y) \subset {}^E R$ such that

$$G^*(z, y_z) = G(z, y_u) = v \tag{9.195}$$

for some $z \in C_v(u) = \mathrm{cg}(u)$. ■

Theorem 9.6.5. *Suppose that $h : S_1 = (M_1, R_1) \to S = (M,R)$ is an S-continuous mapping from a system S_1 into a system S such that S is generally controllable on V' relative to a mapping $G : {}^I R \times {}^E R \to V$. Then S_1 is generally controllable on V' relative to a mapping $G_1 : {}^I R_1 \times {}^E R_1 \to V$.*

Proof: First, we construct a desired mapping $G_1 : {}^I R_1 \times {}^E R_1 \to V$. Let $v_0 \in V$ be a fixed element. For any $x \in {}^I R_1$ and $y \in {}^E R_1$, if there exist relations $r_x, r_y \in R$ such that

$$x \in {}^I(h^{-1}(r_x)) \quad \text{and} \quad y \in {}^E(h^{-1}(r_y)) \tag{9.196}$$

then define

$$G_1(x,y) = G(h(x), h(y)) \tag{9.197}$$

If either r_x or r_y or both do not exist, we define

$$G_1(x,y) = v_0 \tag{9.198}$$

It can be shown that $G_1 : {}^I R_1 \times {}^E R_1 \to V$ is well defined.

Second, we show that S_1 is generally controllable on V' relative to $G_1 : {}^I R_1 \times {}^E R_1 \to V$. Suppose that for each $v \in V'$ there exist sets W_v and $M_v \subset M$ and a mapping $C_v : W_v \to p({}^I R(M_v))$ such that, for each $u \in W_v$, there exists a $y_u \in {}^E R$ such that $G(z, y_u) = v$ for some $z \in C_v(u)$. We now let $h^{-1}(M_v) \subset M_1$ and define a control function ${}^1C_v : W_v \to p({}^I R_1(h^{-1}(M_v)))$ as follows: For each $u \in W_v$,

$$ {}^1C_v(u) = \{x \in {}^I R_1(h^{-1}(M_v)) : h(x) \in C_v(u)\} \tag{9.199} $$

Now it only remains to show that ${}^1C_v$ is a control function of S_1. In fact, for each $u \in W_v$ there exists $y_u \in {}^E R$ such that

$$ G(z, y_u) = v \quad \text{for some } z \in C_v(u) \tag{9.200} $$

so there exists ${}^1y_u \in h^{-1}(y_u) \subset {}^E R_1$ such that

$$ \begin{aligned} G_1(z', {}^1y_u) &= G(h(z'), h({}^1y_u)) \\ &= G(z, y_u) \\ &= v \end{aligned} $$

where $z' \in h^{-1}(z) \subset {}^1C_v(u)$. Therefore, S_1 is generally controllable on V' relative to G_1. ∎

Theorem 9.6.6. *Let $\{S_i : i \in I\}$ be a set of systems, where I is an index set, such that there exists an $i_0 \in I$ such that the system S_{i_0} is generally controllable on V' relative to a mapping $G_{i_0} : {}^I R_{i_0} \times {}^E R_{i_0} \to V$. Then the Cartesian product system $\prod_{cp}\{S_i : i \in I\}$ is also generally controllable on V' (relative to a different mapping $G : {}^I R \times {}^E R \to V$), where it is assumed that $S_{i_0} = (M_{i_0}, R_{i_0})$ and $\prod_{cp}\{S_i : i \in I\} = (M, R)$.*

The proof follows from Theorems 9.3.10 and 9.6.5.

Theorem 9.6.7. *Suppose that $h : S = (M, R) \to S_1 = (M_1, R_1)$ is an embedding mapping from a system S into a system S_1 such that S is generally controllable on V' relative to a mapping $G : {}^I R \times {}^E R \to V$. Then, S_1 is generally controllable on V' relative to a mapping $G_1 : {}^I R_1 \times {}^E R_1 \to V$.*

Proof: We first construct a desired mapping $G_1 : {}^I R_1 \times {}^E R_1 \to V$. Let $v_0 \in V$ be a fixed element. For any $x \in {}^I R_1$ and $y \in {}^E R_1$, if there exist $x^* \in {}^I R$ and $y^* \in {}^E R$ such that $h(x^*) = x$ and $h(y^*) = y$, we define

$$ G_1(x, y) = G(x^*, y^*) \tag{9.201} $$

From the hypothesis that h is 1–1, we see that Eq. (9.201) is well-defined, for the existence of x^* and y^* is unique. If one or both of the elements x^* and y^* do not exist, then we define

$$G_1(x,y) = v_0 \tag{9.202}$$

We then have defined a mapping $G_1 : {}^I R_1 \times {}^E R_1 \to V$.

Second, we show that the system S_1 is generally controllable on V' relative to the mapping G_1. Suppose that for each $v \in V'$ there exist sets W_v and $M_v \subset M$ and a mapping $C_v : W_v \to p({}^I R(M_v))$ such that, for each $u \in W_v$, there exists a $y_u \in {}^E R$ such that $G(z,y_u) = v$ for some $z \in C_v(u)$. We now let $h(M_v) \subset M_1$ and define a control function ${}^1C_v : W_v \to p({}^I R_1(h(M_v)))$ as follows: For each $u \in W_v$,

$${}^1C_v(u) = \{h(z) : z \in C_v(u)\} \tag{9.203}$$

It remains to show that ${}^1C_v$ is indeed a control function for S_1. For each $u \in W_v$ there exists a $y_u \in {}^E R$ such that

$$G(z,y_u) = v \quad \text{for some } z \in C_v(u) \tag{9.204}$$

Therefore,

$$G(h(z),h(y_u)) = G(z,y_u) = v \tag{9.205}$$

where $h(z) \in {}^1C_v(u)$ and $h(y_u) \in {}^E R_1$. ■

Corollary 9.6.1. *If a subsystem $A = (M_A, R_A)$ of a system $S = (M,R)$ is generally controllable on V' relative to a mapping $G_A : {}^I R_A \times {}^E R_A \to V$, then S is generally controllable on V' relative to a mapping $G : {}^I R \times {}^E R \to V$.*

Theorem 9.6.8. *Let $\{S_i = (M_i,R_i) : i \in I\}$ be a set of systems and $\{V_i \subset V : i \in I\}$ a set of nonempty subsets of V. If each system S_i is generally controllable on the subset V_i relative to a mapping $G_i : {}^I R_i \times {}^E R_i \to V$, then $\oplus\{S_i : i \in I\} = (M,R)$ is generally controllable on $\bigcup\{V_i : i \in I\}$ relative to a mapping $G : {}^I R \times {}^E R \to V$, where for each $i \in I$ and any $(x,y) \in {}^I R_i \times {}^E R_i$, $G(x,y) = G_i(x,y)$.*

Proof: From the definition of free sums of systems, it follows that

$${}^I R = \bigcup\{{}^I R_i : i \in I\} \quad \text{and} \quad {}^E R = \bigcup\{{}^E R_i : i \in I\} \tag{9.206}$$

where ${}^I R_i \cap {}^I R_j = \emptyset = {}^E R_i \cap {}^E R_j$ for any distinct $i,j \in I$. Let G be an extension of the mappings $\{G_i : i \in I\}$; i.e., G is a mapping from ${}^I R \times {}^E R$ into V such that for any $(x,y) \in {}^I R \times {}^E R$, if $(x,y) \in {}^I R_i \times {}^E R_i$, then $G(x,y) = G_i(x,y)$.

To show that the free sum $\oplus\{S_i : i \in I\}$ is generally controllable on $\bigcup\{V_i : i \in I\}$ relative to G, we pick an arbitrary $v \in \bigcup\{V_i : i \in I\}$. Then there exists an $i \in I$ such that $v \in V_i$. Therefore, there exist sets W_v and $M_v \subset M_i \subset \bigcup\{M_j : j \in I\}$ and a mapping

$$C_v : W_v \to p({}^IR(M_v)) = p({}^IR_i(M_v)) \tag{9.207}$$

such that for each $u \in W_v$, there exists a $y_u \in {}^ER_i \subset {}^ER$ such that

$$G(z, y_u) = v \quad \text{for some } z \in C_v(u) \tag{9.208}$$

This proves the theorem. ■

In the following, we conclude this section with a discussion of controllabilities of time systems.

Theorem 9.6.9. *Suppose that $\{S, \ell_{ts}, T\}$ is a linked time system. If there exists a $t \in T$ such that*

(a) *S_t is a nondiscrete input–output system,*

(b) *$\ell_{st}(R_t) \cap R_s \neq \emptyset$, for each $s < t$,*

(c) *$\ell_{ts}(R_s) \cap R_t \neq \emptyset$, for each $s > t$,*

then each state S_s of the linked time system has a nondiscrete partial system which is an input–output system.

Proof: Let $S_t = (M_t, R_t)$ and X_t and Y_t be input and output spaces of the state S_t, respectively. If $s < t$, let $\overline{S}_s = (\overline{M}_s, \overline{R}_s)$ be a partial system of the system S_s defined by $\overline{M}_s = \ell_{st}(M_t)$ and $\overline{R}_s = \ell_{st}(R_t) \cap R_s$. Then $\overline{S}_s$ is a nondiscrete input–output system with input space $\ell_{st}(X_t)$ and output space $\ell_{st}(Y_t)$. Finally, if $s > t$, let $\overline{S}_s = (\overline{M}_s, \overline{R}_s)$ be a partial system of S_s defined by $\overline{M}_s = M_s$ and $\overline{R}_s = \{r \in R_s : \ell_{ts}(r) \in R_t\}$. Then $\overline{S}_s$ is a nondiscrete input–output system with input space $X_s = \{x \in M_s : \ell_{ts}(x) \in X_t\}$ and output space $Y_s = \{y \in M_s : \ell_{ts}(y) \in Y_t\}$. ■

Theorem 9.6.10. *Suppose that $\{S, \ell_{ts}, T\}$ is a linked time system such that*

(a) *T has an initial element t_0.*

(b) *Each $\ell_{t_0 s}$ is surjective with $\ell_{t_0 s}(R_s) \cap R_{t_0} \neq \emptyset$.*

(c) *S_{t_0} is a nondiscrete input–output system with input space X_{t_0} and output space Y_{t_0}.*

(d) *S_{t_0} is v-controllable relative to a mapping $G : \mathrm{Dom}(X_{t_0}) \times \mathrm{Ran}(Y_{t_0}) \to V$.*

Then there exists a linked time system $\{\overline{S},\overline{\ell}_{ts},T\}$ such that each state $\overline{S}_t$ is a nondiscrete input–output system and is v-controllable relative to some mapping G_t.

Proof: First we see the construction of the linked time system $\{\overline{S},\overline{\ell}_{ts},T\}$. By AC we well-order the time set T

$$t_0,t_1,\ldots,t_\alpha,\ldots,\alpha<|T| \tag{9.209}$$

where t_0 is the initial element in T. Suppose that the original order on T is $\leq$, and in Eq. (9.209) T is well-ordered by a well-order relation $\leq{}^*$. Generally, $\leq$ and $\leq{}^*$ are different.

Define $\overline{S}_{t_0}=S_{t_0}$ and $\overline{\ell}_{t_0t_0}=\ell_{t_0t_0}$. Suppose that for some $\alpha<|T|$ the system $\overline{S}_{tk}$ and the mappings $\overline{\ell}_{t_jt_r}$ have been defined such that $\overline{\ell}_{t_jt_i}=\overline{\ell}_{t_jt_r}\circ\overline{\ell}_{t_rt_i}$, for each $k<\alpha$ and any $i,r,j<\alpha$ such that $t_i\geq t_r\geq t_j$. Now define $\overline{S}_{t_\alpha}$ and $\{\overline{\ell}_{t_it_\alpha}:i<\alpha$ and $t_\alpha>t_i\}$ and $\{\overline{\ell}_{t_\alpha t_i}:i<\alpha$ and $t_\alpha<t_i\}$ as follows: Let E_{t_α} be the equivalence relation on the object set M_{t_α} defined by $E_{t_\alpha}=\{(x,y)\in M_{t_\alpha}^2:\ell_{t_0t_\alpha}(x)=\ell_{t_0t_\alpha}(y)\}$ and

$$\overline{S}_{t_\alpha}=(S_{t_\alpha})^*/E_{t_\alpha} \tag{9.210}$$

where $(S_{t_\alpha})^*=(M_{t_\alpha},{}^*R_{t_\alpha})$ and ${}^*R_{t_\alpha}=\{r\in R_{t_\alpha}:\ell_{t_0t_\alpha}(r)\in R_{t_0}\}$. For each $\beta<\alpha$, if $t_\alpha>t_\beta$, define

$$\overline{\ell}_{t_\beta t_\alpha}([m_\alpha])=[m_\beta] \tag{9.211}$$

for each $[m_\alpha]\in M_{t_\alpha}/E_{t_\alpha}$, where $\ell_{t_0t_\alpha}(m_\alpha)=\ell_{t_0t_\beta}(m_\beta)$. If $t_\alpha<t_\beta$, we define

$$\overline{\ell}_{t_\alpha t_\beta}([m_\beta])=[m_\alpha] \tag{9.212}$$

for each $[m_\beta]\in M_{t_\beta}/E_{t_\beta}$, where $\ell_{t_0t_\beta}(m_\beta)=\ell_{t_0t_\alpha}(m_\alpha)$. It is then not hard to prove that $\{\overline{\ell}_{t_jt_i}:i,j\leq\alpha(t_i\geq t_j)\}$ is a family of linkage mappings.

By transfinite induction, a linked time system $\{\overline{S},\overline{\ell}_{ts},T\}$ is defined such that $\overline{S}_{t_0}=S_{t_0}$, and each state $\overline{S}_t$ is a nondiscrete input–output system with input space $\ell_{t_0t}^{-1}(X_{t_0})=\{[m_t]\in M_t/E_t:\ell_{t_0t}(m_t)\in X_{t_0}\}$ and output space $\ell_{t_0t}^{-1}(Y_{t_0})=\{[m_t]\in M_t/E_t:\ell_{t_0t}(m_t)\in Y_{t_0}\}$.

Second, we show that each state $\overline{S}_t$ is v-controllable relative to some mapping

$$G_t:\mathrm{Dom}(\ell_{t_0t}^{-1}(X_{t_0}))\times\mathrm{Ran}(\ell_{t_0t}^{-1}(Y_{t_0}))\to V \tag{9.213}$$

Let $\mathrm{Dom}(X_t)=\ell_{t_0t}^{-1}(X_{t_0})$ and $\mathrm{Ran}(Y_t)=\ell_{t_0t}^{-1}(Y_{t_0})$. For any $x_t\in\mathrm{Dom}(X_t)$ there exists $x_{t_0}\in\mathrm{Dom}(X_{t_0})$ such that

$$x_{t_0}=\mathrm{cg}_{t_0}\circ\overline{\ell}_{t_0t}(x_t) \tag{9.214}$$

where cg_{t_0} is a control function of S_{t_0}. Thus, there exists a $y_{t_0} \in \mathrm{Ran}(Y_{t_0})$ such that

$$G(x_{t_0}, y_{t_0}) = v \tag{9.215}$$

Let $\mathrm{cg}_t = \overline{\ell}_{t_0 t}^{-1} \circ \mathrm{cg}_{t_0} \circ \ell_{t_0 t}$ and $y_t = \overline{\ell}_{t_0 t}^{-1}(y_{t_0})$. If we assign $\overline{\ell}_{t_0 t}(x_t, y_t) = (\overline{\ell}_{t_0 t}(x_t), \overline{\ell}_{t_0 t}(y_t))$, then

$$G_t = G_{t_0} \circ \overline{\ell}_{t_0 t} : \mathrm{Dom}(X_t) \times \mathrm{Ran}(Y_t) \to V \tag{9.216}$$

and $G_t(\mathrm{cg}_t(x_t), y_t) = G_{t_0} \circ \overline{\ell}_{t_0 t}(\mathrm{cg}_t(x_t), y_t) = G_{t_0}(x_{t_0}, y_{t_0}) = v$. Therefore, the system $\overline{S}_t$ is v-controllable relative to the mapping G_t. ■

Theorem 9.6.11. *Assume hypotheses (a)–(c) of Theorem 9.6.10 and that S_{t_0} is controllable on $V' \subset V$ relative to a mapping $G : \mathrm{Dom}(X_{t_0}) \times \mathrm{Ran}(Y_{t_0}) \to V$. Then there exists a linked time system $\{\overline{S}, \overline{\ell}_{ts}, T\}$ such that each state $\overline{S}_t$ is a nondiscrete input–output system and is controllable on V' relative to some mapping G_t.*

The proof is a slight modification of that of Theorem 9.6.10 and is omitted.

Theorem 9.6.12. *Assume hypotheses (a)–(c) of Theorem 9.6.10 and that S_{t_0} is collectively controllable on $V^* \subset V$ relative to a mapping $G : \mathrm{Dom}(X_{t_0}) \times \mathrm{Ran}(Y_{t_0}) \to V$. Then there exists a linked time system $\{Q, \overline{\ell}_{ts}, T\}$ such that each state Q_t is a nondiscrete input–output system and is collectively controllable on V^* relative to some mapping G_t.*

The proof is a slight modification of that of Theorem 9.6.10 and is left to the reader.

9.7. Limit Systems

We say that $\mathcal{C}$ is a category if $\mathcal{C}$ consists of a class of objects, denoted by obj$\mathcal{C}$, pairwise disjoint sets of morphisms, denoted by $\mathrm{Hom}_{\mathcal{C}}(A,B)$, for every ordered pair of objects (A,B), and a composition $\mathrm{Hom}_{\mathcal{C}}(A,B) \times \mathrm{Hom}_{\mathcal{C}}(B,C) \to \mathrm{Hom}_{\mathcal{C}}(A,C)$, denoted by $(f,g) \to$ gf, satisfying the following axioms:

(i) For each object A there exists an identity morphism $I_A \in \mathrm{Hom}_{\mathcal{C}}(A,A)$ such that $fI_A = f$, for all $f \in \mathrm{Hom}_{\mathcal{C}}(A,B)$ and $I_A g = g$, for all $g \in \mathrm{Hom}_{\mathcal{C}}(C,A)$.

(ii) Associativity of composition holds whenever possible; that is, if $f \in \mathrm{Hom}_{\mathcal{C}}(A,B)$, $g \in \mathrm{Hom}_{\mathcal{C}}(B,C)$ and $h \in \mathrm{Hom}_{\mathcal{C}}(C,D)$, then

$$h(gf) = (hg)f \tag{9.217}$$

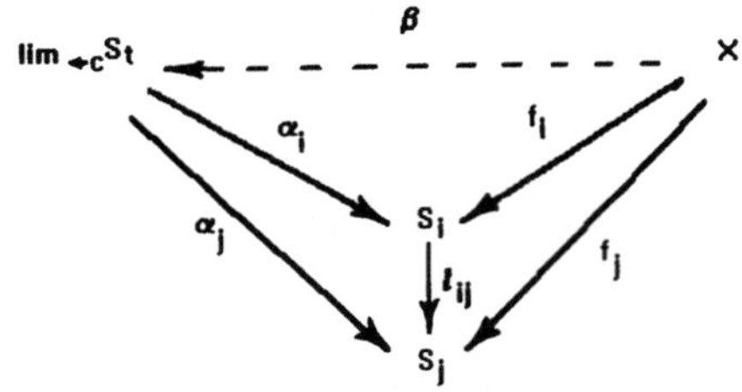

Figure 9.3. Universal mapping property of limit systems.

We give some examples of categories.

a. Suppose $\mathcal{GS}_C$ is the category consisting of all general systems and all S-continuous mappings between systems. Thus, for any two systems S_1 and S_2 the set $\mathrm{Hom}_{\mathcal{GS}_c}(S_1, S_2)$ equals the set of all S-continuous mappings from S_1 into S_2.

b. Suppose $\mathcal{GS}_h$ is the category consisting of all general systems and all homomorphisms between systems. Thus, for two arbitrary systems S_1 and S_2 the set $\mathrm{Hom}_{\mathcal{GS}_h}(S_1, S_2)$ of morphisms equals the set of all homomorphisms from S_1 into S_2.

Let $\{S, \ell_{ts}, T\}$ be an α-type hierarchy of systems in $\mathcal{GS}_C$ (respectively, in $\mathcal{GS}_h$), i.e., each linkage mapping ℓ_{ts} is an S-continuous mapping (respectively, a homomorphism). The inverse limit of this hierarchy, denoted by $\lim_{\leftarrow C}\{S, \ell_{ts}, T\}$ or $\lim_{\leftarrow C} S_t$ (respectively, $\lim_{\leftarrow h}\{S, \ell_{ts}, T\}$ or $\lim_{\leftarrow h} S_t$), is a system and a family of S-continuous mappings $\alpha_t : \lim_{\leftarrow C}\{S, \ell_{ts}, T\} \to S_t$ (respectively, homomorphisms $\alpha_t : \lim_{\leftarrow h}\{S, \ell_{ts}, T\} \to S_t$) with $\alpha_i = \ell_{ij} \circ \alpha_j$ whenever $i \leq j$ satisfying the universal mapping property, contained in Fig. 9.3 (respectively, the universal mapping property, contained in Fig. 9.4) satisfying: For every system X and S-continuous mappings (respectively, homomorphisms) $f_i : X \to S_i$ making the corresponding diagram commute whenever $i \leq j$, there exists a unique S-continuous mapping $\beta : X \to \lim_{\leftarrow C} S_t$ (respectively, a homomorphism $\beta : X \to \lim_{\leftarrow h} S_t$) making the diagram commute.

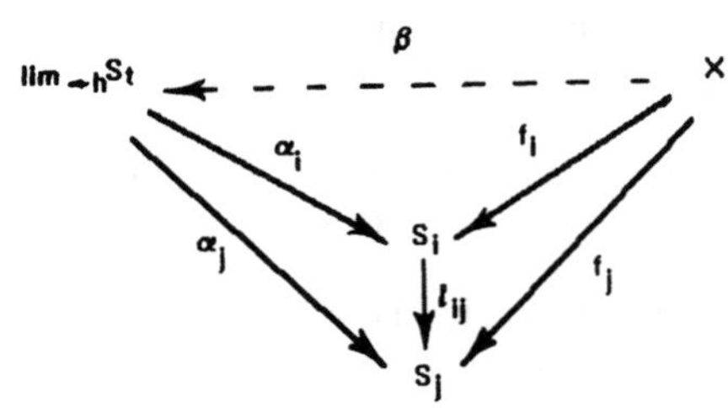

Figure 9.4. Universal mapping property of limit systems.

From the definition of inverse limits, it can be seen that if a linked hierarchy of systems has an inverse limit, the limit is unique up to a similarity.

Theorem 9.7.1. *Let $\{S, \ell_{ts}, T\}$ be a linked α-type hierarchy of systems in the category $\mathcal{GS}_C$. Then the inverse limit $\lim_{\leftarrow C}\{S, \ell_{ts}, T\}$ exists.*

Proof: Let $S_t = (M_t, R_t)$, for each $t \in T$. Define a desired system $\lim_{\leftarrow C} S_t = (M, R)$ as follows:

$$M = \{(m_t)_{t\in T} \in \prod_{t\in T} M_t : \forall s,t \in T(t \leq s \to \ell_{ts}(m_s) = m_t)\} \tag{9.218}$$

and

$$R = \{r : \exists j \in T \exists r_j \in R_j (x \in r \leftrightarrow (x \in M^{n(r_j)} \wedge (x_{jk})_{k<n(r_j)} \in r_j))\} \tag{9.219}$$

and define mappings $\alpha_i : \lim_{\leftarrow C} S_t \to S_i$ by letting

$$\alpha_i((m_t)_{t\in T}) = m_i \tag{9.220}$$

for every object $(m_t)_{t\in T} \in M$ and any $i \in T$. Then it can be checked that the mappings α_t are S-continuous and satisfy $\alpha_t = \ell_{ts} \circ \alpha_s$ whenever $t \leq s$.

We now let $Y = (M_Y, R_Y)$ be a system and $\{f_t : Y \to S_t : t \in T\}$ a family of S-continuous mappings such that $f_t = \ell_{ts} \circ f_s$ whenever $t \leq s$. Define a mapping $\beta : Y \to \lim_{\leftarrow C} S_t$ by letting

$$\beta(y) = (f_t(y))_{t\in T} \tag{9.221}$$

for each $y \in M_Y$. Then from the hypothesis that $f_t = \ell_{ts} \circ f_s$ whenever $t \leq s$, it follows that $\beta : Y \to \lim_{\leftarrow C} S_t$ is well defined.

We now show that $\beta : Y \to \lim_{\leftarrow C} S_t$ is S-continuous. For any relation $r \in R$ there exist $j \in T$ and $r_j \in R_j$ such that $r = \{x \in M^{n(r_j)} : (x_{jk})_{k<n(r_j)} \in r_j\}$. Therefore,

$$\begin{aligned}
\beta^{-1}(r) &= \{\beta^{-1}(x) : x \in M^{n(r_j)} \wedge (x_{jk})_{k<n(r_j)} \in r_j\} \\
&= \{(y_k)_{k<n(r_j)} \in M^{n(r_j)} : f_j((y_k)_{k<n(r_j)}) \in r_j\} \\
&= f_j^{-1}(r_j) \\
&\in R_Y
\end{aligned}$$

From the definition of S-continuity it follows that $\beta : Y \to \lim_{\leftarrow C} S_t$ is S-continuous. The uniqueness of β follows from the commutative diagram in the theorem or Fig. 9.3. ■

Corollary 9.7.1. *Suppose that $\{S, \ell_{ts}, T\}$ is a linked α-type hierarchy of systems such that each linkage mapping ℓ_{ts} is S-continuous. Then $\lim_{\leftarrow C} S_t$ is a partial system of $\prod_{\mathrm{cp}}\{S_t : t \in T\}$.*

The proof follows from the proofs of Theorems 9.7.1 and 9.3.10.

Theorem 9.7.2. *Let $\{S, \ell_{ts}, T\}$ be a linked α-type hierarchy of systems in the category $\mathcal{GS}_h$. Then the inverse limit $\lim_{\leftarrow h}\{S, \ell_{ts}, T\}$ exists.*

Proof: We define a system $\lim_{\leftarrow h} S_t = (M, R)$ as follows:

$$M = \{(m_t)_{t\in T} \in \prod_{t\in T} M_t : \forall t, s \in T(t \le s \to \ell_{ts}(m_s) = m_t)\} \tag{9.222}$$

and

$$R^* = \{(r_t)_{t\in T} \in \prod_{t\in T} R_t : \forall t, s \in T(t \le s \to \ell_{ts}(r_s) = r_t)\} \tag{9.223}$$

and for any $r^* = (r_t)_{t\in T} \in R^*$, define $n = n(r_t)$, for $t \in T$, and

$$r = \{x \in M^n : \exists (x_t)_{t\in T} \in \prod_{t\in T} r_t \forall \alpha < n(x(\alpha) = (x_t(\alpha))_{t\in T})\} \tag{9.224}$$

and

$$R = \{r : r^* \in R^*\} \tag{9.225}$$

Let $\alpha_t : M \to M_t$, for each $t \in T$, be the projection; i.e., $\alpha_t((m_t)_{t\in T}) = m_t$. Then, for any relation $r \in R$, there exists $r^* = (r_t)_{t\in T} \in R^*$ such that r is defined by Eq. (9.224). Thus, $\alpha_t(r) = \{\alpha_t(x) : x \in r\} = r_t$ for each $t \in T$. That implies that each α_t is a homomorphism from (M, R) into S_t.

We now show that the system (M, R) is the inverse limit of the linked hierarchy $\{S, \ell_{ts}, T\}$. Only two things need to be proven: (i) For any system $\overline{S} = (\overline{M}, \overline{R})$ and any set of homomorphisms $\{f_t : \overline{S} \to S_t : t \in T\}$ such that

$$f_t = \ell_{ts} \circ f_s \tag{9.226}$$

whenever $t \le s$, there exists a homomorphism $\beta : \overline{S} \to S$ so that

$$\alpha_t \circ \beta = f_t \tag{9.227}$$

for each $t \in T$. To prove (i), define $\beta : \overline{M} \to M$ by $\beta(\overline{m}) = (f_t(\overline{m}))_{t\in T}$. Then β is a homomorphism from $\overline{S}$ into S. Since for any relation $\overline{r} \in \overline{R}$, $\overline{x} \in \overline{r} \subseteq \overline{M}^{n(\overline{r})}$, then $\beta \circ \overline{x} = \{(f_t(\overline{x}(\alpha)))_{t\in T} : \alpha < n(\overline{r})\}$, so $\beta(\overline{r}) = \{\beta \circ \overline{x} : \overline{x} \in \overline{r}\} \in R$.

Since $\alpha_t \circ \beta(\overline{m}) = \alpha_t((f_t(\overline{m}))_{t\in T}) = f_t(\overline{m})$, for all $\overline{m} \in \overline{M}$, we have $\alpha_\lambda \circ \beta = f_t$, for each $t \in T$, as mappings between sets. On the other hand, for any relation $\overline{r} \in \overline{R}$ and any $\overline{x} \in \overline{r} \subset \overline{M}^{n(\overline{r})}$,

$$\alpha_t(\beta \circ \overline{x}) = \alpha_t(\{f_t(\overline{x}(\alpha)) : \alpha < n(\overline{r})\}_{t\in T}) = \{f_t(\overline{x}(\alpha)) : \alpha < n(\overline{r})\}$$

so

$$\alpha_t \circ \beta(\overline{r}) = \{f_t(\overline{x}(\alpha)) : \alpha < n(\overline{r}) \wedge \overline{x} \in \overline{r}\} = f_t(\overline{r})$$

This implies that $\alpha_t \circ \beta = f_t$ as mappings from the relation set $\overline{R}$ into the relation set R_t for each $t \in T$. Hence, $\alpha_t \circ \beta = f_t$, for each $t \in T$, as homomorphisms from systems into systems.

(ii) The homomorphism β is unique. Suppose $\beta' : \overline{S} \to S$ is another homomorphism satisfying $\alpha_t \circ \beta' = f_t$, for each $t \in T$. Let $\beta'(\overline{m}) = (m_t)_{t\in T} \in M$. Then $m_t = \alpha_t \circ \beta'(\overline{m}) = f_t(\overline{m}) = \alpha_t \circ \beta(\overline{m})$ for all $t \in T$. Thus, $\beta'(\overline{m}) = \beta(\overline{m})$ for all $\overline{m} \in \overline{M}$. This implies that $\beta' = \beta$ on the set $\overline{M}$, so β' and β induce the same homomorphism. Hence, $\beta' = \beta$ as homomorphisms. ∎

Theorem 9.7.3. *Each mapping $\{g, f_{s'}\}$ of a linked hierarchy $H = \{S, \ell_{ts}, T\}$ of systems into a linked hierarchy $H' = \{S', \ell_{t's'}, T'\}$ of systems induces a mapping from* $\lim_{\leftarrow} H$ *into* $\lim_{\leftarrow} H'$, *where* $\lim_{\leftarrow}$ *indicates either* $\lim_{\leftarrow C}$ *or* $\lim_{\leftarrow h}$.

Proof: For an object $x = (x_t)_{t\in T}$ in $\lim_{\leftarrow} H$ and every $s' \in T'$, define

$$y_{s'} = f_{s'}(x_{g(s')}) \tag{9.228}$$

The element $y = \{y_{s'}\}_{s'\in T'}$ thus obtained is an object in $\lim_{\leftarrow} H'$. Indeed, for any $s', t' \in T'$ satisfying $t' \leq s'$, then

$$\begin{aligned}\ell_{t's'}(y_{s'}) &= \ell_{t's'} \circ f_{s'}(x_{g(s')})\\ &= f_{t'} \circ \ell_{g(t')g(s')}(x_{g(s')})\\ &= f_{t'}(x_{g(t')})\\ &= y_{t'}\end{aligned}$$

For the object x in $\lim_{\leftarrow} H$ define $f(x) = y$ in $\lim_{\leftarrow} H'$. Then f is a mapping from $\lim_{\leftarrow} H$ into $\lim_{\leftarrow} H'$. ∎

The mapping $f : \lim_{\leftarrow} H \to \lim_{\leftarrow} H'$ in the proof of Theorem 9.7.3 is called the limit mapping induced by $\{g, f_{s'}\}$ and is denoted by $f = \lim_{\leftarrow}\{g, f_{s'}\}$.

Theorem 9.7.4. *Let $\{g, f_{s'}\}$ be a mapping of a linked hierarchy of systems $H = \{S, \ell_{ts}, T\}$ to a linked hierarchy of systems $H' = \{S', \ell_{t's'}, T'\}$. If all mappings $f_{s'}$ are one-to-one, the limit mapping $f = \lim_{\leftarrow}\{g, f_{s'}\}$ is also one-to-one. If, moreover, all mappings $f_{s'}$ are onto, f is also an onto mapping.*

Proof: Let $x = \{x_s\}_{s \in T}$ and $y = \{y_s\}_{s \in T}$ be two distinct objects of the limit system $\lim_{\leftarrow} H$. Take an $s_0 \in T$ such that $x_{s_0} \neq y_{s_0}$ and an $s' \in T'$ satisfying $s_0 \leq g(s')$. Clearly, $x_{g(s')} \neq y_{g(s')}$, so it follows from the hypothesis that $f_{s'}$ is one-to-one, that is, $f_{s'}(x_{g(s')}) \neq f_{s'}(y_{g(s')})$, which shows that $f(x) \neq f(y)$. Therefore, the limit mapping f is one-to-one.

We now suppose that $f_{s'}$ is a one-to-one mapping from $S_{g(s')}$ onto $S'_{s'}$ for every $s' \in T'$. Take an object $y = \{y_{s'}\}_{s' \in T'}$ from $\lim_{\leftarrow} H'$. It follows that $g(s') = g(t')$ implies $f_{s'}^{-1}(y_{s'}) = f_{t'}^{-1}(y_{t'})$ for any $s', t' \in T'$ satisfying $s' \geq t'$, so that for any $g(s') \in g(T')$, the element

$$z_{g(s')} = f_{s'}^{-1}(y_{s'}) \in M_{g(s')} \tag{9.229}$$

is well defined. For any $s \in T$ choose an element $s' \in T'$ such that $s \leq g(s')$ and define

$$x_s = \ell_{sg(s')}(z_{g(s')}) \in M_s \tag{9.230}$$

It can be verified that x_s does not depend on the choice of s', that $x = \{x_s\}_{s \in T}$ is an element in $\lim_{\leftarrow} H$, and that $f(x) = y$. That shows f is an onto mapping. ■

Theorem 9.7.5. *Suppose that $\{g, f_{s'}\}$ is a mapping of a linked hierarchy of systems $H = \{S, \ell_{ts}, T\}$ into a linked hierarchy of systems $H' = \{S', \ell_{t's'}, T'\}$, where H and H' are in the category $\mathcal{GS}_C$. If all mappings $f_{s'}$ are S-continuous, the limit mapping $f : \lim_{\leftarrow C} H \to \lim_{\leftarrow C} H'$ is also S-continuous.*

Proof: Suppose $\lim_{\leftarrow C} H = (M, R)$ and $\lim_{\leftarrow C} H' = (M', R')$. Theorem 9.7.3 implies that the limit mapping $f : (M, R) \to (M', R')$ is well defined. Let r be a relation in $\lim_{\leftarrow C} H'$. There exist $j \in T'$ and $r_j \in R'_j$ such that

$$r = \{((m_{ik})_{i \in T'})_{k < n(r_j)} \in (M')^{n(r_j)} : (m_{jk})_{k < n(r_j)} \in r_j\} \tag{9.231}$$

Then

$$\begin{aligned} f^{-1}(r) &= \bigcup \{ \prod \{ f^{-1}((m_{ik})_{i \in T'}) : k < n(r_j) \} : \\ &\qquad ((m_{ik})_{i \in T'})_{k < n(r_j)} \in (M')^{n(r_j)} \wedge (m_{jk})_{k < n(r_j)} \in r_j \} \\ &= \{ ((m_{ik})_{i \in T})_{k < n(r_j)} \in M^{n(r_j)} : (m_{jk})_{k < n(r_j)} \in (f_j)^{-1}(r_j) \} \\ &\in R \end{aligned}$$

Thus, from the definition of S-continuity, the limit mapping f is S-continuous. ■

Theorem 9.7.6. *Suppose that* $H = \{S, \ell_{ts}, T\}$ *and* $H' = \{S', \ell_{t's'}, T'\}$ *are two linked hierarchies of systems in the category* $\mathcal{GS}_C$*, and* $\{g, f_{s'}\}$ *is a mapping from* H *into* H'*. If all mappings* $f_{s'}$ *are similarity mappings, the limit mapping* $f : \lim_{\leftarrow C} H \to \lim_{\leftarrow C} H'$ *is also a similarity mapping.*

Proof: Applying Theorem 9.7.4, we know that the limit mapping f is a bijection from $\lim_{\leftarrow C} H$ onto $\lim_{\leftarrow C} H'$. Theorems 9.2.4 and 9.7.5 imply that f is also S-continuous. Now, in order to show that f^{-1} is S-continuous from $\lim_{\leftarrow C} H'$ onto $\lim_{\leftarrow C} H$, it suffices to show that f is a homomorphism from $\lim_{\leftarrow C} H$ into $\lim_{\leftarrow C} H'$.

Let $r \in R$ be a relation such that there exist $j \in T$ and $r_j \in R_j$ satisfying

$$r = \{((m_{ik})_{i\in T})_{k<n(r_j)} \in M^{n(r_j)} : (m_{jk})_{k<n(r_j)} \in r_j\} \tag{9.232}$$

Then

$$\begin{aligned} f(r) &= \{(f[(m_{ik})_{i\in T}])_{k<n(r_j)} : ((m_{ik})_{i\in T})_{k<n(r_j)} \in M^{n(r_j)} \text{ and } (m_{jk})_{k<n(r_j)} \in r_j\} \\ &= \{((f_i(m_{g(i)k}))_{i\in T'})_{k<n(r_j)} : ((m_{ik})_{i\in T})_{k<n(r_j)} \in M^{n(r_j)} \text{ and } (m_{jk})_{k<n(r_j)} \in r_j\} \\ &= \{((m_{ik})_{i\in T'})_{k<n(r_j)} \in (M')^{n(r_j)} : (m_{jk})_{k<n(r_j)} \in f_i(r_j)\} \\ &\in R' \end{aligned}$$

Therefore, the limit mapping f is a similarity mapping from $\lim_{\leftarrow C} H$ onto $\lim_{\leftarrow C} H'$. ■

Theorem 9.7.7. *Suppose that* $H = \{S, \ell_{ts}, T\}$ *and* $H' = \{S', \ell_{t's'}, T'\}$ *are two linked hierarchies of systems in the category* $\mathcal{GS}_h$*, and that* $\{g, f_{s'}\}$ *is a mapping from* H *into* H'*. If all mappings* $f_{s'}$ *are homomorphisms, the limit mapping* $f : \lim_{\leftarrow h} H \to \lim_{\leftarrow h} H'$ *is also a homomorphism.*

Proof: It suffices to show that for any relation r in the system $\lim_{\leftarrow h} H = (M, R)$, $f(r)$ is a relation in the system $\lim_{\leftarrow h} H' = (M', R')$. To this end, let $r^* = (r_t)_{t\in T} \in R^*$, where R^* is defined by Eq. (9.223), such that

$$r = \{x \in M^{n(r^*)} : \exists (x_t)_{t\in T} \in \prod_{t\in T} r_t \forall \alpha < n(r^*)(x(\alpha) = (x_t(\alpha))_{t\in T})\} \tag{9.233}$$

Then

$$\begin{aligned} f(r) = \{x \in (M')^{n(r^*)} : \exists (x_{t'})_{t'\in T'} \in \prod_{t'\in T'} f_{t'}(r_{g(t')}) \\ \forall \alpha < n(r^*)(x(\alpha) = (x_{t'}(\alpha))_{t'\in T'})\} \end{aligned} \tag{9.234}$$

Hence, $f(r) \in R'$. ■

Theorem 9.7.8. *Under the same hypotheses as in Theorem 9.7.7, if all mappings $f_{s'}$, in addition, are similarity mappings, the limit mapping $f : \lim_{\leftarrow h} H \to \lim_{\leftarrow h} H'$ is also a similarity mapping.*

Proof: By Theorems 9.7.4 and 9.7.7, it suffices to show that for any relation r' in the system $\lim_{\leftarrow h} H' = (M', R')$, there exists a relation r in the system $\lim_{\leftarrow h} H = (M, R)$ such that $f(r) = r'$. Pick an element $r'^* = (r_{t'})_{t' \in T'} \in R'^*$, where R'^* is defined by Eq. (9.223) with R_t, T, and ℓ_{ts} replaced by $R'_{t'}, T'$, and $\ell_{t's'}$, respectively, such that

$$r' = \{x \in (M')^{n(r'^*)} : \exists (x_{t'})_{t' \in T'} \in \prod_{t' \in T'} r_{t'}$$
$$\forall \alpha < n(r^*)(x(\alpha) = (x_{t'}(\alpha))_{t' \in T'})\} \tag{9.235}$$

From the hypothesis that each $f_{t'}$ is a similarity mapping from $S_{g(t')}$ onto $S'_{t'}$, it follows that there exists exactly one relation $r_{g(t')} \in R_{g(t')}$ such that

$$f_{t'}(r_{g(t')}) = r_{t'} \tag{9.236}$$

Now for any $t \in T$ choose a $t' \in T'$ such that $g(t') \geq t$. Define a relation $r_t \in R_t$ by

$$r_t = \ell_{tg(t')}(r_{g(t')}) \tag{9.237}$$

It can be shown that the definition of r_t does not depend on the choice of t'. Then an element $r^* = (r_t)_{t \in T} \in R^*$ is well defined, where R^* is defined by Eq. (9.223). The relation $r \in R$, defined by

$$r = \{x \in M^{n(r^*)} : \exists (x_t)_{t \in T} \in \prod_{t \in T} r_t$$
$$\forall \alpha < n(r^*)(x(\alpha) = (x_t(\alpha))_{t \in T})\} \tag{9.238}$$

satisfies $f(r) = r^*$. ■

Corollary 9.7.2. *Let $\{S, \ell_{ts}, T\}$ be a linked α-type hierarchy of systems in category $\mathcal{GS}_C$ or $\mathcal{GS}_h$. T' a cofinal subset of T. Then $\lim_{\leftarrow}\{S, \ell_{ts}, T\}$ is similar to $\lim_{\leftarrow}\{S, \ell_{ts}, T'\}$, where $\lim_{\leftarrow}$ indicates either $\lim_{\leftarrow h}$ or $\lim_{\leftarrow C}$.*

The result follows from Theorems 9.7.6 and 9.7.8.

Corollary 9.7.3. *Let $\{S, \ell_{ts}, T\}$ be a linked α-type hierarchy of systems in category $\mathcal{GS}_C$ or $\mathcal{GS}_h$. If the partially ordered set T has a maximum element $t_0 \in T$, then the systems $\lim_{\leftarrow h}\{S, \ell_{ts}, T\}$, $\lim_{\leftarrow C}\{S, \ell_{ts}, T\}$, and S_{t_0} are similar.*

The result follows from Corollary 9.7.2.

Corollary 9.7.4. *Suppose that* $\{S, \ell_{ts}, T\}$ *is a linked* α*-type hierarchy of systems in the category* $\mathcal{GS}_h$. *Then the limit system* $\lim_{\leftarrow h} S_t$ *is a partial system of* $\prod\{S_t : t \in T\}$.

The result follows from the proofs of Theorems 9.7.2 and 9.3.9.

Theorem 9.7.9. *Let* $\{S, \ell_{ts}, T\}$ *be a linked* α*-type hierarchy of systems in the category* $\mathcal{GS}_C$ *such that one of the states* S_t *is connected. Then the inverse limit system* $\lim_{\leftarrow C}\{S, \ell_{ts}, T\}$ *is connected.*

The result follows from Theorems 9.7.1 and 9.4.3.

Let $\{S, \ell_{ts}, T\}$ be a linked α-type hierarchy of systems such that for every element $m = (m_t)_{t \in T}$ in the thread set Σ, there exists a set $\{\ell_{ts}^m : m_s \to m_t : t, s \in T, s \geq t\}$ of linkage mappings, where each m_s is again a system. The 1-level inverse limit, denoted by $\lim^1_{\leftarrow c}\{S, \ell_{ts}, T\}$ or $\lim^1_{\leftarrow c} S_t$ when all relevant linkage mappings are S-continuous, or $\lim^1_{\leftarrow h}\{S, \ell_{ts}, T\}$ or $\lim^1_{\leftarrow h} S_t$ when all relevant linkage mappings are homomorphisms, is defined by $\lim_{\leftarrow} S_t = (M, R)$, where $\lim_{\leftarrow}$ indicates $\lim^1_{\leftarrow c}$ or $\lim^1_{\leftarrow h}$, and the object set M and the relation set R are defined as follows: There exists a bijection $h : \Sigma \to M$ such that

$$h((m_t)_{t \in T}) = \lim_{\leftarrow} m_t \in M \tag{9.239}$$

and

$$R = \{h(r) : r \text{ is a relation in } \lim_{\leftarrow} S_t\}$$

where $\lim_{\leftarrow}$ indicates either $\lim_{\leftarrow C}$ or $\lim_{\leftarrow h}$.

Theorem 9.7.10. *Let* $\{S, \ell_{ts}, T\}$ *be a linked* α*-type hierarchy of centralizable systems in the category* $\mathcal{GS}_C$. *If the* 1*-level inverse limit system* $\lim^1_{\leftarrow c}\{S, \ell_{ts}, T\}$ *exists, then the system* $\lim^1_{\leftarrow c}\{S, \ell_{ts}, T\}$ *is centralizable.*

Proof: The proof follows from the fact that if C_t is a center of the system S_t, for each $t \in T$, then the inverse limit $\lim_{\leftarrow C} C_t$ is a center of the 1-level inverse limit $\lim^1_{\leftarrow c}\{S, \ell_{ts}, T\}$. ■

Theorem 9.7.11. *Let* $\{S, \ell_{ts}, T\}$ *be a linked* α*-type hierarchy of centralizable systems in the category* $\mathcal{GS}_h$. *If the* 1*-level inverse limit system* $\lim^1_{\leftarrow h}\{S, \ell_{ts}, T\}$ *exists, then the limit system* $\lim^1_{\leftarrow h}\{S, \ell_{ts}, T\}$ *is also centralizable.*

Proof: This theorem follows from the fact that if C_t is a center of the system S_t, for each $t \in T$, then the inverse limit $\lim_{\leftarrow h} C_t$ is a center of the 1-level inverse limit $\lim^1_{\leftarrow h}\{S, \ell_{ts}, T\}$. ■

Theorem 9.7.12. *Suppose that* $\{S, \ell_{ts}, T\}$ *is a linked* α*-type hierarchy of systems in the category* $\mathcal{GS}_C$ *such that for some index* $t_0 \in T$, *the state* $S_{t_0} = (M_{t_0}, R_{t_0})$ *is generally controllable on* V' *relative to a mapping* $G : {}^I R_{t_0} \times {}^E R_{t_0} \to V$. *Then the inverse limit* $\lim_{\leftarrow C}\{S, \ell_{ts}, T\}$ *is also generally controllable (but relative to a different mapping* G^**).*

Proof: From Theorem 9.6.5 it follows that we need only show that the mapping $\alpha_{t_0} : \lim_{\leftarrow C}\{S, \ell_{ts}, T\} \to S_{t_0}$ is S-continuous. Theorem 9.7.1 implies this. ■

9.8. References for Further Study

The search for an ideal definition of general systems has been going on for decades: see von Bertalanffy (1972), Mesarovic and Tahakara (1975), Gratzer (1978), Klir (1985) and Lin (1987). The study of time systems can be found in many publications. For example, Schetzen (1980) studied continuous time systems of multilinearity; Rugh (1981) gave a systematic theory on nonlinear continuous and discrete time systems; a more general setting of the concept of time systems can be found in Mesarovic and Takahara (1975), and their definition was generalized by Ma and Lin (1987).

The constructions of new systems, such as free sum, products and limit systems, were first given by Ma and Lin (1987; 1987; 1988d). Studies on various controllabilities of systems can be found in almost all books on systems. For example, Deng (1985) studied the concept of gray control, Wonham (1979) presented a concept of controllability of linear systems of differential equations, and a study of controllabilities of nonlinear systems can be found in (Casti, 1985). Not all of the concepts of controllabilities are equivalent. The concepts of controllabilities presented here are based on research contained in (Mesarovic and Takahara, 1975; Ma and Lin, 1987). Periodic and order structures of families of systems were studied intensively in (Lin, 1988b; Lin, 1989a). Connectedness and S-continuity were first studied in (Lin, 1990c; Ma and Lin, 1990e). Cornacchio (1972) showed what an important role the concept of continuity played in research of general systems theory and its impacts on engineering, computer science, social and behavioral sciences, and natural sciences.

CHAPTER 10

Systems of Single Relations

As studied in Chapter 9, a (general) system consists of a set of objects and a set of relations between the objects. In practice, many systems, which can be described by one relation, have many complicated structures and properties that we still do not completely understand. In this chapter, we concentrate on systems of one relation and related concepts. Section 10.1 is devoted to the study of two very new concepts: chaos and attractor. The second section contains the general concept of linear systems and characterizations of feedback systems, and the following section introduces important concepts, including MT-time systems, linear time systems, stationary systems, and time invariably realizable systems, and study some feedback-invariant properties. Section 10.4 shows how "whole" systems can be decomposed into factor systems by feedback, and an example is given. Section 10.5 studies the conditions under which "whole" systems can be decomposed.

10.1. Chaos and Attractors

A system $S = (M, \{r\})$ is called an input–output system if there are nonzero ordinal numbers n and m such that

$$\emptyset \neq r \subset M^n \times M^m \tag{10.1}$$

Without loss of generality, we let $X = M^n$ and $Y = M^m$; instead of using the ordered pair $(M, \{r\})$, we will think of S as the binary relation r such that

$$\emptyset \neq S \subset X \times Y \tag{10.2}$$

where the sets X and Y are called the input space and the output space of S, respectively. In the real world, most systems we see are input–output systems. For example, each human being and each factory are input–output systems.

If we let $Z = X \cup Y$, the input–output system S in Eq. (10.2) is a binary relation on Z. Let $D \subset Z$ be an arbitrary subset. If $D^2 \cap S = \emptyset$, then D is called a chaos of

S. Intuitively, the reason why D is known as a chaos is because S has no control over the elements in D.

Theorem 10.1.1. *The necessary and sufficient condition under which an input–output system S over a set Z has a chaotic subset $D \neq \emptyset$, is*

$$S \text{ is not a subset of } I \tag{10.3}$$

where I is the diagonal of the set Z defined by $I = \{(x,x) : x \in Z\}$.

Proof: Necessity. Suppose that the input–output system S over the set Z has a nonempty chaos D. Then $D^2 \cap S = \emptyset$. Therefore, for each $d \in D$, $(d,d) \notin S$. This implies that $S \not\subset I$.

Sufficiency. Suppose that (10.3) holds. Then there exists $d \in Z$ such that $(d,d) \notin S$. Let $D = \{d\}$. Then the nonempty subset D is a chaos of S. ■

Let $S \subset Z^2$ and define

$$S * D = \{x \in Z : \forall y \in D, (x,y) \notin S\} \tag{10.4}$$

where $D \subset Z$.

Theorem 10.1.2. *Let $D \subset Z$. Then D is a chaos of the system S iff $D \subset S * D$.*

Proof: Necessity. Suppose D is a chaos of S. Then for each object $d \in D$ and any $y \in D$,

$$(d,y) \notin S \tag{10.5}$$

Therefore, $d \in S * D$. That is, $D \subset S * D$.

Sufficiency. Suppose D satisfies $D \subset S * D$. Then $D^2 \cap S = \emptyset$. Therefore, D is a chaos of S. ■

Let $\mathrm{COS}(S)$ denote the set of all chaotic subsets on Z of the input–output system S. If $S \not\subset I$, then

$$S_I = \{x \in Z : (x,x) \in S\} \neq Z \tag{10.6}$$

We denote the complement of S_I by $S_I^- = Z - S_I$. Then Theorem 10.1.1 implies that each chaos of S is a certain subset of S_I^-. Therefore,

$$|\mathrm{COS}(S)| \leq 2^{|S_I^-|} \tag{10.7}$$

The following theorem is concerned with determining how many chaotic subsets an input–output system has.

Theorem 10.1.3. *Suppose that an input–output system $S \subset Z^2$ satisfies the following conditions:*

(i) S is symmetric, i.e., $(x,y) \in S$ implies $(y,x) \in S$,

(ii) S is not a subset of I,

(iii) $|S_I^-| = m < \omega_0$, and

(iv) $\forall x \in S_I^-$, there exists at most one $y \in S_I^-$ such that either $(x,y) \in S$ or $(y,x) \in S$.

Let $n = |\{x \in S_I^- : \exists y \in S_I^- ((x,y) \in S)\}|$. Then

$$|\mathrm{COS}(S)| = 2^m \times (3/4)^k \tag{10.8}$$

where $k = n/2$.

Proof: According to the discussion prior to the theorem, it is known that every chaotic subset of Z is a subset of S_I^-. The total number of all subsets of S_I^- is 2^m. Under the assumption of this theorem, a subset of S_I^- is a chaos of S if the subset does not contain two elements of S_I^- which have S-relations. Among the elements of S_I^-, there are $n/2$ pairs of elements with S-relations, and every two pairs do not have common elements. Let $\{x,y\}$ be a pair of elements in S_I^- with an S-relation, then every subset of $S_I^- - \{x,y\} \cup \{x,y\}$ forms a subset of S_I^- not belonging to $\mathrm{COS}(S)$. There are a total of 2^{m-2} of this kind of subset. There are $n/2$ pairs of elements in S_I^- with S-relations, we subtract these $(n/2) \times 2^{m-2}$ subsets from the 2^m subsets of S_I^-. But, some of the $(n/2) \times 2^{m-2}$ subsets, them are in fact the same ones. Naturally, every subset of S_I^-, containing two pairs of elements in S_I^- with S-relations, has been subtracted twice. There are $\binom{k}{2} \times 2^{m-4}$ such sets. We add the number of subsets which have been subtracted twice, and obtain

$$2^m - k \times 2^{m-2} + \binom{k}{2} \times 2^{m-4} \tag{10.9}$$

However, when we add $\binom{k}{2} \times 2^{m-4}$ subsets, the subsets of S_I^- containing three pairs of elements of S_I^- with S-relations, have been added once more than they should have. Therefore, this number should be subtracted. Continuing this process, we have

$$
\begin{aligned}
|\mathrm{COS}(S)| &= 2^m - k \times 2^{m-2} + \binom{k}{2} \times 2^{m-4} - \binom{k}{3} \times 2^{m-6} + \cdots \\
&= \sum_{i=0}^{k} \binom{k}{i} \times 2^{m-2i} \times (-1)^i \\
&= 2^{m-n} \sum_{i=0}^{k} \binom{k}{i} \times 2^{n-2i} \times (-1)^i \\
&= 2^{m-n} \sum_{i=0}^{k} (-1)^i \binom{k}{i} \times 2^{2k-2i} \\
&= 2^{m-n} \sum_{i=0}^{k} (-1)^i \binom{k}{i} \times 4^{k-i} \\
&= 2^{m-n} (4-1)^k \\
&= 2^{m-n} \times 3^k \\
&= 2^m \times (3/4)^k
\end{aligned}
$$

■

By using the same method, the constraint in Theorem 10.1.3 that the system S is symmetric can be relaxed.

Theorem 10.1.4. *Suppose that an input–output system $S \subset Z^2$ satisfies conditions (ii)–(iv) of Theorem 10.1.3. Let $n = |\{x \in S_I^- : \exists y \in S_I^- ((x,y) \in S \text{ or } (y,x) \in S)\}|$ and $k = n/2$. Then*

$$|\mathrm{COS}(S)| = 2^m \times (3/4)^k \tag{10.10}$$

Let S be an input–output system on a set Z. A subset $D \subset Z$ is called an attractor of S if for each $x \in Z - D$, $S(x) \cap D \neq \emptyset$, where $S(x) = \{y \in Z : (x,y) \in S\}$. Figures 10.1 and 10.2 show the geometric meaning of the concept of attractors. When S is not a function, Fig. 10.1 shows that the graph of S outside the vertical bar $D \times Z$ overlaps the horizontal bar $Z \times D$. When S is a function, Figure 10.2 shows that the graph of S outside the vertical bar $D \times Z$ must be contained in the horizontal region $Z \times D$.

Theorem 10.1.5. *Suppose that $S \subset Z^2$ is an input–output system over the set Z and $D \subset Z$. Then D is an attractor of S, iff $D \supset S * D$.*

Proof: Necessity. Suppose that D is an attractor of S. Let $d \in S * D$ be an arbitrary element. By Eq. (10.4), for every $x \in D$, $(d,x) \notin S$. Therefore, $d \in D$. That is, $S * D \subset D$.

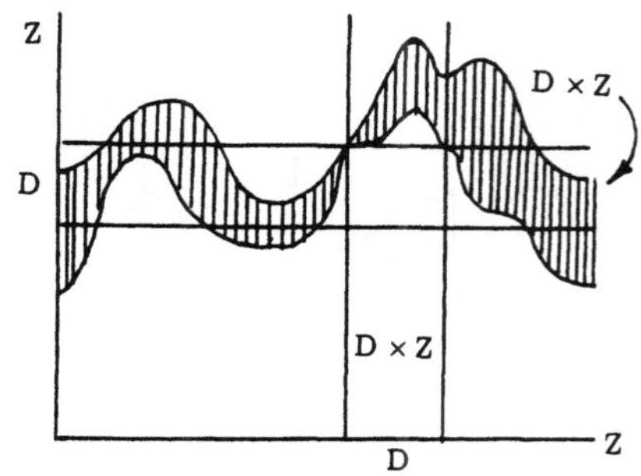

Figure 10.1. D is an attractor of S, and $S(x)$ contains at least one element for each x in $Z - D$.

Sufficiency. Suppose that D satisfies $D \supset S * D$. It follows from Eq. (10.4) that for each object $x \in Z - D$, $x \notin S * D$, so there exists at least one object $d \in D$ such that $(d,x) \in S$. That is, $S(d) \cap D \neq \emptyset$. This implies that D is an attractor of S. ■

Theorem 10.1.6. *The necessary and sufficient condition under which an input–output system $S \subset Z^2$ has an attractor $D \subsetneq Z$ is that S is not a subset of I.*

Proof: Necessity. Suppose S has an attractor D such that $D \subsetneq Z$. By contradiction, suppose $S \subset I$. Then for each $D \subset Z$, $Z - D \subset S * D$. Therefore, when $Z - D \neq \emptyset$, $D \not\subset S * D$; i.e., S, according to Theorem 10.1.5, does not have any attractor not equal to Z. This tells that the hypothesis that S has an attractor D such that $D \subsetneq Z$ implies that $S \not\subset I$.

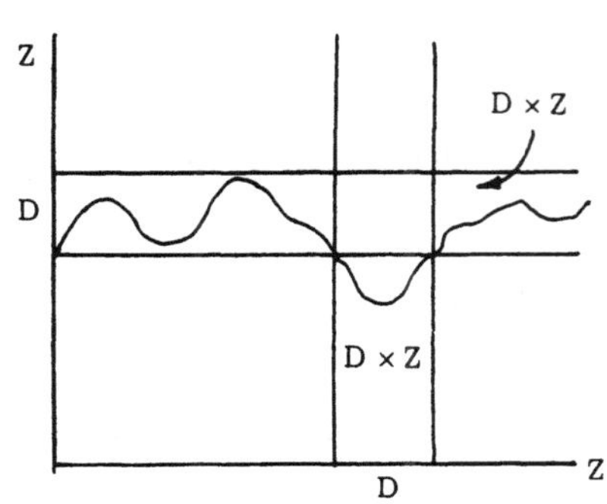

Figure 10.2. D is an attractor of S, and S is a function from Z to Z.

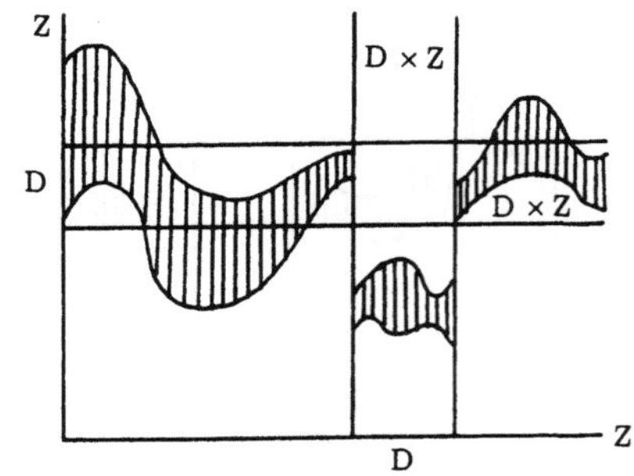

Figure 10.3. D is a strange attractor of the system S.

Sufficiency. Suppose $S \not\subset I$. Choose distinct objects $x,y \in Z$ such that $(x,y) \in S$, and define $D = Z - \{x\}$. Then, according to Eq. (10.4), $D \supset S * D$. Applying Theorem 10.1.5, it follows that S has attractors. ■

Theorem 10.1.7. *Suppose that an input–output system $S \subset Z^2$ satisfies the condition that if $(x,y) \in S$ then, for any $z \in Z$, $(y,z) \notin S$. Let*

$$S_{I^-} = \{x \in Z : \exists y \in Z(y \neq x \quad and \quad (x,y) \in S)\} \tag{10.11}$$

$n = |S_{I^-}|$ and $\mathrm{ATR}(S)$ be the set of all attractors of the system S. Then

$$|\mathrm{ATR}(S)| = 2^n \tag{10.12}$$

Proof: For each subset $A \subset S_{I^-}$, let $D_A = Z - A$. Then $D_A \supset S * D_A$. In fact, for each object $d \in S * D_A$, $(d,x) \notin S$ for every $x \in D_A$. Therefore, $d \notin A$. Otherwise, there exists an element $y \in Z$ such that $y \neq d$ and $(d,y) \in S$. The hypothesis stating that if $(x,y) \in S$ then for any $z \in Z$, $(y,z) \notin S$, implies that the element $y \in Z - A$, contradiction; so $d \in Z - A = D_A$. This says that $D_A \supset S * D_A$. Applying Theorem 10.1.5, it follows that the subset D_A is an attractor of S. Now it can be seen that there are 2^n many such subsets D_A in Z. This completes the proof of the theorem. ■

A subset $D \subset Z$ is a strange attractor of an input–output system $S \subset Z^2$ if D is both a chaos and an attractor of S. Figure 10.3 shows the case when a subset D of Z is a strange attractor of an input–output system S, where the square $D \times D$ is the only portion of the band $Z \times D$ over which the graph of S does not step.

Theorem 10.1.8. *Let $D \subset Z$. Then D is a strange attractor of the input–output system S iff $D = S * D$.*

The proof follows from Theorems 10.1.2 and 10.1.5.

Theorem 10.1.9. *Suppose that $S \subset Z^2$ is an input–output system and $D \subset Z$. Let us denote*

$$S \circ D = S^{-1}(D) = \{x \in Z : \exists y \in D((x,y) \in S)\} \tag{10.13}$$

Then the following statements hold:

(i) D is a chaos of S iff $D \subset Z - S \circ D$.

(ii) D is an attractor of S iff $D \supset Z - S \circ D$.

(iii) D is a strange attractor of S iff $D = Z - S \circ D$.

Proof: It suffices to show that $S * D = Z - S \circ D$. Let $x \in S * D$ be an arbitrary element. Then for any element $y \in D$, $(x,y) \notin D$. Thus, $x \notin S \circ D$; i.e., $x \in Z - S \circ D$. If an element $x \notin S * D$, there exists at least one element $y \in D$ such that $(x,y) \in S$. This implies that $x \in S \circ D$; therefore, $x \notin Z - S \circ D$. This completes the proof that $S * D = Z - S \circ D$. The rest of the proof of the theorem follows from Theorems 10.1.2, 10.1.5, and 10.1.8. ∎

We conclude this section with a discussion of the existence of strange attractors of input–output systems.

Theorem 10.1.10. *Suppose that $S \subset Z^2$ is a symmetric input–output system. Then a necessary and sufficient condition under which there exists a strange attractor of the system S is that there exists a subset $D \subset Z$ such that $D^2 \cap S = \emptyset$ and for each $x \in S_I$, $D \cap S(x) \neq \emptyset$.*

Proof: Necessity. Suppose S has a strange attractor D. From the definition, it follows that $D^2 \cap S = \emptyset$. At the same time, if there exists an element $x \in S_I$ satisfying $D \cap S(x) = \emptyset$ then $x \in S * D$. By Theorem 10.1.8, $x \in D$. However, $x \in S_I$, we must have $x \notin S * \{x\}$, which implies that $x \notin S * D$. This leads to a contradiction.

Sufficiency. Suppose there exists $D \subset Z$ such that $D^2 \cap S = \emptyset$ and for each $x \in S_I$, $D \cap S(x) \neq \emptyset$. Let $\mathcal{A}$ be the collection of subsets of Z defined by

$$\mathcal{A} = \{A \subset Z : A \supset D \quad \text{and} \quad A^2 \cap S = \emptyset\} \tag{10.14}$$

Since $D \in \mathcal{A}$, $\mathcal{A}$ is not empty. It is clear that for each linearly ordered subset $\{A_\xi : \xi < \alpha\}$ of $\mathcal{A}$, ordered by inclusion, $A = \bigcup\{A_\xi : \xi < \alpha\}$ is also contained in $\mathcal{A}$. Therefore, $\mathcal{A}$ contains all maximal elements under the order relation of inclusion.

Let B be a maximal element in $\mathcal{A}$. By Theorem 10.1.2, it follows that $B \subset S * B$. If $B \neq S * B$, pick an element $x \in S * B - B$ and define

$$B^* = B \cup \{x\} \tag{10.15}$$

According to the definition of the element x, for each $y \in B$, $(x,y) \notin S$. From the hypothesis that S is symmetric, it follows that $(y,x) \notin S$ for every $y \in B$. Besides, since for each $z \in S_I$, $A \cap S(z) \neq \emptyset$ for each $A \in \mathcal{A}$, we have $S_I \cap S * B = \emptyset$; i.e., $x \notin S_I$. Therefore, $B^{*2} \cap S_I = \emptyset$. This contradicts the assumption that B is a maximal element in $\mathcal{A}$. ■

We next discuss the conditions under which a composition of input–output systems has strange attractors.

Theorem 10.1.11. *Suppose that f and g are two input–output systems on a set Z such that $g \circ f = f^{-1} \circ g^{-1}$. Then a necessary and sufficient condition under which the composition system $g \circ f$ has a strange attractor is that there exists a subset $D \subset Z$ such that the following conditions hold:*

(i) $(D \times f \circ D) \cap g = \emptyset$.

(ii) For every $x \in (g \circ f)_I$, $f \circ D$ is not a subset of $(Z^2 - g)(x)$.

Proof: Applying Theorem 10.1.10, the composition system $g \circ f$ has a strange attractor if and only if there is a subset $D \subset Z$ such that

(i′) $D^2 \cap g \circ f = \emptyset$.

(ii′) For each $x \in (g \circ f)_I$, $D \cap (g \circ f)(x) \neq \emptyset$.

It now suffices to show that (a) conditions (i) and (i') are equivalent and (b) conditions (ii) and (ii') are equivalent.

Proof of (a). Condition (i') is true iff for any $x,y \in D$, $(x,y) \notin g \circ f$, iff there does not exist a $z \in Z$ such that $(x,z) \in g$ and $(z,y) \in f$, iff for any $x \in D$ and any $z \in f \circ D$, $(x,z) \notin g$, iff $(D \times f \circ D) \cap g = \emptyset$; i.e., (i) holds.

Proof of (b). condition (ii') holds iff for each $x \in (g \circ f)_I$ there exists $d \in D$ such that $(x,d) \in g \circ f$, iff for each $x \in (g \circ f)_I$ there exist $d \in D$ and $z \in Z$ such that $(x,z) \in g$ and $(z,y) \in f$, iff for each $x \in (g \circ f)_I$ there exists $z \in f \circ D$ such that $(x,z) \in g$, iff for each $x \in (g \circ f)_I$ there exists $z \in f \circ D$ such that $(x,z) \notin Z - g$; iff for each $x \in (g \circ f)_I$ $f \circ D \not\subset (Z^2 - g)(x)$; i.e., (ii) is true. ■

10.2. Feedback Transformations

Let $\mathcal{A}$ be a field, X and Y linear spaces over $\mathcal{A}$, and S an input–output system such that

(i) $\emptyset \neq S \subset X \times Y$.

(ii) $s \in S$ and $s' \in S$ imply $s + s' \in S$.

(iii) $s \in S$ and $\alpha \in \mathcal{A}$ imply $\alpha \cdot s \in S$.

Here $+$ and $\cdot$ are addition and scalar multiplication in $X \times Y$, respectively, and defined as follows: For any $(x_1,y_1),(x_2,y_2) \in X \times Y$ and any $\alpha \in \mathcal{A}$,

$$(x_1,y_1)+(x_2,y_2) = (x_1+x_2, y_1+y_2) \tag{10.16}$$

$$\alpha(x_1,y_1) = (\alpha x_1, \alpha y_1) \tag{10.17}$$

The input–output system S is then called a linear system.

Theorem 10.2.1. *Suppose that X and Y are linear spaces over the same field $\mathcal{A}$. Then the following statements are equivalent:*

(1) $S \subset X \times Y$ is a linear system;

(2) There exist a linear space C over $\mathcal{A}$ and a linear mapping $\rho : C \times D(S) \to Y$ such that $(x,y) \in S$ iff there exists a $c \in C$ such that $\rho(c,x) = y$.

(3) There exist a linear space C over $\mathcal{A}$ and linear mappings $R_1 : C \to Y$ and $R_2 : D(S) \to Y$ such that the mapping ρ in (2) is such that $\rho(c,x) = R_1(c) + R_2(x)$ for every $(c,x) \in C \times D(S)$.

Here $D(S)$ is the domain of the input–output system S.

Proof: First, the domain $D(S)$ is a linear subspace of the space X. For any $x_1,x_2 \in D(S)$ and any $\alpha \in \mathcal{A}$, there exist $y_1,y_2 \in Y$ such that $(x_1,y_1),(x_2,y_2) \in S$. From the hypothesis that S is linear, it follows that

$$(x_1+x_2, y_1+y_2) = (x_1,y_1)+(x_2,y_2) \in S \tag{10.18}$$

and

$$\alpha(x_1,y_1) = (\alpha x_1, \alpha y_1) \in S \tag{10.19}$$

and thus $x_1 + x_2$ and $\alpha x_1 \in D(S)$.

The proofs that (3) implies (1) and (2) implies (3) are clear because we can define the desired mappings R_1 and R_2 by $R_1(c) = \rho(c, 0_x)$ and $R_2(x) = \rho(0_c, x)$, where 0_x and 0_c are the additive identities in the linear spaces X and C, respectively.

Proof that (1) implies (2). Suppose S is a linear system and 0_x and 0_y are the additive identities in X and Y, respectively. Define

$$C = \{b \in Y : (0_x, b) \in S\} \tag{10.20}$$

Then $C \neq \emptyset$; in fact, $0_y \in C$ because $(0_x, 0_y) \in S$, and C is a linear subspace over $\mathcal{A}$. Consider the quotient space $Y/C = \{b + C : b \in Y\}$ with addition $+^*$ and scalar multiplication $\cdot^*$ defined by $(b_1 + C) +^* (b_2 + C) = b_1 + b_2 + C$ and $\alpha \cdot^* (b_1 + C) = \alpha b_1 + C$ for any $\alpha \in \mathcal{A}$ and any $(b_1 + C)$ and $(b_2 + C) \in Y/C$.

Define a mapping $\sigma : D(S) \to Y/C$ by letting

$$\sigma(x) = y + C \tag{10.21}$$

if $(x, y) \in S$ for every $x \in D(S)$. Then σ is well defined. In fact, for each $x \in D(S)$ and any $y_1, y_2 \in Y$ satisfying $(x, y_1), (x, y_2) \in S$, $(x, y_1) - (x, y_2) = (0_x, y_1 - y_2) \in S$, and thus $y_1 - y_2 \in C$, so $y_1 + C = y_2 + C$. This means that $\sigma(x) = y_1 + C = y_2 + C$.

Define a linear mapping $\pi : C \times D(S) \to Y/C \oplus C$ as follows:

$$\pi(c, x) = (\sigma(x), c) \tag{10.22}$$

We now choose an isomorphism $I : Y/C \oplus C \to Y$ satisfying $I(y, C) = y$ for every $y \in Y/C$. Now the desired mapping $\rho : C \times D(S) \to Y$ can be defined by $\rho = I \circ \pi$. Then $(x, y) \in S$ if and only if there exists $c \in C$ such that $\rho(c, x) = y$. In fact, if $(x, y) \in S$, there exists $c \in C$ such that $I(\sigma(x), c) = y$, so $\rho(c, x) = I \circ \pi(c, x) = y$. On the other hand, if $(x, y) \in X \times Y$ is such that there exists $c \in C$ with $\rho(c, x) = y$, then $\rho(c, x) = I \circ \pi(c, x) = I(\sigma(x), c) = y$, so $y \in \sigma(x)$. Suppose $\sigma(x) = y_1 + C$. Then $y = y_1 + c_1$ for some $c_1 \in C$; hence,

$$(x, y) = (x, y_1 + c_1) = (x, y_1) + (0_x, c_1) \in S \tag{10.23}$$

This completes the proof of the theorem. ■

Example 10.2.1. Three examples are given to show that the concept of linear systems is an abstraction of different mathematical structures.

(a) Suppose S is a system described by the following linear equations:

$$\begin{aligned} a_{11}x_1 + a_{12}x_2 + \cdots + a_{1n}x_n &= b_1 \\ a_{21}x_1 + a_{22}x_2 + \cdots + a_{2n}x_n &= b_2 \\ &\vdots \\ a_{m1}x_1 + a_{m2}x_2 + \cdots + a_{mn}x_n &= b_m \end{aligned}$$

where x_i, $i = 1,2,\ldots,n$, are n variables, m is the number of equations, a_{ij}, $(i = 1,2,\ldots,m; j = 1,2,\ldots,n)$ are the coefficients of the system, and $b_j, j = 1,2,\ldots,m$, are the constraints of the system. The system S can then be rewritten as a linear system $(M,\{r\})$, where the object set M is the set of all real numbers, and

$$r = \{(x_1,x_2,\ldots,x_n) \in M^n : a_{i1}(x_1+y_1)+a_{i2}(x_2+y_2) + \cdots + a_{in}(x_n+y_n) = b_i \quad \text{for each } i = 1,\ldots,m\}$$

where $(y_1,y_2,\ldots,y_n) \in M^n$ is fixed such that $a_{i1}y_1 + a_{i2}y_2 + \cdots + a_{in}y_n = b_i$ for each $i = 1,2,\ldots,m$.

(b) Suppose a given system S is described by the differential equations

$$\frac{d^n x}{dt^n} + a_1(t)\frac{d^{n-1}x}{dt^{n-1}} + \cdots + a_{n-1}(t)\frac{dx}{dt} + a_n(t)x = f(t)$$

where $a_i(t)$, $i = 1,2,\ldots,n$, and $f(t)$ are continuous functions defined on the interval $[a,b]$. The system S can be rewritten as a linear system $(M,\{r\})$ in the following way: The object set M is the set of all continuous functions defined on $[a,b]$, and the relation r is defined by

$$r = \{x \in M : \frac{d^n(x+y)}{dt^n} + a_1(t)\frac{d^{n-1}(x+y)}{dt^{n-1}} \cdots + a_{n-1}(t)\frac{d(x+y)}{dt} + a_n(t)(x+y) = f(t)\}$$

where $y \in M$ is a fixed function such that

$$\frac{d^n y}{dt^n} + a_1(t)\frac{d^{n-1}y}{dt^{n-1}} + \cdots + a_{n-1}(t)\frac{dy}{dt} + a_n(t)y = f(t)$$

(c) Suppose X is a linear space over a field $\mathcal{A}$, and Y a set of some linear transformations on X. Then Y can be studied as a linear system (X,R), where, for each relation $r \in R$, there exists a transformation $y \in Y$ such that for any $(x_1,x_2) \in X^2$, $(x_1,x_2) \in r$ if and only if $y(x_1) = x_2$.

In Example 10.2.1(b), S is not an input–output linear system. Generally, the concept of general linear systems is the following: Let $\mathcal{A}$ be a field and $S = (M,R)$ a system such that the object set M is a linear space over $\mathcal{A}$. Then S is a linear system if for each relation $r \in R$ there exists an ordinal number $n = n(r)$ such that r is a linear subspace of M^n. We will study only input–output linear systems defined in the beginning of this section.

An input–output system $S \subset X \times Y$ is a functional system if S is a function from the input space X into the output space Y. Let $S \subset X \times Y$ and $S_f : Y \to X$ be a linear system and a linear functional system, respectively. Then the feedback system of S by S_f is defined as the input–output system S' such that

$$(x,y) \in S' \leftrightarrow (\exists z \in X)((x+z,y) \in S \quad \text{and} \quad (y,z) \in S_f) \tag{10.24}$$

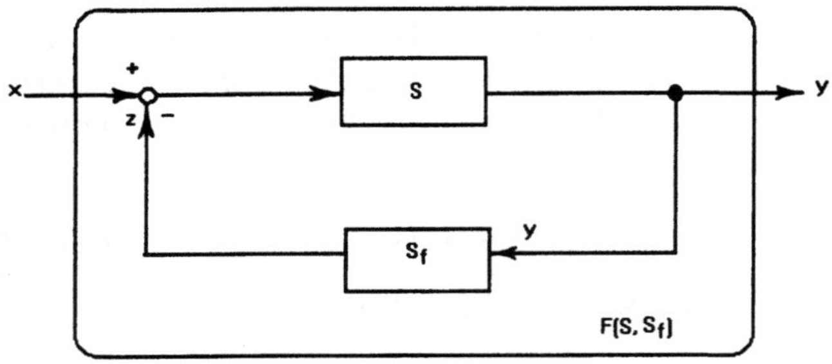

Figure 10.4. Structure of feedback systems.

The systems S and S_f are called an original system and a feedback component system, respectively.

Let

$$\mathcal{S} = \{S \subset X \times Y : S \text{ is a linear system}\} \tag{10.25}$$

$$\mathcal{S}_f = \{S_f : Y \to X : S_f \text{ is a linear functional system}\} \tag{10.26}$$

and

$$\mathcal{S}' = \{S' \subset X \times Y : S' \text{ is a subset}\} \tag{10.27}$$

Then a feedback transformation $F : \mathcal{S} \times \mathcal{S}_f \to \mathcal{S}'$ is defined by Eq. (10.24), which is called the feedback transformation over the linear spaces X and Y. Figure 10.4 shows the geometric meaning of the concept of feedback systems.

The following theorem gives a concrete structure of each feedback system.

Theorem 10.2.2. *Suppose that $S \subset X \times Y$ is a linear system and $S_f : Y \to X$ a linear functional system. Then a subset $S' \subset X \times Y$ is the feedback system of S by S_f, iff*

$$S' = \{(x - S_f(y), y) : (x, y) \in S\} \tag{10.28}$$

Proof: Necessity. Suppose S' is the feedback system of S by S_f. Then, by the definition of feedback systems, for each $(x,y) \in S'$, $(x + S_f(y), y) \in S$. This implies that $(x,y) \in \{(z - S_f(y), y) : (z,y) \in S\}$. Again, the relation in Eq. (10.24) implies that for each element $(x,y) \in S, (x - S_f(y), y) \in S'$. That is, $S' \supset \{(z - S_f(y), y) : (z,y) \in S\}$.

Sufficiency. Clear because Eq. (10.28) implies the relation in Eq. (10.24). ■

Theorem 10.2.3. *Suppose that $S, S' \subset X \times Y$ are linear systems and $S_f : Y \to X$ is a linear functional system such that*

$$F(S, S_f) = S' \tag{10.29}$$

Then $F(S', -S_f) = S$, where $-S_f : Y \to X$ is the linear functional system defined by

$$(-S_f)(y) = -S_f(y) \tag{10.30}$$

for every $y \in Y$.

The proof is straightforward and omitted.

Theorem 10.2.4. *For each linear system $S \subset X \times Y$, there exists a linear functional system $S_f : Y \to X$ such that $F(S, S_f) = S$, that is, S is a feedback system of itself.*

Proof: The desired linear functional system $S_f : Y \to X$ can be defined by

$$S_f(y) = 0_x \tag{10.31}$$

for every $y \in Y$. Then, Eq. (10.28) says that $S = F(S, S_f)$. ■

Theorem 10.2.5 discusses the control problem: A linear structure (say, system) $S \subset X \times Y$ is fixed. According to some predefined targets, we need to produce a new linear structure (say, a linear subspace) $S' \subset X \times Y$. Can we design a feedback process $S_f : Y \to X$ such that $F(S, S_f) = S'$?

Theorem 10.2.5. *Assume the axiom of choice. For each linear system $S \subset X \times Y$ and arbitrary linear subspace $S' \subset X \times Y$, there exists a linear functional system $S_f : Y \to X$ such that $F(S, S_f) = S'$ iff $R(S') = R(S)$ and $N(S') = N(S)$, where $N(W)$ indicates the null space of the input–output system $W \subset X \times Y$, defined by*

$$N(W) = \{x \in X : (x, 0_y) \in W\} \tag{10.32}$$

Proof: Necessity. Suppose there exists a linear functional system $S_f : Y \to X$ such that $F(S, S_f) = S'$. Then Theorem 10.2.2 implies that $R(S') = R(S)$ and that

$$\begin{aligned} N(S') &= \{x \in X : (x, 0_y) \in S'\} \\ &= \{x \in X : (x + S_f(0_y), 0_y) \in S\} \\ &= \{x \in X : (x + 0_x, 0_y) \in S\} \\ &= N(S) \end{aligned}$$

Sufficiency. Suppose that S and S' have the same output space and the null space; i.e., $R(S') = R(S)$ and $N(S') = N(S)$. We define a linear functional system $S_f : Y \to X$ such that $F(S, S_f) = S'$.

For each element $y \in R(S')$, let

$$X'_y = \{x' \in X : (x', y) \in S'\} \tag{10.33}$$

and

$$X_y = \{x \in X : (x, y) \in S\} \tag{10.34}$$

From the Axiom of Choice, it follows that there exist mappings

$$C' : \{X'_y : y \in R(S')\} \to X \tag{10.35}$$

and

$$C : \{X_y : y \in R(S)\} \to X \tag{10.36}$$

such that $C'(X'_y) = C'(y) \in X'_y$ [i.e., $(C'(y), y) \in S'$] and $C(X_y) = C(y) \in X_y$ [i.e., $(C(y), y) \in S$]. We are now ready to check that for each $y \in R(S') = R(S)$,

$$X'_y \neq \emptyset \neq X_y \tag{10.37}$$

Thus, the mappings C and C' are well defined.

Let $\{y_i : i \in I\}$ be a basis in the linear space $R(S')$, where I is a finite or infinite index set. We now define a linear functional system

$$S_f : Y \to X \tag{10.38}$$

satisfying $S_f(y_i) = C(y_i) - C'(y_i)$, for each $i \in I$. Then we make the following claim.

Claim. The systems S, S', and S_f have the property that $F(S, S_f) = S'$.

In fact, for each element $(x, y_i) \in S'$, for an $i \in I$, we have $(x + S_f(y_i), y_i) \in S$ since

$$\begin{aligned}
(x + S_f(y_i), y_i) &= (x + C(y_i) - C'(y_i), y_i) \\
&= (x - C'(y_i), 0_y) + (C(y_i), y_i) \\
&\in N(S') \times \{0_y\} + S \\
&= N(S) \times \{0_y\} + S \\
&= S
\end{aligned} \tag{10.39}$$

Therefore, from Eq. (10.39),

$$(x, y_i) \in S' \to (x + S_f(y_i), y_i) \in S \tag{10.40}$$

Now let $(x,y_i) \in S$ we then have $(x - S_f(y_i),y_i) \in S'$, because

$$\begin{aligned}(x - S_f(y_i),y_i) &= (x - C(y_i) + C'(y_i),y_i) \\ &= (x - C(y_i), 0_y) + (C'(y_i),y_i) \\ &\in N(S) \times \{0_y\} + S' \\ &= N(S') \times \{0_y\} + S' \\ &= S' \end{aligned} \tag{10.41}$$

Hence, we have from Eq. (10.41) the opposite implication of Eq. (10.40). That is, for every $i \in I$,

$$(x,y_i) \in S' \leftrightarrow (x + S_f(y_i),y_i) \in S \tag{10.42}$$

It now remains to show that for each pair $(x,y) \in X \times Y$, the relation in Eq. (10.24) holds. Let (x,y) be an arbitrary element in S'. Then $y = \sum_i c_i \cdot y_i$ for a finite number of y_i's from the basis $\{y_i : i \in I\}$, where each c_i is a nonzero number. Then $(x + S_f(y),y) \in S$. In fact,

$$\begin{aligned}(x + S_f(y),y) &= (x + \sum_i c_i \cdot y_i, y) \\ &= (x + \sum_i c_i(C(y_i) - C'(y_i)), \sum_i c_i \cdot y_i) \\ &= (x - \sum_i c_i \cdot C'(y_i), 0_y) + (\sum_i c_i \cdot C(y_i), \sum_i c_i \cdot y_i) \\ &= [(x,y) - (\sum_i c_i \cdot C'(y_i),y)] + \sum_i c_i \cdot (C(y_i),y_i) \\ &\in N(S') \times \{0_y\} + S \\ &= N(S) \times \{0_y\} + S \\ &= S \end{aligned} \tag{10.43}$$

This gives us the implication from the right to left in Eq. (10.24). The opposite implication can be shown as follows. For each pair $(x,y) \in S$, $(x - S_f(y),y) \in S'$. In fact, let $y = \sum_i c_i \cdot y_i$ for some finite number of nonzero coefficients c_i, where each y_i is from the basis $\{y_i : i \in I\}$. Then

$$\begin{aligned}(x - S_f(y),y) &= (x - \sum_i c_i \cdot S_f(y_i),y) \\ &= (x - \sum_i c_i \cdot (C(y_i) - C'(y_i)), \sum_i c_i \cdot y_i) \\ &= (x - \sum_i c_i \cdot C(y_i), 0_y) + (\sum_i c_i \cdot C'(y_i), \sum_i c_i \cdot y_i) \\ &\in N(S) \times \{0_y\} + S' \\ &= N(S') \times \{0_y\} + S' \\ &= S' \end{aligned} \tag{10.44}$$

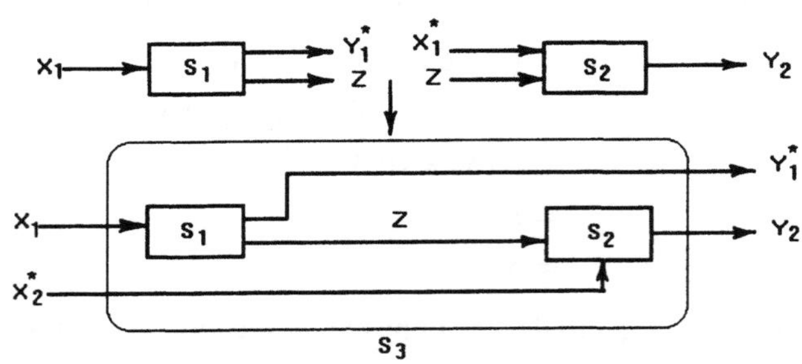

Figure 10.5. Structure of cascade connections of systems.

This completes the proof of Theorem 10.2.5. ■

We conclude this section with a discussion of a more general definition of the concept of feedback systems. Let $*$ be a system connecting operator, termed the cascade (connecting) operation. The operation $*$ is defined as follows: Let S_i, $i = 1,2,3$, be input–output systems such that

$$S_1 \subset X_1 \times (Y_1^* \times Z) \tag{10.45}$$
$$S_2 \subset (X_2^* \times Z) \times Y_2 \tag{10.46}$$
$$S_3 \subset (X_1 \times X_2^*) \times (Y_1^* \times Y_2) \tag{10.47}$$

where it is assumed that no algebraic properties are given on each set involved. The three systems satisfy $S_1 * S_2 = S_3$ if and only if

$$((x_1,x_2),(y_1,y_2)) \in S_3$$
$$\leftrightarrow \exists z((x_1,(y_1,z)) \in S_1 \quad \text{and} \quad (((x_2,z),y_2) \in S_2) \tag{10.48}$$

Figure 10.5 shows the geometric meaning of the cascade connection of systems.

We now let S and S_f be two input–output systems such that

$$S \subset (X \times Z_x) \times (Y \times Z_y) \tag{10.49}$$

and

$$S_f \subset Z_y \times Z_x \tag{10.50}$$

where S_f is called a feedback component of S. The relationship between the two systems is displayed in Fig. 10.6.

The feedback system $F(S,S_f)$ of S by S_f is defined by

$$(x,y) \in F(S,S_f) \leftrightarrow \exists (z_y,z_x) \in S_f(((x,z_x),(y,z_y)) \in S) \tag{10.51}$$

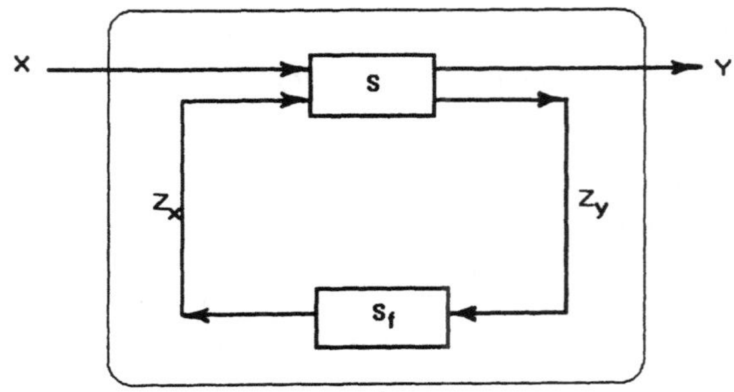

Figure 10.6. Construction of general feedback systems.

In constructing $F(S,S_f)$ step by step, we first cascade S and S_f (i.e., find $S * S_f$) and then input one of the two outputs of the system $S * S_f$.

Without special explanation in this book, by feedback system we always mean the definition given by Eq. (10.24).

10.3. Feedback-Invariant Properties

Some properties of original systems may be kept by their feedback systems and some may not. Those properties of original systems invariant under the feedback transformation F are of special interest in research on feedback systems. Feedback invariance has played a crucial role in characterizing homeostatic or explosive system behaviors in various areas such as biology, engineering, social science, and so on.

A property of a linear system $S \subset X \times Y$ is called feedback invariant with respect to $\mathcal{S}_f$ if the property also holds for the feedback system $F(S,S_f)$ for each linear functional system $S_f \in \mathcal{S}_f$.

Corollary 10.3.1. *Let $S \subset X \times Y$ and $S_f : Y \to X$ be a linear system and a linear functional system, respectively. Then the feedback system $F(S,S_f)$ is a linear system over $X \times Y$ and $R(F(S,S_f)) = R(S)$ and $N(F(S,S_f)) = N(S)$.*

Proof: The linearity of the feedback system $F(S,S_f)$ follows from Theorem 10.2.2. The equations $R(F(S,S_f)) = R(S)$ and $N(F(S,S_f)) = N(S)$ come from Theorem 10.2.5 here the Axiom of Choice is not used. ■

Theorem 10.3.1. *Let $S : X \to Y$ and $S_f : Y \to X$ be linear functional systems such that S is surjective. Then $D(F(S,S_f)) = D(S)$ iff $(I - S \circ S_f) : Y \to Y$ is surjective, where $I : Y \to Y$ indicates the identity mapping; i.e., $I(y) = y$ for each $y \in Y$.*

Proof: Sufficiency. It is clear that $D(F(S,S_f)) \subset D(S)$. Let $x \in D(S)$ be arbitrary. There then exists some $y \in Y$ such that $(x,y) \in S$. Since $(I - S \circ S_f) : Y \to Y$ is surjective, there exists exactly one $\widehat{y} \in Y$ such that

$$(I - S \circ S_f)(\widehat{y}) = y = S(x) \tag{10.52}$$

This implies that

$$S(x + S_f(\widehat{y})) = \widehat{y} \tag{10.53}$$

That is, $(x + S_f(\widehat{y}), \widehat{y}) \in S$. Hence, $x \in D(F(S,S_f))$.

Necessity. Pick an arbitrary $y \in Y$. From the hypothesis that S is a surjective system, it follows that there exists some $x \in X$ such that $S(x) = y$. Since $D(F(S,S_f)) = D(S)$, for this x there exists some $\widehat{y} \in Y$ such that $(x,\widehat{y}) \in F(S,S_f)$; i.e., $(x + S_f(\widehat{y}),\widehat{y}) \in S$. Hence,

$$\begin{aligned}\widehat{y} &= S(x + S_f(\widehat{y})) \\ &= S(x) + S \circ S_f(\widehat{y}) \\ &= y + S \circ S_f(\widehat{y})\end{aligned} \tag{10.54}$$

Thus $\widehat{y} - S \circ S_f(\widehat{y}) = y$, i.e., $(I - S \circ S_f)(\widehat{y}) = y$. This proves that $(I - S \circ S_f) : Y \to Y$ is surjective. ■

Theorem 10.3.2. *Let $S \subset X \times Y$ be a linear system and $S_f : Y \to X$ a linear functional system. Then $D(F(S,S_f)) = X$ iff for each $x \in X$ there exists a $y \in R(S)$ such that $(x + S_f(y),y) \in S$.*

The proof is clear from Eq. (10.24).

Theorem 10.3.3. *Let $S : X \to Y$ be a linear functional system. Then for each arbitrarily fixed $S_f \in \mathcal{S}_f$, the feedback system $F(S,S_f)$ is injective iff the original system S is injective.*

Proof: Sufficiency. Suppose the original system S is injective, and let $S_f \in \mathcal{S}_f$ be arbitrary. Suppose the feedback system $F(S,S_f)$ is not injective. Thus, there exist distinct $x_1,x_2 \in D(F(S,S_f))$ such that

$$(x_1,y), (x_2,y) \in F(S,S_f) \tag{10.55}$$

for some $y \in R(S)$. By Eq. (10.24), we have $(x_1 + S_f(y), y), (x_2 + S_f(y), y) \in S$. Therefore, from the hypothesis that the system $S : X \to Y$ is injective, it follows that $x_1 + S_f(y) = x_2 + S_f(y)$; i.e., $x_1 = x_2$, contradiction.

Necessity. Suppose $F(S, S_f)$ is injective. From Theorem 10.2.3,

$$F(F(S, S_f), -S_f) = S$$

Applying the proof for sufficiency, we know that S is injective. ■

Combining what have been obtained in this section, the following theorem is evident.

Theorem 10.3.4. *Range space, null space, linearity, injectivity, surjectivity, and bijectivity of original systems are feedback invariant with respect to $\mathcal{S}_f$.*

In the rest of this section, we study feedback-invariant properties of MT-time systems. Let T be the time axis defined as the positive half of the real number line; i.e., $T = [0, +\infty)$. Let A and B be two linear spaces over the same field $\mathcal{A}$. We define

$$A^T = \{x : x \text{ is a mapping } T \to A\} \tag{10.56}$$

and

$$B^T = \{x : x \text{ is a mapping } T \to B\} \tag{10.57}$$

Then the sets A^T and B^T can be made linear spaces over $\mathcal{A}$ as follows: For any elements $f, g \in A^T$ (respectively, $\in B^T$), and $\alpha \in \mathcal{A}$,

$$(f + g)(t) = f(t) + g(t) \tag{10.58}$$

and

$$(\alpha f)(t) = \alpha \cdot f(t) \tag{10.59}$$

for each $t \in T$.

Assume the input space X and the output space Y in the preceding discussion are linear subspaces of A^T and B^T, respectively; i.e., $X \subset A^T$ and $Y \subset B^T$.

Proposition 10.3.1. *For any linear system $S \subset X \times Y$ and any linear functional system $S_f : Y \to X$, the feedback system $F(S, S_f)$ is a linear system over $A^T \times B^T$.*

The proof is straightforward and is omitted.

Each input–output system $S \subset A^T \times B^T$ is termed an MT-time system, (Mesarovic–Takahara time system). Therefore, Proposition 10.3.1 implies that the feedback transformation defined in Section 10.2 is well defined on the class of all MT-time systems.

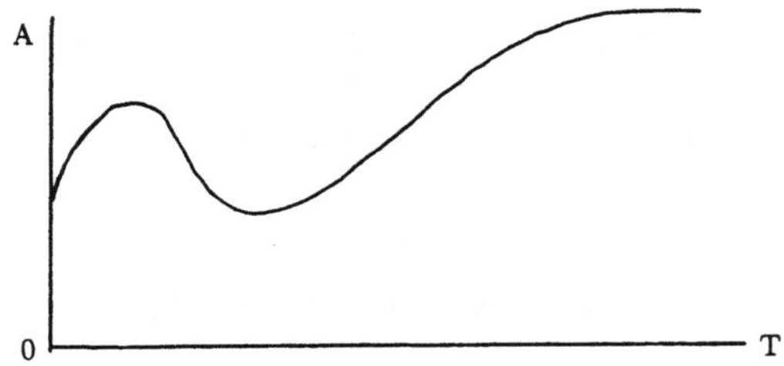

Figure 10.7. The graph of x in the plane $T \times A$.

For any $t, t' \in T$ with $t < t'$, let

$$T^t = [0,t), \quad T_t = [t,+\infty) \tag{10.60}$$

$$T_{tt'} = [t,t'), \quad \overline{T}^t = [0,t], \quad \overline{T}_{tt'} = [t,t'] \tag{10.61}$$

Then the restrictions of a time function $x \in X$ with respect to these time intervals are denoted by

$$x^t = x|T^t, \quad x_t = x|T_t, \quad x_{tt'} = x|T_{tt'} \tag{10.62}$$

and

$$\overline{x}^t = x|\overline{T}^t, \quad \overline{x}_{tt'} = x|\overline{T}_{tt'} \tag{10.63}$$

For each $y \in Y$, or in general, for sets and vectors similar notation will be used; for example, for each $(x,y) \in X \times Y$, $(x,y)^t = (x^t, y^t)$, $(x,y)_t = (x_t, y_t)$, etc.

A subset $S \subset X \times Y \subset A^T \times B^T$ is called a linear time system if S is a linear subspace of $X \times Y$ and the input space $D(S)$ satisfies the condition:

$$\forall x, x' \in X \forall t \in T(x, x' \in D(S) \to x^t \circ x'_t \in D(S)) \tag{10.64}$$

where $x^t \circ x'_t$ is a time function $\in A^T$, called the concatenation of x^t and x'_t, and is defined by the following: For any $s \in T$,

$$(x^t \circ x'_t)(s) = \begin{cases} x(s) & \text{if } s < t \\ x'(s) & \text{if } s \geq t \end{cases} \tag{10.65}$$

Without loss of generality, we will always assume that $D(S) = X$ and $R(S) = Y$.

Let σ^τ be the shift operator defined as follows: For each $x \in X$,

$$\sigma^\tau(x)(\xi) = x(\xi - \tau) \tag{10.66}$$

for each $\xi \in T_\tau$, where $\tau \in \mathbb{R}$ can be any value if $\sigma^\tau(x)$ is meaningful. Figures 10.7–10.9 show the geometric meaning of the concept of shift operators. In Figs. 10.8

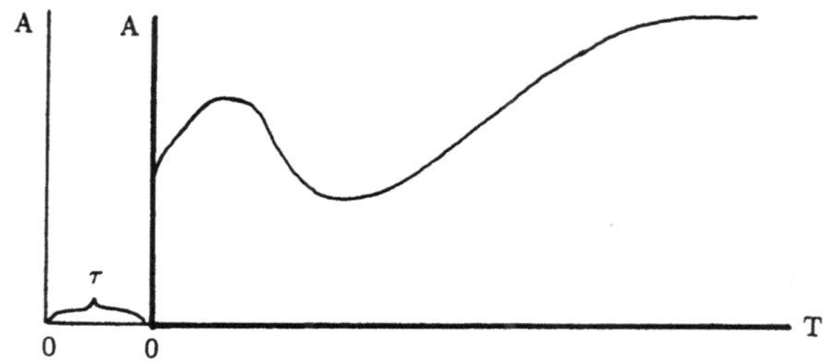

Figure 10.8. The graph of $\sigma^\tau(x)$ with $\tau > 0$.

and 10.9 the dotted $A \times T$ planes indicate the locations of the $A \times T$ plane before the shift operation σ^τ functions. Therefore, when $\tau > 0$, the application of σ^τ implies that we move the graph of x τ units to the right; when $\tau < 0$, the application of σ^τ tells us to consider the portion of the graph of x on the right of the vertical line $t = \tau$.

A linear time system $S \subset X \times Y$ is called strongly stationary if S satisfies the condition

$$\forall t \in T(\sigma^{-t}(S|T_t) = S) \tag{10.67}$$

Theorem 10.3.5. *Let $S \subset X \times Y$ and $S_f : Y \to X$ be a linear time system and a strongly stationary linear functional time system, respectively. Then the feedback system $F(S, S_f)$ is strongly stationary iff the original system S is strongly stationary.*

Proof: Necessity. Suppose that the feedback system $F(S, S_f)$ is strongly stationary. Then

$$S = \{(x + S_f(y), y) : (x, y) \in F(S, S_f)\} \tag{10.68}$$

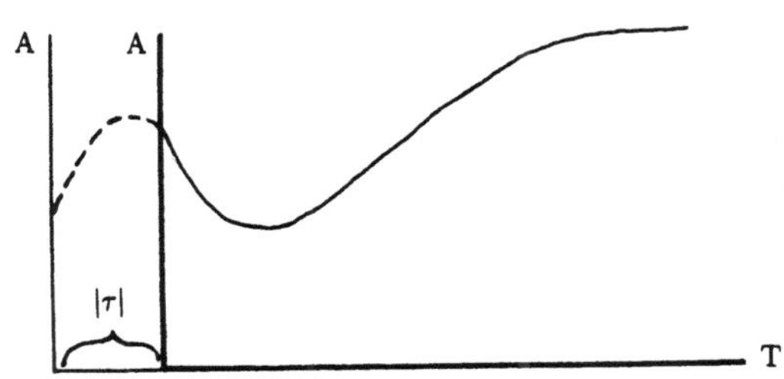

Figure 10.9. The graph of $\sigma^\tau(x)$ with $\tau < 0$.

It suffices to show that for any $(x,y) \in S$, $\sigma^{-t}((x,y)|T_t) \in S$ for every $t \in T$. We now pick an arbitrary $(x,y) \in S$ and $t \in T$. Then

$$\begin{aligned}\sigma^{-t}((x,y)|T_t) &= \sigma^{-t}((x - S_f(y) + S_f(y),y)|T_t) \\ &= (\sigma^{-t}((x - S_f(y))|T_t) + \sigma^{-t}(S_f(y)|T_t), \sigma^{-t}(y|T_t))\end{aligned} \quad (10.69)$$

From the hypothesis that the systems $F(S,S_f)$ and S_f are strongly stationary, it follows that

$$(\sigma^{-t}((x - S_f(y))|T_t), \quad \sigma^{-t}(y|T_t)) \in F(S,S_f) \quad (10.70)$$

and

$$(\sigma^{-t}(y|T_t), \sigma^{-t}(S_f(y)|T_t)) \in S_f \quad (10.71)$$

Let $x' = \sigma^{-t}((x - S_f(y))|T_t)$ and $y' = \sigma^{-t}(y|T_t)$. Then, Eq. (10.71) shows that $S_f(y') = \sigma^{-t}(S_f(y)|T_t)$. Therefore,

$$\sigma^{-t}((x,y)|T_t) = (x' + S_f(y'), y') \quad (10.72)$$

where $(x',y') \in F(S,S_f)$, so Eq. (10.72) implies that $\sigma^{-t}((x,y)|T_t) \in S$.

Sufficiency. This follows from Theorem 10.2.3 and the necessity part of this proof. ■

A linear time system $S \subset X \times Y$ is called precausal if it satisfies the condition:

$$(\forall t \in T)(\forall x \in X)(\bar{x}^t = \bar{0}^t \rightarrow S(x)|\overline{T}^t = S(0)|\overline{T}^t) \quad (10.73)$$

where $S(x) = \{y \in Y : (x,y) \in S\}$. A linear functional time system $S : X \rightarrow Y$ is called causal if it is precausal.

Theorem 10.3.6. *Let $S \subset X \times Y$ be a linear time system. Then S is precausal iff $\forall t \in T \forall x, y \in X(\bar{x}^t = \bar{y}^t \rightarrow S(x)|\overline{T}^t = S(y)|\overline{T}^t)$.*

Proof: Necessity. Suppose S is precausal. Then for any $x,y \in X$, $(\overline{x-y})^t = \bar{0}^t$ implies that $S(x-y)|\overline{T}^t = S(0)|\overline{T}^t$. This shows that when $\bar{x}^t = \bar{y}^t$, $S(x)|\overline{T}^t = S(y)|\overline{T}^t$, because $(\overline{x-y})^t = \bar{0}^t$ is the same as $\bar{x}^t - \bar{y}^t = \bar{0}^t$. Hence, $\bar{x}^t = \bar{y}^t$, and $S(x-y)|\overline{T}^t = S(0)|\overline{T}^t$ implies that $S(x)|\overline{T}^t - S(y)|\overline{T}^t = S(0)|\overline{T}^t$, so $S(x)|\overline{T}^t = S(y)|\overline{T}^t + S(0)|\overline{T}^t = S(y)|\overline{T}^t$.

Sufficiency. Suppose $\forall t \in T \forall x, y \in X(\bar{x}^t = \bar{y}^t \rightarrow S(x)|\overline{T}^t = S(y)|\overline{T}^t)$. This implies that $\forall t \in T \forall x \in X(\bar{x}^t = \bar{0}^t \rightarrow S(x)|\overline{T}^t = S(0)|\overline{T}^t)$. ■

Theorem 10.3.7. *Let $S : X \to Y$ and $S_f : Y \to X$ be linear functional causal time systems. Then the feedback system $F(S,S_f)$ is causal iff the composition time system $S \circ S_f \circ F(S,S_f) : D(F(S,S_f)) \to Y$ is causal.*

Proof: The argument of the necessity part follows from the fact that each composition of causal functional time systems is causal.

Sufficiency. Suppose that the composition system $S \circ S_f \circ F(S,S_f)$ is causal. That is, for any $t \in T$ and any $x \in D(F(S,S_f))$, $\overline{x}^t = \overline{0}^t$ implies that

$$S \circ S_f \circ F(S,S_f)(x)|\overline{T}^t = \overline{0}^t \tag{10.74}$$

Let $y = F(S,S_f)(x)$. We need to show that $\overline{y}^t = \overline{0}^t$. By Eq. (10.24),

$$S(x + S_f(y)) = y \tag{10.75}$$

Therefore, $S(x) + S \circ S_f(y) = y$, where $S(x)|\overline{T}^t = \overline{0}^t$ and $S \circ S_f(y)|\overline{T}^t = S \circ S_f \circ F(S,S_f)(x)|\overline{T}^t = \overline{0}^t$. Thus,

$$\begin{aligned}\overline{y}^t &= [S(x) + S \circ S_f(y)]|\overline{T}^t \\ &= S(x)|\overline{T}^t + S \circ S_f(y)|\overline{T}^t \\ &= \overline{0}^t + \overline{0}^t \\ &= \overline{0}^t\end{aligned}$$

This implies that the feedback system $F(S,S_f)$ is causal. ■

A linear functional time system $S : X \to Y$ is called time invariably realizable if it satisfies the condition

$$\forall t \in T \forall x \in X(\lambda^t S(0^t \circ \sigma^t(x)) = S(x)) \tag{10.76}$$

where $\lambda^t(\cdot) = \sigma^{-t}(\cdot|T_t)$.

Theorem 10.3.8. *A linear functional time system $S : X \to Y$ is time invariably realizable iff for any $t \in T$ and any $x \in X$*

$$S(0^t \circ \sigma^t(x))|T_t = \sigma^t S(x) \tag{10.77}$$

Proof: Necessity. Suppose S is time invariably realizable. Then S satisfies the relation in Eq. (10.76). It suffices to show that Eq. (10.77) is derivable from the relation in Eq. (10.76). For any $t \in T$ and any $x \in X$, let $y = S(0^t \circ \sigma^t(x))$. Then

$$\lambda^t S(0^t \circ \sigma^t(x)) = \lambda^t(y) = \sigma^{-t}(y|T_t) = S(x) \tag{10.78}$$

The last equality of Eq. (10.78) shows that

$$\begin{aligned}\sigma^t \circ \sigma^{-t}(y|T_t) &= y|T_t \\ &= S(0^t \circ \sigma^t(x))|T_t \\ &= \sigma^t S(x)\end{aligned} \tag{10.79}$$

Sufficiency. Suppose S satisfies Eq. (10.77) for any $t \in T$ and any $x \in X$. Then

$$\sigma^{-t} S(0^t \circ \sigma^t(x))|\overline{T}^t = \sigma^{-t}\sigma^t S(x) = S(x) \tag{10.80}$$

Comparing Eq. (10.80) and the relation in Eq. (10.76), we see that S is time invariably realizable. ■

Theorem 10.3.9. *Suppose that $S : X \to Y$ and $S_f : Y \to X$ are two linear functional time systems such that S and S_f are time invariably realizable. Then the feedback system $F(S, S_f)$ is time invariably realizable iff for any $t \in T$ and any $x \in \dot{D}(F(S, S_f))$,*

$$S \circ S_f \circ F(S,S_f)(0^t \circ \sigma^t(x))|T_t = \sigma^t S \circ S_f \circ F(S,S_f)(x) \tag{10.81}$$

i.e., the system $S \circ S_f \circ F(S, S_f)$ is time invariably realizable.

Proof: Sufficiency. Suppose the system $S \circ S_f \circ F(S, S_f)$ is time invariably realizable. Then for any $t \in T$ and any $x \in D(F(S, S_f))$, Eq. (10.81) holds. From the hypothesis that S is time invariably realizable, it follows from Theorem 10.3.8 that Eq. (10.77) holds. Therefore,

$$\begin{aligned}S(0^t \circ \sigma^t(x))|T_t + S \circ S_f(F(S,S_f)(0^t \circ \sigma^t(x)))|T_t \\ = \sigma^t S(x) + \sigma^t S \circ S_f \circ F(S,S_f)(x)\end{aligned} \tag{10.82}$$

This implies that

$$\begin{aligned}S[0^t \circ \sigma^t(x) + S_f(F(S,S_f)(0^t \circ \sigma^t(x)))]|T_t \\ = \sigma^t S(x + S_f \circ F(S,S_f)(x))\end{aligned} \tag{10.83}$$

This equation is the same as

$$F(S,S_f)(0^t \circ \sigma^t(x))|T_t = \sigma^t F(S,S_f)(x) \tag{10.84}$$

Therefore, applying Theorem 10.3.8, we have showed that the feedback system $F(S, S_f)$ is time invariably realizable.

Necessity. Suppose $F(S, S_f)$ is time invariably realizable. Then Theorem 10.3.8 says that $\forall t \in T$ and $\forall x \in D(F(S, S_f))$,

$$F(S,S_f)(0^t \circ \sigma^t(x))|T_t = \sigma^t F(S,S_f)(x) \tag{10.85}$$

Let $y = F(S,S_f)(0^t \circ \sigma^t(x))$ and $y' = F(S,S_f)(x)$. Then from Eq. (10.85) we have

$$S(0^t \circ \sigma^t(x) + S_f(y))|T_t = \sigma^t S(x + S_f(y')) \tag{10.86}$$

$$S(0^t \circ \sigma^t(x))|T_t + S \circ S_f(y)|T_t = \sigma^t S(x) + \sigma^t S \circ S_f(y') \tag{10.87}$$

From the hypothesis that S is time invariably realizable and Theorem 10.3.8, it follows that Eq. (10.87) can be written as

$$S \circ S_f(y)|T_t = \sigma^t S(S_f(y')) \tag{10.88}$$

Equation (10.88) is the same as Eq. (10.81). This ends the proof. ■

We conclude this section with some discussion of chaos, attractor, and feedback transformation.

Theorem 10.3.10. *For each linear system $S \subset X \times X$ there exists a feedback component system $S_f : X \to X$, such that $D \subset X$ is a chaos, attractor, or strange attractor, respectively, of S, iff D is a chaos, attractor, or strange attractor of $F(S,S_f)$, respectively.*

The proof follows from Theorem 10.2.4.

Theorem 10.3.11. *Suppose that $S \subset X \times X$ and $S_f : X \to X$ are a linear system and a functional linear system, respectively. Then a subset $D \subset X$ is a chaos of S [iff D is a chaos of the feedback system $F(S,S_f)$], if*

$$(X - D) \pm S_f(D) \subset X - D \tag{10.89}$$

Proof: Necessity. Suppose $D \subset X$ is a chaos of S but not a chaos of $F(S,S_f)$. Pick an element $(x,y) \in D^2 \cap F(S,S_f)$. Then $(x + S_f(y), y) \in S$, so $x' = x + S_f(y) \notin D$. Now $x = x' - S_f(y)$. This contradicts relation (10.89). Therefore, D must be a chaos of $F(S,S_f)$.

Sufficiency. Suppose $D \subset X$ is a chaos of $F(S,S_f)$ but not a chaos of S. Pick an element $(x,y) \in D^2 \cap S$. Then $(x - S_f(y), y) \in F(S,S_f)$. Therefore, it must be that $x' = x - S_f(y) \notin D$. Now $x = x' + S_f(y)$, which contradicts relation (10.89). So D must be a chaos of S. ■

Theorem 10.3.12. *Suppose that $S \subset X \times X$ and $S_f : X \to X$ are a linear system and a functional linear system, respectively. Then $D \subset X$ is an attractor of S (iff D is an attractor of the feedback system $F(S,S_f)$), if*

$$X - D \subset \{x - S_f(y) : (x,y) \in S \cap ((X - D) \times D)\} \tag{10.90}$$

and

$$X - D \subset \{x + S_f(y) : (x,y) \in F(S,S_f) \cap ((X - D) \times D)\} \tag{10.91}$$

Proof: Necessity. Suppose $D \subset X$, satisfying condition (10.90), is an attractor of S but not of $F(S, S_f)$. Then there exists an element $x \in X - D$ such that

$$F(S,S_f)(x) \cap D = \emptyset \tag{10.92}$$

From condition (10.90) it follows that there exists $(z,y) \in S \cap ((X-D) \times D)$ such that $z - S_f(y) = x$. Thus, $(x,y) \in F(S,S_f)$; i.e., $y \in F(S,S_f)(x) \cap D$, contradiction.

Sufficiency. Suppose $D \subset X$, satisfying condition (10.91), is an attractor of $F(S,S_f)$, but not of the original system S. There then exists an element $x \in X - D$ such that

$$S(x) \cap D = \emptyset \tag{10.93}$$

From condition (10.91) it follows that there exists $(z,y) \in F(S,S_f) \cap ((X-D) \times D)$ such that $z + S_f(y) = x$. Therefore, $(x,y) \in S$; i.e., $y \in S(x) \cap D$, contradiction. ■

10.4. Decoupling of Single-Relation Systems

Let $S \subset X \times Y$ be an input–output system. We assume that the input space X and the output space Y can be represented in terms of the factor sets; i.e., there are sets $\{X_i : i \in I\}$ and $\{Y_i : i \in I\}$, where I is a fixed index set, such that

$$X = \prod \{X_i : i \in I\} \tag{10.94}$$

and

$$Y = \prod \{Y_i : i \in I\} \tag{10.95}$$

For each index $i \in I$, let

$$\prod_i = (\prod_{ix}, \prod_{iy}) : X \times Y \to X_i \times Y_i \tag{10.96}$$

be the projection defined by letting $\prod_{ix} = p_i : X \to X_i$ and $\prod_{iy} = p_i : Y \to Y_i$ such that $\prod_{ix}((x_j)_{j\in I}) = x_i$, $\prod_{iy}((y_j)_{j\in I}) = y_i$, and $\prod_i((x_j)_{j\in I}, (y_j)_{j\in I}) = (\prod_{ix}((x_j)_{j\in I}), \prod_{iy}((y_j)_{j\in I})) = (x_i, y_i)$. We now can represent factor systems of S on the factor sets X_i and Y_i; i.e., for each $i \in I$, the factor system S_i of S can be defined by

$$S_i = \prod_i (S) \subset X_i \times Y_i \tag{10.97}$$

The system S can now be viewed as decomposed into a family of factor systems $\overline{S} = \{S_i : i \in I\}$, and S can be represented as a relation among its factor systems (i.e., $S \subset \prod\{S_i : i \in I\}$) by identifying $((x_i)_{i\in I}, (y_i)_{i\in I}) \in S \subset X \times Y$ with $((x_i, y_i)_{i\in I}) \in S \subset \prod\{S_i : i \in I\}$. The systems S and S_i are called the overall system and the factor (or component) system, respectively, and $S \subset \prod\{S_i : i \in I\}$ is called a complex system representation over $\{S_i : i \in I\}$.

Generally, there are many interactions among subsystems, so S is a proper subset of the product of its factor systems $\prod\{S_i : i \in I\}$. If $S = \prod\{S_i : i \in I\}$, S is called noninteracted

We now consider what properties are transferable from the overall system S to its component systems S_i or from the component systems S_i to an overall system S.

Proposition 10.4.1. *Let $S \subset \prod\{S_i : i \in I\}$ be an input–output system which has been decomposed into a family of factor systems $\overline{S} = \{S_i : i \in I\}$. If S is a linear system over a field $\mathcal{A}$, each component system S_i is then also a linear system over the field $\mathcal{A}$.*

Proof: From the hypothesis that S is a linear system over the field $\mathcal{A}$, it follows that, for each $i \in I$, S_i is a nonempty system on $X_i \times Y_i$. We need to show two things: (1) For any $(x_1,y_1),(x_2,y_2) \in S_i, (x_1,y_1)+(x_2,y_2) \in S_i$; (2) for any $(x,y) \in S_i$ and $\alpha \in \mathcal{A}$, $\alpha(x,y) \in S_i$.

The argument of (1). For any (a_i,c_i) and $(b_i,d_i) \in S_i$, there are (x_1,y_1) and $(x_2,y_2) \in S$ such that $\prod_i(x_1,y_1) = (a_i,c_i)$ and $\prod_i(x_2,y_2) = (b_i,d_i)$. Therefore,

$$\begin{aligned}(a_i,c_i)+(b_i,d_i) &= (a_i+b_i, c_i+d_i)\\ &= \prod_i(x_1+x_2, y_1+y_2) \in S_i\end{aligned} \tag{10.98}$$

because $(x_1+x_2, y_1+y_2) \in S$.

The argument of (2). For any $(a_i,c_i) \in S_i$ and any $\alpha \in \mathcal{A}$, there exists a pair $(x,y) \in S$ such that $\prod_i(x,y) = (a_i,c_i)$. Therefore,

$$\begin{aligned}\alpha(a_i,c_i) &= (\alpha a_i, \alpha c_i)\\ &= \prod_i(\alpha x, \alpha y)\\ &= \prod_i(\alpha(x,y))\\ &\in S_i\end{aligned} \tag{10.99}$$

because $\alpha(x,y) \in S$. ■

Proposition 10.4.2. *Suppose that $S \subset X \times Y$ is a linear time system that has been decomposed into a family of component systems $\{S_i \subset X_i \times Y_i : i \in I\}$; i.e., $S \subset \prod\{S_i : i \in I\}$. Then each component system S_i is also a linear time system.*

Proof: It suffices from Proposition 10.4.1 to show that for any $x,y \in D(S_i)$ and any $t \in T$, the concatenation of x,y, $x^t \circ y_t \in D(S_i)$ for each $i \in I$. In fact, for any $x,y \in D(S_i)$ and any $t \in T$, there exist $u,v \in D(S)$ such that

$$\prod_{ix}(u) = x \quad \text{and} \quad \prod_{ix}(v) = y \tag{10.100}$$

Therefore, $x^t \circ y_t = \prod_{ix}(u^t \circ v_t) \in D(S_i)$, because $u^t \circ v_t \in D(S)$. ■

Proposition 10.4.3. *Under the same assumption as Proposition 10.4.2 and if the system S is strongly stationary, then each component system S_i is also strongly stationary.*

Proof: Let $(x_t,y_t) \in S_i|T_t$ be arbitrary. There then exists some element $(x_i,y_i) \in S_i$ such that $(x_t,y_t) = (x_{it}, y_{it})$. Furthermore, there exists some $(x,y) \in S$ such that $\prod_i(x,y) = (x_i,\ y_i)$. Since $\lambda^t S \subset S$, $\lambda^t(x,y) = (\lambda^t(x), \lambda^t(y)) \in S$, where $\prod_i(\lambda^t x, \lambda^t y) = (\lambda^t x_i, \lambda^t y_i)$, which implies that $(\lambda^t x_i,\ \lambda^t y_i) \in S_i$. Hence, $(x_t,y_t) = (x_{it}, y_{it}) \in \sigma^t(S_i)$. Consequently, $S_i|T_t \subset \sigma^t(S_i)$.

Conversely, let $(x_t,y_t) \in \sigma^t(S_i)$ be arbitrary. There then exists some $(x_i,y_i) \in S_i$ such that $(x_t,y_t) = \sigma^t(x_i,y_i)$. Furthermore, there exists some $(x,y) \in S$ such that $\prod_i(x,y) = (x_i,y_i)$. Since $\sigma^t(S) \subset S|T_t$, we have $\sigma^t(x,y) = (\sigma^t(x), \sigma^t(y)) \in \sigma^t S \subset S|T_t$. That is, there exists some $(x',y') \in S$ such that $(\sigma^t(x), \sigma^t(y)) = (x',y')|T_t$. Hence,

$$\begin{aligned}(x_t,y_t) &= \sigma^t(x_i,y_i) \\ &= \sigma^t \prod_i(x,y) \\ &= \prod_i(\sigma^t(x), \sigma^t(y)) \\ &= \prod_i[(x',y')|T_t]\end{aligned}$$

That is, $\sigma^t(S_i) \subset S_i|T_t$.

Combining these arguments, we obtain the following: For each index $i \in I$ and any $t \in T$, $S_i|T_t = \sigma^t(S_i)$; i.e., $\sigma^{-t}(S_i|T_t) = S_i$. From Eq. (10.66), we can show that the component system S_i is strongly stationary. ■

Proposition 10.4.4. *Under the same assumption as in Proposition 10.4.2 and if the system S is precausal with $D(S) = X$, each component system S_i is also precausal.*

Proof: Let $i \in I$ be arbitrarily fixed. It can then be seen that $D(S_i) = X_i$. In fact, if $D(S_i) \neq X_i$, we can pick an element $x_i \in X_i - D(S_i)$. This implies that $D(S) \neq X$, because no $x \in D(S)$ satisfies $\prod_{ix}(x) = x_i$, contradiction.

Let $x_i, x_i' \in X_i$, $t \in T$ arbitrary, and assume $\overline{x}_i^t = \overline{x}_i'^t$. Since it is assumed that S is precausal, for any $x,y \in X$, $\overline{x}^t = \overline{y}^t$ implies that $S(x)|\overline{T}^t = S(y)|\overline{T}^t$, from Theorem 10.3.6. Therefore, when the elements $x,y \in X$ satisfy $\prod_{ix}(x) = x_i$, $\prod_{ix}(y) = x_i'$ and $\prod_j(x) = \prod_j(y)$, for every $j \in I$ with $j \neq i$, we have $\overline{x}^t = \overline{y}^t$. Thus

$$\begin{aligned} S_i(x_i)|\overline{T}^t &= \prod_{iy}(S(x))|\overline{T}^t \\ &= \prod_{iy}(S(y))|\overline{T}^t \\ &= S_i(x_i')|\overline{T}^t \end{aligned} \tag{10.101}$$

Applying Theorem 10.3.6 we have shown that each component system S_i is precausal. ■

Example 10.4.1. We will construct an example to show that even though an overall system S is functional, none of the component systems S_i is functional.

Suppose the index set $I = \{1,2\}$ and the sets X_1, X_2, Y_1, and Y_2 are all the same, namely the set $\mathbb{Z}$ of all integers. Let S be a functional linear system from $X_1 \times X_2$ into $Y_1 \times Y_2$ defined by

$$S(n,m) = (m,n) \tag{10.102}$$

for any $(n,m) \in X_1 \times X_2$. Then, $S_1 = \{(n,m) : (n,m) \in X_1 \times Y_1\}$ and $S_2 = \{(m,n) : (m,n) \in X_2 \times Y_2\}$. Then it can be easily seen that neither S_1 nor S_2 is functional.

Proposition 10.4.5. *Suppose that $S \subset X \times Y$ is a functional system that has been decomposed into a family of component systems $\{S_i \subset X_i \times Y_i : i \in I\}$ with $S = \prod\{S_i : i \in I\}$. Then each component system S_i is also functional. If, moreover, the overall system S is bijective, each component system S_i is also bijective.*

Proof: For any fixed $i \in I$, let $(x_i,y_i), (x_i,z_i) \in S_i$ be arbitrary. Since $S = \prod\{S_i : i \in I\}$ due to the assumption, there exist elements $(x,y), (x,z) \in S$ such that $\prod_i(x,y) = (x_i, y_i)$ and $\prod_i(x,z) = (x_i, z_i)$. Because S is functional, $y = z$, so, $y_i = \prod_{iy}(y) = \prod_{iy}(z) = z_i$. That is, the component system S_i is functional.

Suppose S is a bijective function from X onto Y. For any $y_i \in Y_i$ there exists an element $y \in Y$ such that $\prod_{iy}(y) = y_i$. From the hypothesis that S is bijective, it follows that there exists an element $x \in X$ such that $S(x) = y$. Therefore, for $x_i = \prod_{ix}(x)$, $S_i(x_i) = y_i$. That is, S_i is surjective. Let (x_i,y_i) and $(w_i,y_i) \in S_i$ be arbitrary. Then there exist (x,y) and $(w,y) \in S$ such that $\prod_i(x,y) = (x_i,y_i)$ and

$\prod_i(w,y) = (w_i,y_i)$. From the hypothesis that S is injective, it follows that $x = w$, that is, $x_i = \prod_{ix}(x) = \prod_{ix}(w) = w_i$. Hence, S_i is bijective. ■

Proposition 10.4.6. *Suppose that $S \subset X \times Y$ is a functional linear time system that S has been decomposed into a family of component systems $\{S_i \subset X_i \times Y_i : i \in I\}$ such that S is noninteracted; i.e., $S = \prod\{S_i : i \in I\}$. If S is time invariably realizable, each component system S_i is then also time invariably realizable.*

Proof: From the hypothesis that S is a functional system from X into Y, it follows that $D(S) \subset X$. Thus, from the proof of Proposition 10.4.4, for each fixed $i \in I, D(S_i) = X_i$.

We now let $t \in T$ and $x_i \in X_i$ be arbitrary. There exists an element $x_0 \in X$ such that $\prod_{ix}(x_0) = x_i$. Since S is time invariably realizable, from Eq. (10.76) it follows that for $t \in T$,

$$\lambda^t S(0^t \circ \sigma^t(x_0)) = S(x_0) \tag{10.103}$$

Let $A = \{x \in X : \prod_{ix}(x) = x_i\}$. Then $x_0 \in A \neq \emptyset$ and each $x \in A$ satisfies Eq. (10.103), which implies that

$$\bigcup\{\lambda^t S(0^t \circ \sigma^t(x)) : x \in A\} = \bigcup\{S(x) : x \in A\} \tag{10.104}$$

Therefore,

$$\begin{aligned} S_i(x_i) &= \prod_{ix}(\bigcup\{S(x) : x \in A\}) \\ &= \prod_{ix}(\bigcup\{\lambda^t S(0^t \circ \sigma^t(y)) : y \in A\}) \\ &= \bigcup\{\prod_{ix} \lambda^t S(0^t \circ \sigma^t(y)) : y \in A\} \\ &= \bigcup\{\lambda^t \prod_{ix} S(0^t \circ \sigma^t(y)) : y \in A\} \\ &= \bigcup\{\lambda^t S_i[\prod_{ix}(0^t \circ \sigma^t(y))] : y \in A\} \\ &= \bigcup\{\lambda^t S_i(0^t \circ \prod_{ix} \sigma^t(y)) : y \in A\} \\ &= \bigcup\{\lambda^t S_i(0^t \circ \sigma^t(\prod_{ix}(y))) : y \in A\} \\ &= \lambda^t S_i(0^t \circ \sigma^t(x_i)) \end{aligned} \tag{10.105}$$

Applying Eq. (10.76), we complete the proof that S_i is time invariably realizable.■

The following result combines Propositions 10.4.1–10.4.6.

Theorem 10.4.1. *Let $S \subset X \times Y$ be an input–output system which has been decomposed into a family of factor systems $\overline{\mathbb{S}} = \{S_i : i \in I\}$ such that $S = \prod\{S_i : i \in I\}$. Then the following statements hold:*

(1) S is a linear system if and only if each S_i is a linear system.

(2) S is a linear time system if and only if each S_i is a linear time system.

(3) S is strongly stationary if and only if each S_i is strongly stationary.

(4) S is causal if and only if each S_i is causal.

(5) S is bijective if and only if each S_i is bijective.

(6) The functional system S is time invariably realizable if and only if each S_i is functional and time invariably realizable.

The proof is not hard and is left to the reader.

Theorem 10.4.1 shows that the theoretical properties of many important systems are either hereditary from the overall system S to each component system S_i or able to be lifted up from S_i to S. In practice, we always hope to study S by studying each component system S_i.

Let $S \subset X \times Y$ be a linear system. We say that S can be decoupled by feedback if there exists some functional system $S_f \in \mathbb{S}_f$ such that

$$F(S,S_f) = \prod\{\prod_i(F(S,S_f)) : i \in I\} \tag{10.106}$$

where it is assumed that $X = \prod\{X_i : i \in I\}$, $Y = \prod\{Y_i : i \in I\}$ and $\prod_i : X \times Y \to X_i \times Y_i$ is the projection defined by relation (10.96).

The concept of decoupling by feedback shows the following actions: The original system S may not be noninteracted, so there might be some properties which cannot be lifted up from component systems to the overall system S. If S is decoupled by feedback, after applying appropriate feedback transformation, S is transformed to a system which has a noninteracted complex systems representation. We now can use that representation to analyze and control S.

Theorem 10.4.2. *Suppose that $S \subset X \times Y$ and $S_f : Y \to X$ are a linear system and a functional linear system, respectively, such that S is decomposed into $\overline{\mathbb{S}} = \{S_i : i \in I\}$ and S_f is decomposed into $\overline{\mathbb{S}}_f = \{S_{fi} : i \in I\}$. Then, for each $i \in I$,*

$$\prod_i(F(S,S_f)) = F(S_i,S_{fi}) \tag{10.107}$$

Proof: For each $(x_i,y_i) \in \prod_i(F(S,S_f))$, there exists $(x,y) \in F(S,S_f)$ such that $\prod_i(x,y) = (x_i,y_i)$. Since, $\prod_{ix}(S_f(y)) = S_{fi}(y_i)$, it follows that

$$\begin{aligned}(x_i + S_{fi}(y_i),y_i) &= (\prod_{ix}(x+S_f(y)),\prod_{iy}(y)) \\ &= \prod_i(x+S_f(y),y) \\ &\in \prod_i(S) \\ &= S_i \end{aligned} \tag{10.108}$$

Equation (10.108) implies that $(x_i,y_i) \in F(S_i,S_{fi})$. That is, $\prod_i(F(S,S_f)) \subset F(S_i,S_{fi})$.

Let $(x_i,y_i) \in F(S_i,S_{fi})$ be arbitrary. There then exists $(x,y) \in S$ such that

$$\prod_i(x,y) = (x_i + S_{fi}(y_i),y_i) \tag{10.109}$$

Therefore, $(x - S_f(y),y) \in F(S,S_f)$. From this, we have

$$\begin{aligned}(x_i,y_i) &= (x_i + S_{fi}(y_i) - S_{fi}(y_i),y_i) \\ &= (\prod_{ix}(x - S_f(y)),\prod_{iy}(y)) \\ &= \prod_i(x - S_f(y),y) \\ &\in \prod_i(F(S,S_f)) \end{aligned} \tag{10.110}$$

Equation (10.110) gives $F(S_i,S_{fi}) \subset \prod_i(F(S,S_f))$. ■

Theorem 10.4.3. *Assume the Axiom of Choice. Suppose that $S \subset X \times Y$ is a linear system that can be decomposed into a family of component systems $\overline{S} = \{S_i : i \in I\}$. Then S can be decoupled by feedback if and only if $R(S) = \prod\{R(S_i) : i \in I\}$ and $N(S) = \prod\{N(S_i) : i \in I\}$.*

Proof: Necessity. Suppose S can be decoupled by feedback. By the definition of decoupling by feedback, there exists a functional linear system $S_f \in \mathcal{S}_f$ such that

$$F(S,S_f) = \prod\{\prod_i(F(S,S_f)) : i \in I\} \tag{10.111}$$

Therefore, by Theorem 10.2.5,

$$\begin{aligned} R(S) &= R(F(S, S_f)) \\ &= \prod\{R(\prod_i(F(S, S_f))) : i \in I\} \\ &= \prod\{R(F(S_i, S_{fi})) : i \in I\} \\ &= \prod\{R(S_i) : i \in I\} \end{aligned} \tag{10.112}$$

and

$$\begin{aligned} N(S) &= N(F(S, S_f)) \\ &= \prod\{N(\prod_i(F(S, S_f))) : i \in I\} \\ &= \prod\{N(F(S_i, S_{fi})) : i \in I\} \\ &= \prod\{N(S_i) : i \in I\} \end{aligned} \tag{10.113}$$

Sufficiency. Let $S' \subset X \times Y$ be the system defined by

$$S' = \prod\{S_i : i \in I\} \tag{10.114}$$

Since $R(S) = \prod\{R(S_i) : i \in I\}$ and $N(S) = \prod\{N(S_i) : i \in I\}$ due to our assumption, we see that $R(S) = R(S')$ and $N(S) = N(S')$. By applying Theorem 10.2.5, we find there exists a functional linear system $S_f \in \mathcal{S}_f$ such that

$$\begin{aligned} F(S, S_f) &= S' \\ &= \prod\{S_i : i \in I\} \\ &= \prod\{\prod_i(F(S, S_f)) : i \in I\} \end{aligned} \tag{10.115}$$

That is, S can be decoupled by feedback. ■

Example 10.4.2. As an example of decoupling by feedback, consider the multivariable system S defined by

$$\begin{aligned} \dot{z} &= Az + Bx \\ y &= Cz + Dx \\ z(0) &= 0 \end{aligned} \tag{10.116}$$

where z is an $m \times 1$ variable vector, A an $m \times m$ constant matrix, B an $m \times n$ constant matrix, C an $n \times m$ constant matrix, D an $n \times n$ nonsingular constant matrix, and

$$X = \{x : [0, \infty) \to \mathbb{R}^n : x \text{ is piecewise continuous}\} \tag{10.117}$$

is the input space and

$$Y = \{y : [0,\infty) \to \mathbb{R}^n : y \text{ is piecewise continuous}\} \tag{10.118}$$

is the output space. For each $i = 1,2,\ldots,n$, define the system S_i by

$$\begin{aligned} \dot{z} &= Az + B_i x_i \\ y_i &= C_i z + D_i x_i \\ z(0) &= 0 \end{aligned} \tag{10.119}$$

where B_i is an $m \times 1$ constant matrix such that $B = [B_1 B_2 \cdots B_n]$, C_i is a $1 \times m$ constant matrix such that

$$C = \begin{bmatrix} C_1 \\ C_2 \\ \vdots \\ C_n \end{bmatrix} \tag{10.120}$$

D_i is a nonzero constant, and

$$X_i = \{x_i : [0,\infty) \to \mathbb{R} : x_i \text{ is piecewise continuous}\} \tag{10.121}$$

is the input space and

$$Y_i = \{y_i : [0,\infty) \to \mathbb{R} : y_i \text{ is piecewise continuous}\} \tag{10.122}$$

is the output space. Then the systems S and S_i, $i = 1,2,\ldots,n$, satisfy

$$R(S) = \prod\{R(S_i) : i \in I\} \quad \text{and} \quad N(S) = \prod\{N(S_i) : i \in I\} \tag{10.123}$$

In fact, for any $x \in X$, the solution of the differential equation (10.116) in S is given by

$$\phi(t,0,x) = \int_0^t e^{(t-s)A} \cdot B \cdot x(s)\, ds \tag{10.124}$$

Thus, $D(S) = X$. At the same time, since the matrix D is nonsingular, the inverse system S^{-1} of S is obtained as follows:

$$\begin{aligned} \dot{z} &= (A - BD^{-1}C)z + BD^{-1}y \\ x &= -D^{-1}Cz + D^{-1}y \\ z(0) &= 0 \end{aligned} \tag{10.125}$$

Therefore, a similar argument shows that $R(S) = D(S^{-1}) = Y$. For the same reason, we can show that for each $i = 1,2,\ldots,n$, $R(S_i) = Y_i$. That is,

$$\begin{aligned} R(S) &= Y \\ &= \prod\{Y_i : i = 1,2,\ldots,n\} \\ &= \prod\{R(S_i) : i = 1,2,\ldots,n\} \end{aligned} \tag{10.126}$$

For the null spaces of the systems S and S_i, $i = 1,2,\ldots,n$,

$$N(S) = \{-D^{-1}Cz : \dot{z} = (A - BD^{-1}C)z\} \tag{10.127}$$

and

$$N(S_i) = \{-D_i^{-1}C_iz : \dot{z} = (A - B_iD_i^{-1}C_i)z\} \tag{10.128}$$

Therefore, $N(S) \doteq \prod\{N(S_i) : i = 1,2,\ldots,n\}$. By Theorem 10.4.3, S can be decoupled by feedback. In fact, define

$$\alpha = \begin{bmatrix} A & & & & 0 \\ & A & & & \\ & & \cdot & & \\ & & & \cdot & \\ & & & & \cdot \\ 0 & & & & A \end{bmatrix}$$

$$\beta = \begin{bmatrix} B_1 & & & & 0 \\ & B_2 & & & \\ & & \cdot & & \\ & & & \cdot & \\ & & & & \cdot \\ 0 & & & & B_n \end{bmatrix}$$

$$\gamma = \begin{bmatrix} C_1 & & & & 0 \\ & C_2 & & & \\ & & \cdot & & \\ & & & \cdot & \\ & & & & \cdot \\ 0 & & & & C_n \end{bmatrix}$$

$$\delta = \begin{bmatrix} D_1 & & & & 0 \\ & D_2 & & & \\ & & \cdot & & \\ & & & \cdot & \\ & & & & \cdot \\ 0 & & & & D_n \end{bmatrix}$$

Then the system $S_d = S_1 \times S_2 \times \cdots \times S_n$ is represented by

$$\begin{aligned} \dot{z} &= \alpha z + \beta x \\ y &= \gamma z + \delta x \\ z(0) &= 0 \end{aligned} \tag{10.129}$$

Since δ is nonsingular, S_d^{-1} is obtained as follows:

$$\begin{aligned} \dot{z} &= (\alpha - \beta\delta^{-1}\gamma)z + \beta\delta^{-1}y \\ x &= -\delta^{-1}\gamma z + \delta^{-1}y \\ z(0) &= 0 \end{aligned} \tag{10.130}$$

If we define a functional linear system $S_f : Y \to X$ as follows:

$$\begin{aligned} \begin{bmatrix} \dot{z} \\ \dot{z}' \end{bmatrix} &= \begin{bmatrix} A - BD^{-1}C & 0 \\ 0 & \alpha - \beta\delta^{-1}\gamma \end{bmatrix} \cdot \begin{bmatrix} z \\ z' \end{bmatrix} + \begin{bmatrix} BD^{-1} \\ \beta\delta^{-1} \end{bmatrix} \cdot y \\ x &= \begin{bmatrix} -D^{-1}C & \delta^{-1}\gamma \end{bmatrix} \cdot \begin{bmatrix} z \\ z' \end{bmatrix} + (D^{-1} - \delta^{-1}) \cdot y \\ \begin{bmatrix} z \\ z' \end{bmatrix}(0) &= 0 \end{aligned} \tag{10.131}$$

Then the original system S can be transferred into S_d by this feedback component system S_f. In fact, for any pair $(x,y) \in X \times Y$, $(x,y) \in S_d$ iff the input x and the output y satisfy the systems representation in Eq. (10.130) iff $(x + S_f(y), y) \in S$. We show the last "if and only if" condition as follows.

First, suppose that the input x and the output y satisfy the systems representation in Eq. (10.130). Let z' be the solution of the differential equation in the representation in Eq. (10.130), and let z be the solution of the differential equation in the systems representation in Eq. (10.125). Then $[z\ z']^T$ is the solution of the differential equation in the systems representation in Eq. (10.131). From the definition of S_f, it follows that

$$\begin{aligned} S_f(y) &= -D^{-1}Cz + \delta^{-1}\gamma z' + D^{-1}y - \delta^{-1}y \\ &= (-D^{-1}Cz + D^{-1}y) - (-\delta^{-1}\gamma z' + \delta^{-1}y) \end{aligned} \tag{10.132}$$

Therefore,

$$(x + S_f(y), y) = (-D^{-1}Cz + D^{-1}y, y) \in S \tag{10.133}$$

where the membership relation comes from the systems representation in Eq. (10.125) of the inverse system S^{-1}.

Second, suppose that a pair $(x,y) \in X \times Y$ satisfies $(x + S_f(y), y) \in S$. From the systems representations in Eqs. (10.125) and (10.131), we have

$$x + S_f(y) = -D^{-1}Cz + D^{-1}y \tag{10.134}$$

That is,

$$\begin{aligned} x &= (-D^{-1}Cz + D^{-1}y) - S_f(y) \\ &= (-D^{-1}Cz + D^{-1}y) - (-D^{-1}Cz + \delta^{-1}\gamma z' + D^{-1}y - \delta^{-1}y) \\ &= -\delta^{-1}\gamma z' + \delta^{-1}y \end{aligned} \tag{10.135}$$

Combining Eq. (10.135) with the systems representation in Eq. (10.130) proves that the pair (x,y) satisfies the systems representation in Eq. (10.130).

10.5. Decomposability Conditions

Intuitively speaking, a complex system consists of several subsystems (see, for example, Sections 9.3 and 9.6). Let us first consider the following example. Suppose that

$$S_i \subset X_i \times Y_i \tag{10.136}$$

for $i = 1,2,3$, are three input–output systems such that $X_1 = X_2 = X_3 = Y_1 = Y_2 = Y_3 = \mathbb{R}$, $(x_i, y_i) \in S_i$, $i = 1,2,3$ if and only if $y_1 = x_1$, $y_2 = 2x_2$, and $y_3 = 2x_3$. Suppose these three systems are connected in series; that is, $x_2 = y_1$ and $x_3 = y_2$ hold; furthermore, suppose $\widehat{X} = X_1$ and $\widehat{Y} = Y_3$ are the input space and the output space of a new system $\widehat{S}$, defined as

$$\widehat{S} = \{(x,y) \in \widehat{X} \times \widehat{Y} : y = 4x\} \tag{10.137}$$

Then $\widehat{S}$ is the composition of S_1, S_2, and S_3. However, in Eq. (10.137), the component systems S_1, S_2, and S_3 completely disappeared.

Let us now consider another construction of a new system S. We let

$$X = X_1 \times X_2 \times X_3 \quad \text{and} \quad Y = Y_1 \times Y_2 \times Y_3 \tag{10.138}$$

and the new system $S \subset X \times Y$ is defined by

$$\begin{aligned} S = \{((x_1,x_2,x_3),(y_1,y_2,y_3)) \in X \times Y :& \\ (x_i,y_i) \in S_i \text{ and } x_2 = y_1 \text{ and } x_3 = y_2\}& \end{aligned} \tag{10.139}$$

Figures 10.10 and 10.11 show the geometric meanings of $\widehat{S}$ and S. The system S is a representation of the whole system, which explicitly reflects the fact that

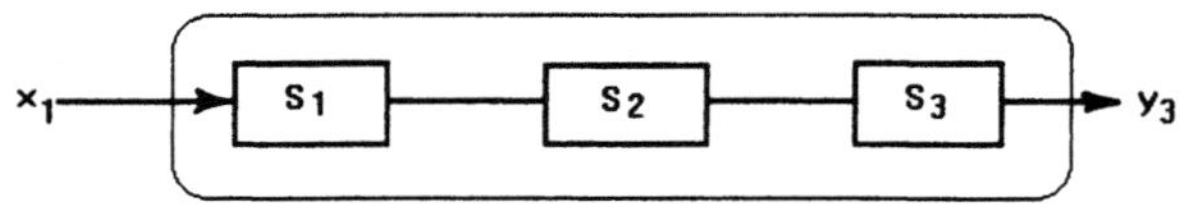

Figure 10.10. Structure of the system $\widehat{S}$.

the whole system is composed of three component systems S_i, $i = 1,2,3$. The structure of $\widehat{S}$ is called a global system representation of the whole system, while S is a complex system representation of the whole system. In fact, S is the product system of S_i, $i = 1,2,3$.

The complex system S is defined over X and Y, but to emphasize fact that S consists of the component systems S_1, S_2, and S_3, we identify $((x_1,x_2,x_3),(y_1,y_2,y_3)) \in X \times Y$ with $((x_1,y_1),(x_2,y_2),(x_3,y_3)) \in S_1 \times S_2 \times S_3$ and denote the complex system S by $S \subset S_1 \times S_2 \times S_3$. For the global system $\widehat{S}$, we emphasize the structure by writing $\widehat{S} = S_3 \circ S_2 \circ S_1$.

The global system $\widehat{S}$ and the complex system S are intimately related to each other. In fact, if we define two mappings $h_1 : X \to \widehat{X}$ and $h_2 : Y \to \widehat{Y}$ by $h_1(x_1, x_2, x_3) = x_1$ and $h_2(y_1,y_2,y_3) = y_3$, then

$$\forall(x,y) \in X \times Y((x,y) \in S \to (h_1(x), h_2(y)) \in \widehat{S}) \tag{10.140}$$

and

$$\forall(\widehat{x},\widehat{y}) \in \widehat{X} \times \widehat{Y}((\widehat{x},\widehat{y}) \in \widehat{S} \to \exists(x,y) \in S((h_1(x),h_2(y)) = (\widehat{x},\widehat{y}) \in \widehat{S})) \tag{10.141}$$

If we define two mappings $k_1 : \widehat{X} \to X$ and $k_2 : \widehat{Y} \to Y$ by $k_1(\widehat{x}) = (\widehat{x},\widehat{x},2\widehat{x})$ and $k_2(\widehat{y}) = (\widehat{y}/4,\widehat{y}/2,\widehat{y})$, then k_1 and k_2 satisfy the relations

$$\forall(\widehat{x},\widehat{y}) \in \widehat{X} \times \widehat{Y}((\widehat{x},\widehat{y}) \in \widehat{S} \to (k_1(\widehat{x}),k_2(\widehat{y})) \in S) \tag{10.142}$$

and

$$\begin{aligned}\forall(\widehat{x},\widehat{y}),(\widehat{x}',\widehat{y}') \in \widehat{X} \times \widehat{Y}&(k_1(\widehat{x}),k_2(\widehat{y})) \\ &= (k_1(\widehat{x}'),k_2(\widehat{y}')) \to (\widehat{x},\widehat{y}) = (\widehat{x}',\widehat{y}')\end{aligned} \tag{10.143}$$

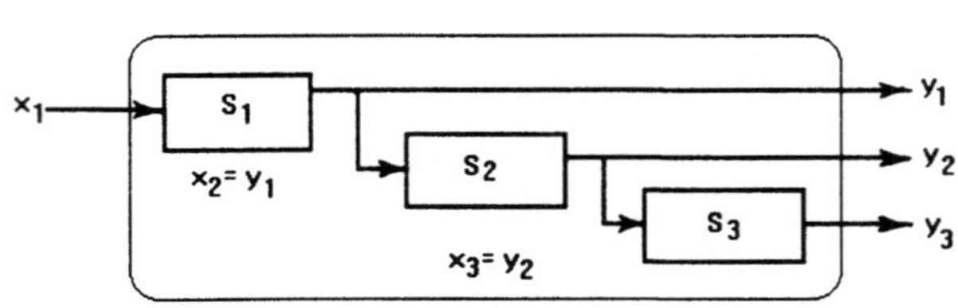

Figure 10.11. Structure of the system S.

Combining relations (10.140)–(10.143), we can see that for any $(x,y) \in X \times Y$,

$$(k_1,k_2) \circ (h_1,h_2)(x,y) = (k_1,k_2)(h_1(x),h_2(y))$$
$$= (k_1h_1(x),k_2h_2(y)) = (x,y) \qquad (10.144)$$

and for any $(\widehat{x},\widehat{y}) \in \widehat{X} \times \widehat{Y}$,

$$(h_1,h_2) \circ (k_1,k_2)(\widehat{x},\widehat{y}) = (h_1,h_2)(k_1(\widehat{x}),k_2(\widehat{y}))$$
$$= (h_1k_1(\widehat{x}),h_2k_2(\widehat{y}))$$
$$= (\widehat{x},\widehat{y}) \qquad (10.145)$$

Therefore, S and $\widehat{S}$ are similar.

Generally, let $S \subset X \times Y$ and $S' \subset X' \times Y'$ be two input–output systems, and h_i, $i = 1,2$, be mappings such that

$$h_1 : X \to X' \quad \text{and} \quad h_2 : Y \to Y' \qquad (10.146)$$

satisfying the condition

$$\forall(x,y) \in X \times Y((x,y) \in S \to (h_1(x),h_2(y)) \in S') \qquad (10.147)$$

Then $h = (h_1,h_2)$ is called a modeling relation from S to S' and is denoted by $(h_1,h_2) : S \to S'$ or $S \xrightarrow{h} S'$. Notice that h is actually a mapping from S into S'.

Proposition 10.5.1. *Suppose that $h = (h_1,h_2)$ and $k = (k_1,k_2)$ are modeling relations from system S_1 to system S_2 and from S_2 to system S_3, respectively. Then the composition $k \circ h = (k_1,k_2) \circ (h_1,h_2) = (k_1 \circ h_1, k_2 \circ h_2)$ is a modeling relation from S_1 to S_3.*

The proof is straightforward and is omitted.

Suppose $\{X_i : i \in I\}$ and $\{Y_i : i \in I\}$ are families of sets and that $\{f_i : X_i \to Y_i : i \in I\}$ is a family of mappings. The product mapping $f = \prod_{i\in I} f_i : \prod\{X_i : i \in I\} \to \prod\{Y_i : i \in I\}$ is defined by

$$f((x_i)_{i\in I}) = (f_i(x_i))_{i\in I} \qquad (10.148)$$

for any $(x_i)_{i\in I} \in \prod\{X_i : i \in I\}$. When $I = \{1,2,\ldots,n\}$ is finite, we denote the product mapping f by $f_1 \times f_2 \times \cdots \times f_n$. The following result shows the relation between the concept of modeling relations and that of product mappings.

Proposition 10.5.2. *Suppose that $h = (h_1,h_2)$ and $k = (k_1,k_2)$ are two modeling relations from system S to system S'. Then the restriction of the product mappings $h_1 \times h_2$ and $k_1 \times k_2$ on S are mappings from S to S'. Furthermore, if $D(S) = X$ and $R(S) = Y$, then $h_1 \times h_2|S = k_1 \times k_2|S$ if and only if $h_i = k_i$, $i = 1,2$.*

The proof is straightforward and is omitted.

Due to Proposition 10.5.2, we still use h for the mapping $h_1 \times h_2|S$ and the modeling relation $(h_1,h_2) : S \to S'$, whenever no confusion occurs.

A modeling relation $h = (h_1,h_2) : S \to S'$ is called an epimorphism from the system S into the system S' if for any fixed system S'' and modeling relations h' and h'' from S' to S'', $h' \circ h = h'' \circ h$ implies $h' = h''$.

Theorem 10.5.1. *Let $h = (h_1,h_2)$ be a modeling relation from a system S to a system S' such that $D(S) = X$, $R(S) = Y$, $D(S') = X'$, and $R(S') = Y'$. Then h is an epimorphism from S to S' if and only if h_1 and h_2 are surjective.*

Proof: Necessity. Suppose h is an epimorphism from S to S' and, without loss of generality, the mapping h_1 is not surjective. That is, $h_1(X) \neq D(S') = X'$. There exist mappings h_1' and h_1'' such that

$$h_1'|h_1(X) = h_1''|h_1(X) \quad \text{and} \quad h_1' \neq h_1'' \tag{10.149}$$

where, for each arbitrarily fixed system $S'' \subset X'' \times Y''$, h_1' and h_1'' are from X' to X''. Pick mappings h_2' and $h_2'' : Y' \to Y''$ such that $h_2' = h_2''$. Then we $h' \circ h = h'' \circ h$, but $h' \neq h''$. This contradicts the fact that h is an epimorphism from S to S'.

Sufficiency. Suppose h_1 and h_2 are surjective. Let S'' be an arbitrary system over $X'' \times Y''$ and $h' = (h_1', h_2')$ and $h'' = (h_1'', h_2'')$ modeling relations from S' to S'' such that $h' \circ h = h'' \circ h$. Then

$$\begin{aligned} h' \circ h = h'' \circ h &\to h' \circ h(x,y) = h'' \circ h(x,y) \\ &\to h_1' \circ h_1(x) = h_1'' \circ h_1(x) \\ &\to h_1' = h_1'' \end{aligned} \tag{10.150}$$

The last implication follows because $h_1 : X \to X'$ is surjective and because $D(S) = X$. A similar argument gives $h_2' = h_2''$. Hence, applying Proposition 10.5.2, we have $h' = h''$. That is, h is an epimorphism from S to S'. ■

Theorem 10.5.2. *Under the same assumption as in Theorem 10.5.1, if the modeling relation h is also surjective as a mapping (that is, $h_1 \times h_2|S$ is surjective), then h is an epimorphism. The converse is not true.*

Proof: It can be clearly seen from the assumptions that the mappings h_1 and h_2 are surjective. Applying Theorem 10.5.1, it follows that h is an epimorphism.

We now see an example when the converse is not true. Let $X = Y$, $S = \{(x,x) : x \in X\}$, $S' = X \times X$, and $h_1 = h_2 = \mathrm{id}_X$. Then $h = (h_1,h_2)$ is an epimorphism but not surjective. ■

Theorem 10.5.3. *Under the same assumption as in Theorem 10.5.1, if the modeling relation h is also a relation (that is, h has a right cancellation), then h is surjective as a mapping. The converse is not true.*

Proof: Let $k : S' \to S$ be a right cancellation of h; i.e., $h \circ k = (\mathrm{id}_X, \mathrm{id}_Y)$. Let $(x', y') \in S'$ be arbitrary. Then $h \circ k(x', y') = h(k(x', y')) = (x', y')$. That is, h is surjective.

We now construct an example to show the converse may not be true. Suppose that $S \subset X \times Y$ and $S' \subset X' \times Y'$ are systems defined by

$$S = \{(x_1, y_1), (x_2, y_2), (x_3, y_2)\} \tag{10.151}$$

and

$$S' = \{(x'_1, y'_1), (x'_1, y'_2), (x'_2, y'_2)\} \tag{10.152}$$

Let $h_1 : D(S) \to D(S')$ and $h_2 : R(S) \to R(S')$ be two mappings defined by $h_1(x_1) = h_1(x_2) = x'_1$, $h_1(x_3) = x'_2$, $h_2(y_1) = y'_1$, $h_2(y_2) = y'_2$. If $h = (h_1, h_2)$ has a right cancellation $k = (k_1, k_2)$, we have $k_1(x_1) = x_1$ or x_2, $k_2(y'_1) = y_1$, and $k_2(y'_2) = y_2$. *Case 1:* Suppose that $k_1(x'_1) = x_1$. It then follows that $(k_1, k_2)(x'_1, y'_2) = (x_1, y_2) \notin S$, which contradicts the hypothesis that k is a modeling relation from S' to S. *Case 2:* Suppose $k_1(x'_1) = x_2$. We then have that $(k_1, k_2)(x'_1, y'_1) = (x_2, y_1) \notin S$, which contradicts the hypothesis that k is a modeling relation from S' to S. The contradictions imply that the modeling relation h does not have a right cancellation; i.e., h is not a retraction. ■

A modeling relation h from a system S to a system S' is an isomorphism from S onto S' if h has both right and left cancellation. In this case, S is isomorphic to S'.

Theorem 10.5.4. *A modeling relation $h : S \to S'$ is an isomorphism from S onto S' if and only if h is bijective from S onto S' as a mapping.*

Proof: Necessity. Suppose $h : S \to S'$ is an isomorphism. From Theorem 10.5.3 it follows that $h : S \to S'$ is surjective as a mapping. It remains for us to show that h is injective. Let $p : S' \to S$ be a left cancellation of h. That is, $p \circ h = (\mathrm{id}_X, \mathrm{id}_Y)$. By contradiction, suppose $h : S \to S'$ is not injective as a mapping. There then exist distinct (x, y) and $(u, v) \in S$ such that $h(x, y) = h(u, v)$. So, $p \circ h(x, y) = p \circ h(u, v)$, which contradicts the assumption that $p \circ h = (\mathrm{id}_X, \mathrm{id}_Y)$ is the identity mapping on S. ■

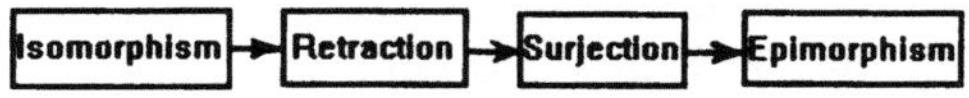

Figure 10.12. Properties of the mapping h from S to S' with $D(S) = X$ and $R(S) = Y$.

Example 10.5.1. We show that retraction does not guarantee an isomorphism. Let $S \subset X \times Y$ and $S' \subset X' \times Y'$ be two systems defined by

$$S = \{(x_1,y_1),(x_1,y_2),(x_2,y_2),(x_3,y_2)\} \tag{10.153}$$

and

$$S' = \{(x_1',y_1'),(x_1',y_2'),(x_2',y_2')\} \tag{10.154}$$

Define a modeling relation $h = (h_1,h_2)$ from S to S' by letting $h_1(x_1) = h_1(x_2) = x_1'$ and $h_1(x_3) = x_2'$, and $h_2(y_1) = y_1'$ and $h_2(y_2) = y_2'$. Let $k = (k_1,k_2)$ from $D(S') \times R(S') \to D(S) \times R(S)$ be defined by

$$k_1(x_1') = x_1, \quad k_1(x_2') = x_3, \quad k_2(y_1') = y_1, \quad k_2(y_2') = y_2 \tag{10.155}$$

Then k is a modeling relation from S' to S and $h \circ k = (\mathrm{id}_{X'}, \mathrm{id}_{Y'})$. But, according to Theorem 10.5.4, S is not isomorphic to S' because $|S| \neq |S'|$.

Combining Theorems 10.5.2 and 10.5.3 we obtain Fig. 10.12.

A modeling relation $h = (h_1,h_2)$ from S to S' is termed as a monomorphism from S to S' if h is injective as a mapping from S to S'. Then, similar to the previous results, we have the following.

Theorem 10.5.5. *Under the same assumption as in Theorem 10.5.1, the following hold:*

(i) If both mappings h_1 and h_2 are injective, then h is a monomorphism, but the converse is not true.

(ii) If the modeling relation h is a section (that is, h has a left cancellation), then h_1 and h_2 are injective, but the converse is not true.

(iii) If h is an isomorphism, then h is a section, but the converse is not true.

Proof: (i) If h_1 and h_2 are injective, then the modeling relation h is injective as a mapping from S to S'. Therefore, h is a monomorphism. Conversely, define two systems $S \subset X \times Y$ and $S' \subset X' \times Y'$ such that

$$S = \{(x_1,y_1),(x_2,y_3),(x_3,y_2),(x_3,y_3)\} \tag{10.156}$$

and

$$S' = \{(x_1', y_1'), (x_1', y_2'), (x_2', y_1'), (x_2', y_2')\} \tag{10.157}$$

and define $h = (h_1, h_2) : S \to S'$ by letting $h_1(x_1) = h_1(x_2) = x_1'$, $h_1(x_3) = x_2'$, $h_2(y_1) = h_2(y_2) = y_1'$ and $h_2(y_3) = y_2'$. Then $h : S \to S'$ is an injective modeling relation, but none of the mappings h_1 nor h_2 is injective.

(ii) Let $k = (k_1, k_2) : S' \to S$ be a left cancellation of h; i.e., $k \circ h = (k_1 \circ h_1, k_2 \circ h_2) = (\mathrm{id}_{X'}, \mathrm{id}_{Y'})$. Then h_1 and h_2 are injective. Conversely, define systems $S \subset X \times Y$ and $S' \subset X' \times Y'$ such that

$$S = \{(x_1, y_1), (x_2, y_2)\} \tag{10.158}$$

and

$$S' = \{(x_1', y_1'), (x_2', y_2'), (x_3', y_1'), (x_3', y_2')\} \tag{10.159}$$

and define mappings h_1 and h_2 by $h_1(x_1) = x_1'$, $h_1(x_2) = x_2'$, $h_2(y_1) = y_1'$ and $h_2(y_2) = y_2'$. Then h_1 and h_2 are both injective. Now if the modeling relation h has a left cancellation $k = (k_1, k_2)$, then either $k_1(x_3') = x_1$ or $k_1(x_3') = x_2$. If $k_1(x_3') = x_1$, then

$$k(x_3', y_2') = (x_1, y_2) \notin S \tag{10.160}$$

a contradiction. If $k_1(x_3') = x_2$, then

$$k(x_3', y_1') = (x_2, y_1) \notin S \tag{10.161}$$

This contradicts the hypothesis that k is a modeling relation from S' to S.

(iii) We need only define two systems $S \subset X \times Y$ and $S' \subset X' \times Y'$ and a section $h : S \to S'$ such that h is not an isomorphism. Let

$$S = \{(x_1, y_1), (x_2, y_2), (x_1, y_2)\} \tag{10.162}$$

and

$$S' = \{(x_1', y_1'), (x_1', y_2'), (x_2', y_2'), (x_3', y_1'), (x_3', x_2')\} \tag{10.163}$$

and let the modeling relation h be defined as in the proof of (ii). Define a left cancellation $k = (k_1, k_2)$ of h as follows: $k_1(x_1') = k_1(x_3') = x_1$, $k_1(x_2') = x_2$, $k_2(y_1') = y_1$ and $k_2(y_2') = y_2$. But h cannot be an isomorphism from S to S' by Theorem 10.5.4 because $|S| \neq |S'|$. ■

Theorem 10.5.6. *A modeling relation $h = (h_1, h_2) : S \to S'$ is an isomorphism if and only if $h_1 : D(S) \to D(S')$ and $h_2 : R(S) \to R(S')$ are bijective.*

Proof: Necessity. Suppose h is an isomorphism. Then Theorems 10.5.3 and 10.5.5 imply that the mappings $h_1 : D(S) \to D(S')$ and $h_2 : R(S) \to R(S')$ are surjective and injective.

Sufficiency follows from Theorem 10.5.4. ■

Theorem 10.5.7. *Let $h_1 : X \to X'$ and $h_2 : Y \to Y'$ be two mappings, and $S \subset X \times Y$ and $S' \subset X' \times Y'$ be two systems. Then the product mapping $h_1 \times h_2$ reduces a modeling relation from S to S'; i.e., $h_1 \times h_2|S$ is a mapping from S to S' if and only if for any $x \in D(S)$, $h_2[S(x)] \subset S'(h_1(x))$.*

Proof: Necessity. Suppose $h_1 \times h_2|S$ is a mapping from S to S'. Then for any $x \in D(S)$ and any $y \in S(x)$, we have $h_1 \times h_2|S(x,y) = (h_1(x), h_2(x)) \in S'$. That is, $h_2(y) \in S'(h_1(x))$. This gives $h_2[S(x)] \subset S'(h_1(x))$.

Sufficiency. Suppose, for any $x \in D(S)$, $h_2[S(x)] \subset S'(h_1(x))$. Then, for each $(x,y) \in S$, $h_2(y) \in S'(h_1(x))$; i.e., $(h_1(x), h_2(y)) \in S'$. Therefore, $h = (h_1, h_2)$ is a modeling relation from S to S'. ■

From Theorem 10.5.6, it can be seen that if a system S is isomorphic to a system S', then S' is also isomorphic to S. In fact, let $h = (h_1, h_2)$ be an isomorphism from S to S'. Then there are mappings $k_1 : X' \to X$ and $k_2 : Y' \to Y$ such that

$$k_1|D(S') = h_1^{-1} \quad \text{and} \quad k_2|R(S') = h_2^{-1} \tag{10.164}$$

Applying Theorem 10.5.6, it follows that $k = (k_1, k_2)$ is an isomorphism from S' to S. Therefore, in the future we can speak about isomorphic systems.

Suppose $S \subset X \times Y$ is an input–output system such that $D(S) = X$ and $R(S) = Y$, and $\overline{S} = \{S_\alpha \subset S : S_\alpha \neq \emptyset, \alpha \in I\}$, where I is an index set, is a covering of S; i.e., $\overline{S}$ satisfies $\bigcup \overline{S} = \bigcup\{S_\alpha : \alpha \in I\} = S$. The collection $\overline{S}$ is referred to as an input–output system covering of S if $\overline{S}$ satisfies the following two conditions:

(i) There are coverings $\overline{X} = \{X_\alpha : \alpha \in I_x\}$ and $\overline{Y} = \{Y_\alpha : \alpha \in I_y\}$ of the input space X and the output space Y, respectively.

(ii) There is a relation $I_s \subset I_x \times I_y$ with $D(I_s) = I_x$ and $R(I_s) = I_y$ such that $\overline{S}$ and $\{X_i \times Y_j : (i,j) \in I_s\}$ are related by a one-to-one correspondence $\psi : I_s \to I$ with

$$(X_i \times Y_j) \cap S = S_{\psi(i,j)} \tag{10.165}$$

In particular, if the collections $\overline{S}$, $\overline{X}$, and $\overline{Y}$ are partitions of the sets S, X, and Y, the corresponding input–output system covering of S will be referred to as an input–output system partition. The system covering will be denoted by $\{\overline{S}, \overline{X}, \overline{Y}, I_s\}$.

A system covering will be used when viewing a system macroscopically by ignoring its detailed structures. Generally, not every covering is compatible with an input–output model structure. When it is compatible, it can generate an approximation of the original input–output system. Here, the relation I_s is a representation of the approximation and will be called an associated input–output system of the covering.

Theorem 10.5.8. *Each input–output system $S \subset X \times Y$ has an input–output system covering, which is also an input–output system partition.*

Proof: We define the collections $\overline{S}$, $\overline{X}$, and $\overline{Y}$ as follows:

$$\overline{S} = \{S\}, \quad \overline{X} = \{X\}, \quad \overline{Y} = \{Y\} \tag{10.166}$$

The rest of the proof is clear. ■

Example 10.5.2. An input–output system covering (respectively, partition)

$$\{\overline{S}, \overline{X}, \overline{Y}, I_s\}$$

is termed a trivial covering (respectively, partition) if $|I_x| = 1$ or $|I_y| = 1$. We now construct an input–output system $S \subset X \times Y$ which does not have a nontrivial partition. Let $S \subset X \times Y$ be the system defined by

$$S = \{(x_1,y_1),(x_2,y_2)\} \tag{10.167}$$

where $x_1,x_2 \in X$ and $y_1,y_2 \in Y$ are distinct elements in X and Y, respectively. Then S may not have a nontrivial partition. In fact, by contradiction, suppose S has a nontrivial input–output system partition $\{\overline{S}, \overline{X}, \overline{Y}, I_s\}$. Then the associated input–output system $I_s \subset I_x \times I_y$ must satisfy $|I_x| = 2 = |I_y|$. In fact, if $|I_x| > 2$, there is an $i \in I_x$ such that $X_i \in \overline{X}$ contains neither x_1 nor x_2. That is, for each $j \in I_y$,

$$(X_i \times Y_j) \cap S = \emptyset \tag{10.168}$$

Equation (10.168) contradicts the condition that $(X_i \times Y_i) \cap S \in \overline{S}$ and each element in $\overline{S}$ is nonempty. A similar argument shows that $|I_y|$ has to be 2.

Suppose $\overline{X} = \{X_1, X_2\}$ and $\overline{Y} = \{Y_1, Y_2\}$. Then for each $i = 1,2$,

$$X_i \cap D(S) \neq \emptyset \neq Y_i \cap R(S) \tag{10.169}$$

To guarantee that for each $i = 1,2$ and $j = 1,2$,

$$(X_i \times Y_j) \cap S = S_{\psi(i,j)} \in \overline{S} \tag{10.170}$$

the sets X_i and Y_j must contain $\{x_1,x_2\}$ and $\{y_1,y_2\}$, respectively. This fact implies that if the input space $X = \{x_1,x_2\}$ and the output space $Y = \{y_1,y_2\}$, then the system covering $\{\overline{S}, \overline{X}, \overline{Y}, I_s\}$ would be a trivial covering, contradiction.

When a system covering is a partition, the associated input–output system I_s is characterized by the following results.

Theorem 10.5.9. *Let $\overline{S} = \{S_i : S_i \subset S$ and $i \in I\}$ be an input–output system partition of a system $S \subset X \times Y$. Then the following holds:*

$$\begin{aligned}(i,j) \in I_s &\leftrightarrow (\exists(x,y) \in S)(x \in X_i \text{ and } y \in Y_i) \\ &\leftrightarrow (X_i \times Y_j) \cap S \neq \emptyset\end{aligned} \tag{10.171}$$

where I_s is the associated input–output system of the partition, $X_i \in \overline{X}$ and $Y_j \in \overline{Y}$. In particular, if the system I_s exists, I_s is unique for the given sets of $\overline{S}$, $\overline{X}$, and $\overline{Y}$.

Proof: The following is clear:

$$(\exists(x,y) \in S)(x \in X_i \text{ and } y \in Y_j) \leftrightarrow (X_i \times Y_j) \cap S \neq \emptyset \tag{10.172}$$

Suppose $(i,j) \in I_s$. The definition of a system covering implies that $(X_i \times Y_j) \cap S = S_{\psi(i,j)} \neq \emptyset$, so there is $(x,y) \in S_{\psi(i,j)} = (X_i \times Y_j) \cap S$. Consequently, $(\exists(x,y) \in S)(x \in X_i$ and $y \in Y_j)$. Conversely, suppose there exists $(x,y) \in S$ such that $x \in X_i$ and $y \in Y_j$. Since $(x,y) \in (X_i \times Y_j) \cap S \subset S = \bigcup_{\alpha \in I} S_\alpha$, we have

$$(x,y) \in S_k \quad \text{for some } k \in I \tag{10.173}$$

Let $\psi^{-1}(k) = (i',j')$. Since $(X_{i'} \times Y_{j'}) \cap S = S_{\psi(i',j')} = S_k$ and $(x,y) \in S_k$ hold, we have $x \in X_{i'}$ and $y \in Y_{j'}$. Since $\overline{X}$ and $\overline{Y}$ are partitions, we must have $X_i = X_{i'}$ and $Y_j = Y_{j'}$. The uniqueness of existence of the associated input–output system I_s comes from the fact that $(i,j) \in I_s \leftrightarrow (X_i \times Y_j) \cap S \neq \emptyset$. ■

Let $f : X \to Y$ be a mapping from a set X to a set Y. An equivalence relation E_f can then be defined on X as follows:

$$xE_f y \leftrightarrow f(x) = f(y) \tag{10.174}$$

The quotient set X/E_f will be denoted by X/f.

Theorem 10.5.10. *Suppose that $h = (h_1, h_2) : S \to S'$ is a modeling relation with $D(S) = X$ and $R(S) = Y$. There then exists an input–output system partition $\{\overline{S}, \overline{X}, \overline{Y}, I_s\}$ of S such that $\overline{X} = X/h_1$, $\overline{Y} = Y/h_2$, and $I_s \subset (X/h_1) \times (Y/h_2)$ is defined by*

$$([x],[y]) \in I_s \leftrightarrow ([x] \times [y]) \cap S \neq \emptyset \tag{10.175}$$

A one-to-one correspondence $\psi : I_s \to S/h_1 \times h_2$ is given by

$$\psi([x],[y]) = [(x,y)] \tag{10.176}$$

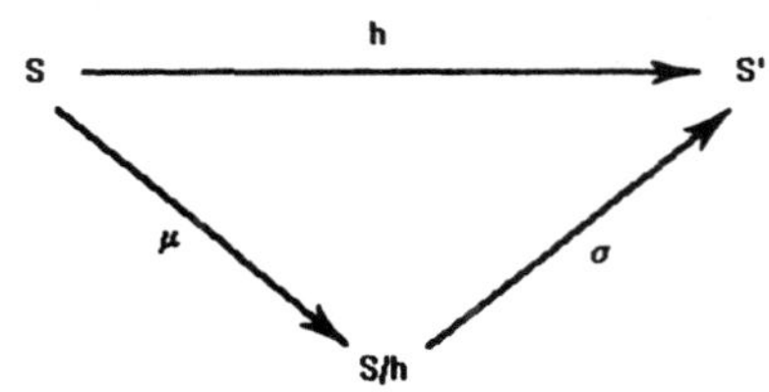

Figure 10.13. Epi–mono decomposition of a modeling relation h.

Proof: We first show that the correspondence ψ is well defined. Let $([x],[y]) \in I_s$ and $(\hat{x}, \hat{y}) \in X \times Y$ be arbitrary such that $[x] = [\hat{x}]$ and $[y] = [\hat{y}]$. It suffices to show that

$$[(x,y)] = [(\hat{x},\hat{y})] \tag{10.177}$$

From the definition of equivalence class, it follows that $[(x,y)] = \{(\hat{x},\hat{y}) \in X \times Y : h(\hat{x},\hat{y}) = (x,y)\} = [(\hat{x},\hat{y})]$; that is, ψ is a mapping from I_s into $X \times Y/h_1 \times h_2$. We now need to show that for each $([x],[y]) \in I_s$, $\psi([x],[y]) \in S/h_1 \times h_2$; i.e., there exists $(\hat{x}, \hat{y}) \in S$ such that $[(x,y)] = [(\hat{x},\hat{y})]$. The last equation follows from the fact that $([x] \times [y]) \cap S \neq \emptyset$.

Second, we show that $\psi : I_s \to S/h_1 \times h_2$ is a one-to-one correspondence. Let $[(\hat{x},\hat{y})] \in S/h_1 \times h_2$ be arbitrary. Since $\psi([\hat{x}],[\hat{y}]) = [(\hat{x},\hat{y})]$ holds, ψ is surjective. Suppose $\psi([x],[y]) = \psi([\hat{x}],[\hat{y}])$. Then $[x] = [\hat{x}]$ and $[y] = [\hat{y}]$; i.e., the mapping ψ is injective. ■

Theorem 10.5.11. *Suppose that $S \subset X \times Y$ and $S' \subset X' \times Y'$ are input–output systems such that $D(S) = X$, $R(S) = Y$, $D(S') = X'$, and $R(S') = Y'$, and $h = (h_1,h_2) : S \to S'$ is a modeling relation. Let $\mu_1 : X \to X/h_1$ and $\mu_2 : Y \to Y/h_2$ be canonical mappings. Then the following statements hold:*

(i) $\mu = (\mu_1, \mu_2)$ is an epimorphism from S to S/h. In particular, μ is surjective.

(ii) A modeling relation $\sigma = (\sigma_1, \sigma_2) : S/h \to S'$ can be defined by $\sigma([x],[y]) = (h_1(x),h_2(y))$ such that σ_i, $i = 1,2$, are injective.

(iii) The diagram in Fig. 10.13 is commutative; that is, there is an epi–mono decomposition of h.

(iv) If h_1 and h_2 are injective, then the modeling relation μ is an isomorphism.

(v) If h is surjective, then σ is an isomorphism.

Proof: (i) It is clear that the canonical mappings $\mu_1 : X \to X/h_1$ and $\mu_2 : Y \to Y/h_2$ are surjective when defined by $\mu_i(m) = [m]$ for $i = 1,2$. If $(x,y) \in S$, $(x,y) \in ([x] \times [y]) \cap S$, and thus $\mu(x,y) = (\mu_1(x), \mu_2(y)) = ([x],[y]) \in S/h$. By Theorem 10.5.1, the modeling relation μ is an epimorphism. Since there is $(x',y') \in S$ for any $([x],[y]) \in S/h$ such that $[x'] = [x]$ and $[y'] = [y]$, μ is surjective.

(ii) Since $[x] = [x']$ implies $h_1(x) = h_1(x')$, $\sigma_1([x]) = h_1(x)$ is well defined. The mapping σ_2 is also well defined. Suppose $([x],[y]) \in S/h$. Then the definition of S/h implies there exists $(\widehat{x},\widehat{y})$ such that $(\widehat{x},\widehat{y}) \in ([x] \times [y]) \cap S$. Consequently, $h_1(x) = h_1(\widehat{x})$, $h_2(y) = h_2(\widehat{y})$, and $(\widehat{x},\widehat{y}) \in S$; i.e., $\sigma([x],[y]) = (h_1(x), h_2(y)) \in S'$. Hence, σ is a modeling relation, and σ_1 and σ_2 are clearly injective.

(iii) The commutativity of the diagram in Fig. 10.13 is clear from the definitions of μ and σ.

(iv) If h_1 and h_2 are injective, $h_1 = \sigma_1 \circ \mu_1$ says that μ_1 is injective. Similarly, the mapping μ_2 is also injective. From statement (i) and Theorem 10.5.6, we know that μ is an isomorphism.

(v) The desired result can be shown as in the proof of (iv). ■

Suppose $\widehat{S} \subset \widehat{X} \times \widehat{Y}$ is a system, $\{S_i \subset X_i \times Y_i : i = 1,2,\ldots,n\}$ is a collection of systems, and $S \subset X \times Y$ satisfies the conditions

$$X \subset X_1 \times X_2 \times \cdots \times X_n \quad \text{and} \quad Y \subset Y_1 \times Y_2 \times \cdots \times Y_n \tag{10.178}$$

and

$$S \subset S_1 \times S_2 \times \cdots \times S_n \tag{10.179}$$

where we identify each element $((x_1,x_2,\ldots,x_n),(y_1,y_2,\ldots,y_n)) \in S$ with $((x_1,y_1),(x_2,y_2),\ldots,(x_n,y_n)) \in S_1 \times S_2 \times \cdots \times S_n$.

The system S is called a surjective complex system representation of $\widehat{S}$ with order n if there exist mappings $h_1 : X \to \widehat{X}$ and $h_2 : Y \to \widehat{Y}$ such that $h_1 \times h_2(S) = \widehat{S}$. S is called an injective complex system representation of $\widehat{S}$ with order n if there exist mappings $k_1 : \widehat{X} \to X$ and $k_2 : \widehat{Y} \to Y$ such that $k_1 \times k_2(\widehat{S}) \subset S$ and k_1 and k_2 are injective. S is called an isomorphic complex system representation of $\widehat{S}$ with order n if there exist mappings $h_1 : X \to \widehat{X}$ and $h_2 : Y \to \widehat{Y}$ such that $h = (h_1, h_2)$ is an isomorphism from S onto S'.

Proposition 10.5.3. *For each system $\widehat{S} \subset \widehat{X} \times \widehat{Y}$ and each natural number $n > 0$, $\widehat{S}$ has a surjective, an injective, and an isomorphic complex system representation with order n.*

Proof: Let $X \subset \widehat{X}^n$ and $Y \subset \widehat{Y}^n$ be the diagonals of the product sets, respectively. That is,

$$X = \{\underbrace{(x,x,\ldots,x)}_{n \text{ times}} : x \in \widehat{X}\} \quad \text{and} \quad Y = \{\underbrace{(y,y,\ldots,y)}_{n \text{ times}} : y \in \widehat{Y}\} \tag{10.180}$$

We now define $S \subset X \times Y$ by letting

$$((x,x,\dots,x),(y,y,\dots,y)) \in S \leftrightarrow (x,y) \in \widehat{S} \tag{10.181}$$

and define mappings $h_1 : X \to \widehat{X}$ and $h_2 : Y \to \widehat{Y}$ by $h_1(x,x,\dots,x) = x$ and $h_2(y,y,\dots,y) = y$. Then we have the desired results with respect to (h_1,h_2). ■

Theorem 10.5.12. *A system $\widehat{S} \subset \widehat{X} \times \widehat{Y}$ has an injective complex system representation S with order n over a class of systems $\{S_i \subset X_i \times Y_i : i = 1,2,\dots,n\}$ iff there exist n partitions of $\widehat{S}$, $P_i = \{\widehat{S}_{ij} \subset \widehat{S} : j \in I_i\}$, for each $i = 1,2,\dots,n$, such that the following conditions hold:*

(i) P_i is an input–output system partition of $\widehat{S}$, for each $i = 1,2,\dots,n$.

(ii) There exists an embedding $\sigma_i = (\sigma_{1i},\sigma_{2i}) : \widehat{S}_i \to S_i$, where $\widehat{S}_i$ is an associated input–output system of P_i with σ_{1i} and σ_{2i} injective.

(iii) Let $\mathcal{X}_i = \{[x_j]_i : x_j \in \widehat{X}\}$ and $\mathcal{Y}_i = \{[y_j]_i : y_j \in \widehat{Y}\}$ be the input and output spaces of $\widehat{S}_i$, respectively, where $\mathcal{X}_i$ and $\mathcal{Y}_i$ are partitions by definition and $[x_j]_i$ and $[y_j]_i$ are equivalence classes of $\mathcal{X}_i$ and $\mathcal{Y}_i$, respectively. Then, for arbitrary $[x_i]_i \in \mathcal{X}_i$ and $[y_i]_i \in \mathcal{Y}_i$, for each $i = 1,2,\dots,n$,

$$\bigcap_{i=1}^{n}[x_i]_i \neq \emptyset \to (\exists \widehat{x} \in \widehat{X})(\bigcap_{i=1}^{n}[x_i]_i = \{\widehat{x}\}) \tag{10.182}$$

and

$$\bigcap_{i=1}^{n}[y_i]_i \neq \emptyset \to (\exists \widehat{y} \in \widehat{Y})(\bigcap_{i=1}^{n}[y_i]_i = \{\widehat{y}\}) \tag{10.183}$$

Proof: Necessity. Let $S \subset X \times Y$ be an injective complex system representation with order n, where $X \subset X_1 \times X_2 \times \cdots \times X_n$ and $Y \subset Y_1 \times Y_2 \times \cdots \times Y_n$. Let $\prod_i = (\prod_{ix},\prod_{iy}) : X \times Y \to X_i \times Y_i$ be the projection defined by Eq. (10.96). Then $\prod_i : S \to S_i$ is a modeling relation. Let $e = (e_1,e_2) : \widehat{S} \to S$ be an injective modeling relation. Since $\prod_i \circ e = (\prod_{ix} \circ e_1, \prod_{iy} \circ e_2) : \widehat{S} \to S_i$ is a modeling relation, we can define a partition P_i of $\widehat{S}$ by

$$P_i = \widehat{S}/(\prod_{ix} \circ e_1) \times (\prod_{iy} \circ e_2) \tag{10.184}$$

We now check the conditions (i)–(iii) for these partitions P_i, $i = 1,2,\dots,n$. Since Theorem 10.5.10 implies that P_i is an input–output system partition, condition (i) holds. Since the associated input–output system $\widehat{S}_i$ is given by $\widehat{S}_i = \widehat{S}/\prod_i \circ e$, there is an embedding mapping $\sigma : \widehat{S}_i \to S_i$ due to Theorem 10.5.11(ii). Hence, condition (ii) also holds. We will check condition (iii). In the present case,

$$\mathcal{X}_i = \widehat{X} / \prod_{ix} \circ e_1 \quad \text{and} \quad \mathcal{Y}_i = \widehat{Y} / \prod_{iy} \circ e_2 \tag{10.185}$$

Choose an equivalence class $[x_i]_i$ from each $\mathcal{X}_i$ for $i = 1, 2, \ldots, n$. The definition of $\mathcal{X}_i$ implies $\prod_{ix} \circ e_1([x_i]_i) = \{\widehat{x}_i\}$ for some $\widehat{x}_i \in X_i$, so

$$e_1([x_i]_i) \subset \prod_{ix}^{-1}(\{\widehat{x}_i\}) \tag{10.186}$$

Suppose $\widehat{x} \in \bigcap_{i=1}^{n} \prod_{ix}^{-1}(\{\widehat{x}_i\})$. Since $\widehat{x} \in \prod_{ix}^{-1}(\{\widehat{x}_i\})$ for each i,

$$(\forall i = 1, 2, \ldots, n)(\prod_{ix}(\widehat{x}) = \widehat{x}_i) \tag{10.187}$$

Hence, $\widehat{x} = (\widehat{x}_1, \widehat{x}_2, \ldots, \widehat{x}_n)$. Consequently,

$$\bigcap_{i=1}^{n} \prod_{ix}^{-1}(\{\widehat{x}_i\}) \neq \emptyset \to \bigcap_{i=1}^{n} \prod_{i=1}^{n} \prod_{ix}^{-1}(\{\widehat{x}_i\}) = \{(\widehat{x}_1, \ldots, \widehat{x}_n)\} \tag{10.188}$$

Since, in general, $e_1(\bigcap_{i=1}^{n}[x_i]_i) \subseteq \bigcap_{i=1}^{n} e_1([x_i]_i)$, $e_1([x_i]_i) \subset \prod_{ix}^{-1}(\{\widehat{x}_i\})$ implies

$$e_1(\bigcap_{i=1}^{n}[x_i]_i) \neq \emptyset \to e_1(\bigcap_{i=1}^{n}[x_i]_i) = \{(\widehat{x}_1, \ldots, \widehat{x}_n)\} \tag{10.189}$$

Since e_1 is injective, $\bigcap_{i=1}^{n}[x_i]_i$ is empty or singleton. A similar argument works for the collection $\mathcal{Y}_i$. Hence, we have showed condition (iii).

Sufficiency. Suppose $\widehat{S}$ satisfies conditions (i)–(iii). Let $S' \subset \widehat{S}_1 \times \cdots \times \widehat{S}_n$ be defined by

$$(([x_1]_1, \cdots, [x_n]_n), ([y_1]_1, \ldots, [y_n]_n)) \in S'$$

$$\leftrightarrow (\forall i = 1, \ldots, n)(([x_i]_i, [y_i]_i) \in \widehat{S}_i) \quad \text{and} \quad (\bigcap_{i=1}^{n}[x_i]_i, \bigcap_{i=1}^{n}[y_i]_i) \in \widehat{S} \tag{10.190}$$

where we identify singleton sets with the elements. Let $X' = D(S')$ and $Y' = R(S')$ and mappings $k_1 : \widehat{X} \to X'$ and $k_2 : \widehat{Y} \to Y'$ be defined by

$$k_1(\widehat{x}) = ([\widehat{x}]_1, \ldots, [\widehat{x}]_n) \quad \text{and} \quad k_2(\widehat{y}) = ([\widehat{y}]_1, \ldots, [\widehat{y}]_n) \tag{10.191}$$

for each $\widehat{x} \in \widehat{X}$ and $\widehat{y} \in \widehat{Y}$.

Claim. The mapping $k = (k_1, k_2)$ is an isomorphism from $\widehat{S}$ to S'.

In fact, let $(\widehat{x}, \widehat{y}) \in \widehat{S}$ be arbitrary. Theorem 10.5.9 implies that $([\widehat{x}]_i, [\widehat{y}]_i) \in \widehat{S}_i$. Since condition (iii) implies that

$$(\bigcap_{i=1}^{n}[\widehat{x}]_i, \bigcap_{i=1}^{n}[\widehat{y}]_i) = (\widehat{x},\widehat{y}) \in \widehat{S} \tag{10.192}$$

we have $k_1 \times k_2(\widehat{x}, \widehat{y}) \in S'$. That is, k is a modeling relation from $\widehat{S}$ to S'. To show k is one-to-one, we let $k_1(\widehat{x}) = k_1(\widehat{x}')$. It follows from the definition of k_1 that $\forall i([\widehat{x}]_i = [\widehat{x}']_i)$. The condition (iii) implies that

$$\{\widehat{x}\} = \bigcap_{i=1}^{n}[\widehat{x}]_i = \bigcap_{i=1}^{n}[\widehat{x}']_i = \{\widehat{x}'\} \tag{10.193}$$

Hence, k_1 is injective. The same argument can be used to show that k_2 is also injective.

To show that k is surjective from $\widehat{S}$ onto S', let

$$(([x_1]_1, \ldots, [x_n]_n), ([y_1]_1, \ldots, [y_n]_n)) \in S'$$

be arbitrary. Then the definition of S' implies that there exists $(\widehat{x}, \widehat{y}) \in \widehat{S}$ such that

$$\{\widehat{x}\} = \bigcap_{i=1}^{n}[x_i]_i \quad \text{and} \quad \{\widehat{y}\} = \bigcap_{i=1}^{n}[y_i]_i \tag{10.194}$$

Consequently, for any i, $\widehat{x} \in [x_i]_i$ and $\widehat{y} \in [y_i]_i$; i.e., $[\widehat{x}]_i = [x_i]_i$ and $[\widehat{y}]_i = [y_i]_i$. Therefore,

$$k_1 \times k_2(\widehat{x},\widehat{y}) = (([x_1]_1, \ldots, [x_n]_n), ([y_1]_1, \ldots, [x_n]_n)) \tag{10.195}$$

This completes the proof that k is an isomorphism from $\widehat{S}$ to S'.

We now let $S = S_1 \times S_2 \times \cdots \times S_n$. Since there is an embedding σ_i from $\widehat{S}_i$ to S_i, an injective modeling relation from $\widehat{S}$ to S is

$$((\sigma_{11} \times \cdots \times \sigma_{1n}) \circ k_1, (\sigma_{21} \times \cdots \times \sigma_{2n}) \circ k_2) \tag{10.196}$$

This ends the proof of Theorem 10.5.12. ■

Theorem 10.5.13. *If a system $\widehat{S} \subset \widehat{X} \times \widehat{Y}$ has a surjective complex system representation S with order n over a class $\{S_i \subset X_i \times Y_i : i = 1,2,\ldots,n\}$, then the following conditions hold:*

(i) $\widehat{S}$ has n input–output coverings

$$C_i = \{\widehat{S}^i_j \subset \widehat{S} : j \in I_i\}, \quad i = 1,2,\ldots,n \tag{10.197}$$

(ii) *There is a surjective mapping from S_i to an associated input–output system $\widehat{S}_i$ of C_i.*

Proof: Let $h = (h_1, h_2) : S \to \widehat{S}$ be a surjective modeling relation, and $\prod_i = (\prod_{ix}, \prod_{iy}) : S \to S_i$ the projection. For each i let

$$C_i' = \{(h_1(\prod_{ix})^{-1}(j) \times h_2(\prod_{iy})^{-1}(k)) \cap \widehat{S} : (j,k) \in S_i\} \tag{10.198}$$

Then C_i' is a covering of $\widehat{S}$. In fact, let $(\widehat{x},\widehat{y}) \in \widehat{S}$ be arbitrary. Since h is surjective, $(\widehat{x},\widehat{y}) = h_1 \times h_2(x,y)$ for some $(x,y) \in S$. If $\prod_i(x,y) = (j,k) \in S_i$,

$$(x,y) \in \prod_{ix}^{-1}(j) \times \prod_{iy}^{-1}(k) \tag{10.199}$$

That is,

$$\begin{aligned}(\widehat{x},\widehat{y}) = h(x,y) &\in h(\prod_{ix}^{-1}(j) \times \prod_{iy}^{-1}(k)) \\ &= h_1 \circ \prod_{ix}^{-1}(j) \times h_2 \circ \prod_{iy}^{-1}(k)\end{aligned} \tag{10.200}$$

Next let C_i be the subset of C_i' defined by deleting empty sets and duplicated sets and write

$$C_i = \{(h_1 \circ \prod_{ix}^{-1}(j) \times h_2 \circ \prod_{iy}^{-1}(k)) \cap \widehat{S} : (j,k) \in I_s^i \subset S_i\} \tag{10.201}$$

Then C_i is an input–output system covering. In fact, let $I_x^i = D(I_s^i)$ and $I_y^i = R(I_s^i)$ and

$$\mathcal{X}_i = \{h_1 \circ \prod_{ix}^{-1}(j) : j \in I_x^i\} \tag{10.202}$$

and

$$\mathcal{Y}_i = \{h_2 \circ \prod_{iy}^{-1}(k) : k \in I_y^i\} \tag{10.203}$$

Then $\mathcal{X}_i$ and $\mathcal{Y}_i$ are coverings of the sets $\widehat{X}$ and $\widehat{Y}$, respectively. Next let the desired relation between I_x^i and I_y^i be I_s^i, and let ψ be the identity mapping on I_s^i. Then (ii) holds. ∎

Theorem 10.5.14. *Suppose that* $\widehat{S} \subset \widehat{X} \times \widehat{Y}$ *is a system and* $\{S_i \subset X_i \times Y_i : i = 1,2,\dots,n\}$ *is a family of systems. If the following conditions are true,* $\widehat{S}$ *has a surjective complex system representation with order n,* $S \subset X \times Y$*, where*

$$X = X_1 \times X_2 \times \cdots \times X_n \quad \text{and} \quad Y = Y_1 \times Y_2 \times \cdots \times Y_n \tag{10.204}$$

and

$$S \subset S_1 \times S_2 \times \cdots \times S_n \tag{10.205}$$

(i) $\widehat{S}$ *has n input–output coverings*

$$C_i = \{\widehat{S}^i_j \subset \widehat{S} : j \in I^i\}, \quad i = 1,2,\dots,n \tag{10.206}$$

(ii) There is a surjective modeling relation $\sigma_i = (\sigma_{1i}, \sigma_{2i})$ *from* S_i *to an associated input–output system* $\widehat{S}_i$ *of* C_i.

(iii) Let $\mathcal{X}_i = \{\widehat{X}^i_j \subset \widehat{X} : j \in I^i_x\}$ *and* $\mathcal{Y}_i = \{\widehat{Y}^i_k \subset \widehat{Y} : k \in I^i_y\}$ *be the input and output spaces of* $\widehat{S}_i$*, respectively. Then for any* $\widehat{X}^i_i \in \mathcal{X}_i$ *and* $\widehat{Y}^i_i \in \mathcal{Y}_i$,

$$\bigcap_{i=1}^{n} \widehat{X}^i_i \neq \emptyset \rightarrow \left| \bigcap_{i=1}^{n} \widehat{X}^i_i \right| = 1 \tag{10.207}$$

and

$$\bigcap_{i=1}^{n} \widehat{Y}^i_i \neq \emptyset \rightarrow \bigcap_{i=1}^{n} \left| \widehat{Y}^i_i \right| = 1 \tag{10.208}$$

Proof: Suppose (i)–(iii) hold and I^i_s is an associated input–output system of C_i. Define a system $S' \subset I^1_s \times \cdots \times I^n_s$ as follows:

$$(j_1,\dots j_n, k_1,\dots,k_n) \in S' \leftrightarrow$$
$$\forall i((j_i,k_i) \in I^i_s) \quad \text{and} \quad (\bigcap_{i=1}^{n} \widehat{X}^i_{j_i}, \bigcap_{i=1}^{n} \widehat{Y}^i_{j_i}) \in \widehat{S} \tag{10.209}$$

where each singleton set is identified with its element. Let $X' = D(S')$ and $Y' = R(S')$ and $h_1 : X' \to \widehat{X}$ and $h_2 : Y' \to \widehat{Y}$ be the mappings defined by

$$h_1(j_1,\dots j_n) = \bigcap_{i=1}^{n} \widehat{X}^i_{j_i} \tag{10.210}$$

and

$$h_2(k_1,\ldots,k_n) = \bigcap_{i=1}^{n} \widehat{Y}^i_{k_i}$$

Condition (iii) and the definition of S' ensure that h_1 and h_2 are well defined. We now show that $h = (h_1, h_2)$ is a surjective modeling relation from S to S'. If $(x,y) \in S'$, then the definitions of S' and h imply that $(h_1(x), h_2(y)) \in \widehat{S}$ and hence h is a modeling relation. Suppose $(\widehat{x},\widehat{y}) \in \widehat{S}$ is arbitrary. For each i,

$$\widehat{S} \subset \bigcup_{j \in I^i} \widehat{S}^i_j \rightarrow \exists r \in I^i((\widehat{x},\widehat{y}) \in \widehat{S}^i_r) \tag{10.211}$$

The definition of an input–output system covering implies that for some $(j_i, k_i) \in I^i_s$,

$$(\widehat{X}^i_{j_i} \times \widehat{Y}^i_{k_i}) \cap \widehat{S} = \widehat{S}^i_r \tag{10.212}$$

Consequently, $\widehat{x} \in \widehat{X}^i_{j_i}$ and $\widehat{y} \in \widehat{Y}^i_{k_i}$; that is,

$$h_1 \times h_2(j_1,\ldots,j_n,k_1,\ldots,k_n) = (\widehat{x},\widehat{y}) \tag{10.213}$$

The modeling relation h is surjective. We now define $S \subset S_1 \times S_2 \times \cdots \times S_n$ and $\sigma : S \rightarrow S'$ by

$$(x,y) \in S \leftrightarrow \forall i((x_i,y_i) \in S_i) \quad \text{and}$$
$$\sigma(x,y) = (\sigma_{11}(x_1),\ldots,\sigma_{1n}(x_n),\sigma_{21}(y_1),\ldots,\sigma_{2n}(y_n)) \in S' \tag{10.214}$$

Then $h \circ \sigma$ is a surjective modeling relation from S to $\widehat{S}$. ■

Theorem 10.5.15. *A system $\widehat{S} \subset \widehat{X} \times \widehat{Y}$ has an isomorphic complex system representation S of order n over a family $\{S_i \subset X_i \times Y_i : i = 1, 2, \ldots, n\}$ of systems if and only if there are partitions P_i, $i = 1,2,\ldots,n$, of $\widehat{S}$ such that the following conditions hold:*

(i) P_i is an input–output system partition.

(ii) There is an isomorphism σ_i from an associated input–output system $\widehat{S}_i$ of P_i to S'_i, where $S'_i \subset S_i$ and $S \subset S'_1 \times \cdots \times S'_n$.

(iii) Condition (iii) in Theorem 10.5.14 is true.

The proof is similar to those of Theorems 10.5.12 and 10.5.14 and is left to the reader.

10.6. Some References for Further Study

The concepts of chaos and attractors of single-relation systems were introduced by Zhu and Wu (1987). They named the concepts panchaos and panattractor, and systematically studied them under the name pansystems theory. The theory appeared in the late 1970s. For more information consult Wu's work (1981b; 1982a; 1982b; 1982c; 1983a; 1983b; 1984b; 1984c; 1984d; 1985).

The concept of feedback systems has been a major research topic in systems theory and has been studied in many different ways by, for example, Mesarovic and Takahara (1975), Bunge (1979), Rugh (1981), Deng (1985), Klir (1985), and Saito and Mesarovic (1985). The definition of feedback systems, given in this chapter was introduced by Saito and Mesarovic (1985), and some results are from Lin and Ma (1990).

General systems of a single relation originated with Mesarovic, who began to study this concept in 1964 (Mesarovic, 1964). Its final form was contained in the book by Mesarovic and Takahara (1975), in which the concept of general linear systems was studied in detail. Theorem 10.2.1 is a combination of a result in (Lin, 1989b) and a fundamental theorem given by Mesarovic and Takahara (1975).

Feedback transformation has been a really hot topic in recent years. Because all living and many nonliving systems are feedback systems, the method of studying these systems is important. The concepts of feedback transformation and feedback decoupling shed some light on the matter, because the concepts tell when the system under consideration can be divided into smaller systems. For recent research, see Takahara and Asahi (1985) and Saito (1986; 1987).

The study of decomposability conditions enlightens the research of relations between "parts" and "whole." Section 10.5 is mainly from Takahara's excellent article (1982). The concept of decomposition of systems shows its power in Section 10.4.

CHAPTER 11

Calculus of Generalized Numbers

As shown in Chapter 6, it is necessary to develop a more mature theory of mathematics of levels. In this chapter the system of generalized numbers is treated rigorously, such that the concepts of limit, continuity, differential calculus, integral calculus, and their applications in the well-known Schwartz distribution theory are introduced. Further research topics are also listed.

As before, boldface letters are used for **GNS**-related concepts and lightface letters for real-valued concepts. For example, $f(x)$ represents a real-valued function, and $\mathbf{f}(\mathbf{x})$ a function from a subset of **GNS** to a subset of **GNS**. Without loss of generality, we will always assume that each real-valued function is from $\mathbb{R}$ into $\mathbb{R}$, and each generalized number-valued function is from **GNS** into **GNS**, where $\mathbb{R}$ is the set of all real numbers and **GNS** stands for the set of all generalized numbers. A comprehensive understanding of mathematical analysis is needed to understand this chapter.

11.1. Introduction

Let $\mathbb{Z}$ be the set of all integers, $\mathbb{R}$ the set of all real numbers, and $\mathbb{N}$ the set of all natural numbers. A generalized number $\mathbf{x}$ is a function from $\mathbb{Z}$ into $\mathbb{R}$ such that there exists a number $k = k(\mathbf{x}) \in \mathbb{Z}$, called the level index of $\mathbf{x}$, satisfying the condition

$$\mathbf{x}(t) = 0 \quad \text{for all } t < k(\mathbf{x}) \quad \text{and} \quad \mathbf{x}(k(\mathbf{x})) \neq 0 \tag{11.1}$$

For convenience, the zero function, denoted by $\mathbf{0}$, from $\mathbb{Z}$ into $\mathbb{R}$ will also be considered a generalized number, and each such $\mathbf{x}$ is expressed as

$$\begin{aligned} \mathbf{x} &\equiv \sum_{t \geq k(\mathbf{x})} \mathbf{x}(t)\mathbf{1}_{(t)} \\ &\equiv (\ldots, 0, \underset{\substack{\uparrow \\ k(x)\text{th level}}}{\mathbf{x}(k(\mathbf{x}))}, \mathbf{x}(k(\mathbf{x})+1), \ldots) \\ &\equiv (\ldots, 0, x_{k(\mathbf{x})}, x_{k(\mathbf{x})+1}, \ldots) \end{aligned} \tag{11.2}$$

where $\mathbf{1}_{(t)}$ stands for the function from $\mathbb{Z}$ into $\mathbb{R}$ such that

$$\mathbf{1}_{(t)}(s) = \begin{cases} 0 & \text{if } s \neq t \\ 1 & \text{if } s = t \end{cases} \tag{11.3}$$

and $\mathbf{x}(t) = x_t$. The set of all generalized numbers will be written as **GNS**. The operations of addition, subtraction, scalar multiplication, multiplication, division, and ordering relation of the generalized numbers are defined as follows:

(1) Addition and subtraction. Let $\mathbf{x}$ and $\mathbf{y}$ be arbitrary from **GNS**. Define

$$\mathbf{x} \pm \mathbf{y} = \sum_t [\mathbf{x}(t) \pm \mathbf{y}(t)] \tag{11.4}$$

(2) Ordering. The ordering relation between $\mathbf{x}, \mathbf{y} \in \mathbf{GNS}$ is the lexicographical ordering; i.e., $\mathbf{x} < \mathbf{y}$ iff there is an integer k_0 such that $\mathbf{x}(t) = \mathbf{y}(t)$, if $t < k_0$, and $\mathbf{x}(k_0) < \mathbf{y}(k_0)$.

(3) Scalar multiplication. Let $c \in \mathbb{R}$ and $\mathbf{x} \in \mathbf{GNS}$, and define

$$c\mathbf{x} = \sum_t c\mathbf{x}(t)\mathbf{1}_{(t)} \tag{11.5}$$

(4) Multiplication. Let $\mathbf{x}$, $\mathbf{y} \in \mathbf{GNS}$. The product of $\mathbf{x}$ and $\mathbf{y}$, denoted $\mathbf{x} \times \mathbf{y}$ or $\mathbf{xy}$, is defined as

$$\mathbf{x} \times \mathbf{y} = \sum_t \sum_{n+m=t} [\mathbf{x}(m)\mathbf{y}(n)]\mathbf{1}_{(\cdot)} \tag{11.6}$$

(5) Division. For $\mathbf{x}, \mathbf{y} \in \mathbf{GNS}$, if there exists a $\mathbf{z} \in \mathbf{GNS}$ such that

$$\mathbf{y} \times \mathbf{z} = \mathbf{x} \tag{11.7}$$

then the generalized number $\mathbf{z}$ is called the quotient of $\mathbf{x}$ divided by $\mathbf{y}$, denoted by $\mathbf{x} \div \mathbf{y}$ or $\mathbf{x}/\mathbf{y}$. As shown in Chapter 6, if $\mathbf{y} \neq \mathbf{0} \in \mathbf{GNS}$, then $\mathbf{x} \div \mathbf{y}$ always uniquely exists.

Theorem 11.1.1. $(\mathbf{GNS}, +, \times, <)$ *forms a non-Archimedean field.*

Proof: It suffices to show that $(\mathbf{GNS}, +, \times, <)$ satisfies the following axioms:

A1 $\mathbf{x} + \mathbf{y} = \mathbf{y} + \mathbf{x}$, for any $\mathbf{x}, \mathbf{y} \in \mathbf{GNS}$.

A2 $\mathbf{x} + (\mathbf{y} + \mathbf{z}) = (\mathbf{x} + \mathbf{y}) + \mathbf{z}$, for any $\mathbf{x}, \mathbf{y}, \mathbf{z} \in \mathbf{GNS}$.

A3 There exists $\mathbf{0} \in \mathbf{GNS}$ such that $\mathbf{x} + \mathbf{0} = \mathbf{x}$, for any $\mathbf{x} \in \mathbf{GNS}$.

A4 For any $\mathbf{x} \in \mathbf{GNS}$, there exists $\mathbf{y} \in \mathbf{GNS}$ such that $\mathbf{x}+\mathbf{y}=\mathbf{0}$.

M1 $\mathbf{x} \times \mathbf{y} = \mathbf{y} \times \mathbf{x}$, for any $\mathbf{x},\mathbf{y} \in \mathbf{GNS}$.

M2 $\mathbf{x} \times (\mathbf{y} \times \mathbf{z}) = (\mathbf{x} \times \mathbf{y}) \times \mathbf{z}$, for any $\mathbf{x},\mathbf{y},\mathbf{z} \in \mathbf{GNS}$.

M3 There exists $\mathbf{1} \in \mathbf{GNS}$ such that $\mathbf{x} \times \mathbf{1} = \mathbf{x}$, for any $\mathbf{x} \in \mathbf{GNS}$.

M4 For any $\mathbf{0} \neq \mathbf{x} \in \mathbf{GNS}$, there exists $\mathbf{y} \in \mathbf{GNS}$ such that $\mathbf{x} \times \mathbf{y} = 1$.

D1 $\mathbf{0} \neq \mathbf{1}$.

D2 $\mathbf{x} \times (\mathbf{y}+\mathbf{z}) = (\mathbf{x} \times \mathbf{y}) + (\mathbf{x} \times \mathbf{z})$, for any $\mathbf{x},\mathbf{y},\mathbf{z} \in \mathbf{GNS}$.

R1 $\mathbf{x} < \mathbf{x}$ can never be true, for any $\mathbf{x} \in \mathbf{GNS}$.

R2 For any $\mathbf{x},\mathbf{y},\mathbf{z} \in \mathbf{GNS}$, $\mathbf{x} < \mathbf{y}$ and $\mathbf{y} < \mathbf{z}$ imply $\mathbf{x} < \mathbf{z}$.

R3 For any $\mathbf{x},\mathbf{y} \in \mathbf{GNS}$, one of the following must be true:

$$\mathbf{x} < \mathbf{y}, \quad \mathbf{x} = \mathbf{y}, \quad \mathbf{x} > \mathbf{y}$$

R4 For any $\mathbf{x},\mathbf{y},\mathbf{z} \in \mathbf{GNS}$, $\mathbf{x} < \mathbf{y}$ implies $\mathbf{x}+\mathbf{z} < \mathbf{y}+\mathbf{z}$.

R5 For any $\mathbf{x},\mathbf{y},\mathbf{z} \in \mathbf{GNS}$, $\mathbf{x} < \mathbf{y}$ and $\mathbf{z} > \mathbf{0}$ imply $\mathbf{x} \times \mathbf{z} < \mathbf{y} \times \mathbf{z}$.

N1 For each natural number $n \in \mathbb{N}$, there exists $\mathbf{x} \in \mathbf{GNS}$ such that $n < \mathbf{x}$.

It is straightforward to check all these axioms except, M4 and R5, if one observes the following facts: In **GNS** $\mathbf{0}$ is the zero function from $\mathbb{Z}$ to $\mathbb{R}$, defined by $\mathbf{0}(t) = 0$ for all $t \in \mathbb{Z}$; in **GNS** $\mathbf{1}$ is the unit element $\mathbf{1}_{(0)}$ of the zeroth level, and for each natural number $n \in \mathbb{N}$, treat n as the generalized number $\mathbf{n} = (\ldots, 0, n, 0, \ldots)$ such that $k(\mathbf{n}) = 0$.

Proof of Axiom M4: Assume that the fixed and unknown generalized numbers $\mathbf{x}$ and $\mathbf{y}$ have expressions

$$\mathbf{x} = \sum_{t \geq k(\mathbf{x})} x_t \mathbf{1}_{(t)} \quad \text{and} \quad \mathbf{y} = \sum_{t \geq k(\mathbf{y})} y_t \mathbf{1}_{(t)}$$

Then the equation $\mathbf{x} \times \mathbf{y} = \mathbf{1}$ to be satisfied by $\mathbf{x}$ and $\mathbf{y}$ is equivalent to

$$\sum_{t+s=u} x_t y_t = \mathbf{1}_{(u)} \tag{11.8}$$

where $\mathbf{1} = \mathbf{1}_{(0)}$. That is, $\mathbf{1}(u) = 0$, if $u \neq 0$, and $\mathbf{1}(0) = 1$. If $u = 0$ in Eq. (11.8), we see that $k(\mathbf{y})$ can be defined as $-k(\mathbf{x})$, and

$$y_{k(\mathbf{y})} = 1/x_{k(\mathbf{x})}$$

If $u = 1$, Eq. (11.8) becomes

$$x_{k(\mathbf{x})}y_{k(\mathbf{y})+1} + x_{k(\mathbf{x})+1}y_{k(\mathbf{y})} = 0$$

Hence, $y_{k(\mathbf{y})+1}$ can be defined by $-x_{k(\mathbf{x})+1}/x^2_{k(\mathbf{x})}$. By mathematical induction, the generalized number $\mathbf{y}$ is defined.

It can also be shown that $\mathbf{y}$ is unique. To this end, it suffices to show that there is no other way to define the level index $k(\mathbf{y})$. By contradiction, assume that $k(\mathbf{y}) = -k(\mathbf{x}) - d$, for some $d \in \mathbb{N}$. Replacing u by $-d$, in Eq. (11.8), we get

$$x_{k(\mathbf{x})} \times y_{k(\mathbf{y})} = \mathbf{1}_{(-d)} = 0$$

This contradicts the fact that $x_{k(\mathbf{x})} \times y_{k(\mathbf{y})} \neq 0$. This ends the proof of Axiom M4.

Proof of Axiom R5: The assumption that $\mathbf{x} < \mathbf{y}$ and $\mathbf{z} > \mathbf{0}$ implies that there is $t_0 \in \mathbb{Z}$ such that $\mathbf{x}(t) = \mathbf{y}(t)$ for all $t < t_0$, $\mathbf{x}(t_0) < \mathbf{y}(t_0)$ and $\mathbf{z}(k(\mathbf{z})) > 0$. To compare $\mathbf{x} \times \mathbf{z}$ and $\mathbf{y} \times \mathbf{z}$, let $k_0 = t_0 + k(\mathbf{z})$. It then suffices to show that for all $s < k_0$, $(\mathbf{x} \times \mathbf{z})(s) = (\mathbf{y} \times \mathbf{z})(s)$ and $(\mathbf{x} \times \mathbf{z})(k_0) < (\mathbf{y} \times \mathbf{z})(k_0)$. To this end, for $s < k_0$, one has

$$\begin{aligned}(\mathbf{x} \times \mathbf{z})(s) &= \sum_{u+v=s} \mathbf{x}(u)\mathbf{z}(v) \\ &= \sum_{u+v=s} \mathbf{y}(u)\mathbf{z}(v) \\ &= (\mathbf{y} \times \mathbf{z})(s)\end{aligned}$$

since $v \geq k(\mathbf{z})$ implies $u < t_0$ and $\mathbf{x}(u) = \mathbf{y}(u)$. If $s = k_0 = t_0 + k(\mathbf{z})$, $\mathbf{x}(t_0) < \mathbf{y}(t_0)$ implies

$$\begin{aligned}(\mathbf{x} \times \mathbf{z})(k_0) &= \sum_{u+v=k_0} \mathbf{x}(u)\mathbf{z}(v) \\ &< \sum_{u+v=k_0} \mathbf{y}(u)\mathbf{z}(v) \\ &= (\mathbf{y} \times \mathbf{z})(k_0)\end{aligned}$$

since if $v = k(\mathbf{z})$, u has to equal t_0, so

$$\mathbf{x}(t_0) \times \mathbf{z}(k(\mathbf{z})) < \mathbf{y}(t_0) \times \mathbf{z}(k(\mathbf{z}))$$

while other terms in the summation $\sum_{u+v=k_0} \mathbf{x}(u)\mathbf{z}(v)$ are the same as the corresponding terms in $\sum_{u+v=k_0} \mathbf{y}(u)\mathbf{z}(v)$. This ends the proof of axiom R5. ■

Theorem 11.1.2. *Let $I_{(k)} = \{\mathbf{x} = x_k \times \mathbf{1}_{(k)} : x_k \in \mathbb{R}\}$. If $h, k \in \mathbb{Z}$ satisfy $k < h$, then the generalized numbers in $I_{(k)}$ are infinities compared with those in $I_{(h)}$, and the generalized numbers in $I_{(h)}$ are infinitesimals compared with those in $I_{(k)}$.*

The proof is straightforward and omitted.

Theorem 11.1.3. *Let* $\mathbf{x}, \mathbf{y} \in \mathbf{GNS}$ *such that* $\mathbf{y} \neq \mathbf{0} \neq \mathbf{x}$. *Then*

(1) $k(\mathbf{x}+\mathbf{y}) \geq \min\{k(\mathbf{x}), k(\mathbf{y})\}$.

(2) $k(\mathbf{x} \times \mathbf{y}) = k(\mathbf{x}) + k(\mathbf{y})$.

(3) $k(\mathbf{x}/\mathbf{y}) = k(\mathbf{x}) - k(\mathbf{y})$.

Proof: The argument for (1) is straightforward and is left to the reader. The proof of (2). Assume that $\mathbf{z} = \mathbf{x} \times \mathbf{y}$ and that

$$\mathbf{y} = \sum_{m \geq k(y)} \mathbf{y}(m)\mathbf{1}_{(m)}$$
$$\mathbf{x} = \sum_{m \geq k(x)} \mathbf{x}(m)\mathbf{1}_{(m)}$$
$$\mathbf{z} = \sum_{m \geq k(z)} \mathbf{z}(m)\mathbf{1}_{(m)}$$

Hence, $\mathbf{x} \times \mathbf{y} = \mathbf{z}$ is equivalent to, for any integer s,

$$\sum_{m+n=s} \mathbf{x}(m)\mathbf{y}(n) = \mathbf{z}(s) \tag{11.9}$$

where $k(\mathbf{x}) \leq m \leq s - k(\mathbf{y})$ and $k(\mathbf{y}) \leq n \leq s - k(\mathbf{z})$. We can now define $k(\mathbf{z}) = k(\mathbf{x}) + k(\mathbf{y})$. When $s = k(\mathbf{z})$, Eq. (11.9) becomes

$$\mathbf{x}(k(\mathbf{x})) \times \mathbf{y}(k(\mathbf{y})) = \mathbf{z}(k(\mathbf{z}))$$

When $s = k(\mathbf{z}) + 1$, Eq. (11.9) becomes

$$\mathbf{x}(k(\mathbf{x})) \times \mathbf{y}(k(\mathbf{y})+1) + \mathbf{x}(k(\mathbf{x})+1) \times \mathbf{y}(k(\mathbf{y})) = \mathbf{z}(k(\mathbf{z})+1)$$

By applying induction on the index $k(\mathbf{z}) + i$, the generalized number $\mathbf{z}$ can be well defined.

To show the uniqueness of the index number $k(\mathbf{z})$, it suffices to see that $k(\mathbf{z})$ cannot be less than $k(\mathbf{x}) + k(\mathbf{y})$, as defined. To this end, assume $k(\mathbf{z}) = k(\mathbf{x}) + k(\mathbf{y}) - n$, for some nonzero whole number n. Replacing s by $k(\mathbf{x}) + k(\mathbf{y}) - n$ in Eq. (11.9) gives

$$0 = \sum_{m+n=k(\mathbf{x})+k(\mathbf{y})-n} \mathbf{x}(m)\mathbf{y}(n) = \mathbf{z}(k(\mathbf{x}) + k(\mathbf{y}) - n) \neq 0$$

contradiction. That is, the definition of $k(\mathbf{z})$ is unique. This ends the proof of (2).

Proof of (3). Let $\mathbf{z} = \mathbf{x}/\mathbf{y}$. Assume that

$$\mathbf{y} = \sum_{m \geq k(y)} \mathbf{y}(m)\mathbf{1}_{(m)}$$

$$\mathbf{x} = \sum_{m \geq k(x)} \mathbf{x}(m)\mathbf{1}_{(m)}$$

$$\mathbf{z} = \sum_{m \geq k(z)} \mathbf{z}(m)\mathbf{1}_{(m)}$$

Then these numbers satisfy

$$\sum_{u+v=s} \mathbf{y}(u) \times \mathbf{z}(v) = \mathbf{x}(s) \tag{11.10}$$

with $k(\mathbf{y}) \leq u \leq s - k(\mathbf{z})$ and $k(\mathbf{z}) \leq v \leq s - k(\mathbf{y})$. This second inequality implies that we can define $k(\mathbf{z}) = k(\mathbf{x}) - k(\mathbf{y})$. Hence, replacing s by $k(\mathbf{x})$ in Eq. (11.10), one has

$$\mathbf{y}(k(\mathbf{y})) \times \mathbf{z}(k(\mathbf{x}) - k(\mathbf{y})) = x(k(\mathbf{x}))$$

That is,

$$\mathbf{y}(k(\mathbf{y})) \times \mathbf{z}(k(\mathbf{z})) = \mathbf{x}(k(\mathbf{x}))$$

Thus, $\mathbf{z}(k(\mathbf{z})) = \mathbf{x}(k(\mathbf{x}))/\mathbf{y}(k(\mathbf{y}))$. When s is replaced by $k(\mathbf{x}) + 1$ in Eq. (11.10), one has

$$\mathbf{y}(k(\mathbf{y})) \times \mathbf{z}(k(\mathbf{z}) + 1) + \mathbf{y}(k(\mathbf{y}) + 1) \times \mathbf{z}(k(\mathbf{z})) = \mathbf{x}(k(\mathbf{x}) + 1)$$

Therefore,

$$\mathbf{z}(k(\mathbf{z}) + 1) = \{\mathbf{x}(k(\mathbf{x}) + 1) - \mathbf{y}(k(\mathbf{y}) + 1) \times \mathbf{z}(k(\mathbf{z}))\}/\mathbf{y}(k(\mathbf{y}))$$

By induction, we can continue this process and define each $\mathbf{z}(m)$ for $m \geq k(\mathbf{z}) = k(\mathbf{x}) - k(\mathbf{y})$. Thus, the relation between the generalized numbers $\mathbf{z}$, $\mathbf{x}$, and $\mathbf{y}$ can be found.

The uniqueness of $k(\mathbf{z})$ can be shown as follows: Assume that $k(\mathbf{z}) = k(\mathbf{x}) - k(\mathbf{y}) - n$ for some nonzero whole number n. Then replacing s by $k(\mathbf{x}) - n$ in Eq. (11.10) gives

$$0 \neq \mathbf{y}(k(\mathbf{y})) \times \mathbf{z}(k(\mathbf{z})) = \mathbf{x}(k(\mathbf{x}) - n) = 0$$

contradiction. The whole number n must therefore be zero. That is, $k(\mathbf{z})$ must be unique. ■

11.2. Continuity

Let E be a subset of **GNS** and $\mathbf{f}: E \to \mathbf{GNS}$ a function. For a fixed $\mathbf{a} \in E$ and fixed $m, n \in \mathbb{Z}$, if for any real number $\varepsilon > 0$ there exists a real number $\delta > 0$ such that, for any $\mathbf{x} \in E$

$$k(\mathbf{x}-\mathbf{a}) = m \quad \text{and} \quad |(\mathbf{x}-\mathbf{a})_m| < \delta$$

imply that

$$k(\mathbf{f}(\mathbf{x})-\mathbf{f}(\mathbf{a})) \geq n \quad \text{and} \quad |(\mathbf{f}(\mathbf{x})-\mathbf{f}(\mathbf{a}))_n| < \varepsilon$$

then $\mathbf{f}(\mathbf{x})$ is quasi-(m,n) continuous at point $\mathbf{a}$ (Wang, 1985). A special case of quasi-(m,n) continuity is quasi-$(0,0)$ continuity, which is the ordinary definition of continuity of real-valued functions. If $n < m$, Theorem 11.1.2 implies that when the function value is infinity, one can still study the concept of continuity. That is, the concept of continuity is expanded to include the study of continuity of infinities.

Theorem 11.2.1. *If* $\mathbf{f}(\mathbf{x})$ *and* $\mathbf{g}(\mathbf{x})$ *are quasi-*(m,n) *continuous at a point* $\mathbf{a}$, *then so are the functions* $\mathbf{h}(\mathbf{x}) = \mathbf{f}(\mathbf{x})+\mathbf{g}(\mathbf{x})$ *and* $\mathbf{k}(\mathbf{x}) = \mathbf{f}(\mathbf{x})-\mathbf{g}(\mathbf{x})$.

Proof: Since $\mathbf{f}(\mathbf{x})$ and $\mathbf{g}(\mathbf{x})$ are quasi-(m,n) continuous at point $\mathbf{a}$, by the definition of quasi-(m,n) continuity, for any real number $\varepsilon > 0$, there exist real numbers $\delta_f > 0$ and $\delta_g > 0$ such that for any $\mathbf{x} \in \mathbf{GNS}$,

$$k(\mathbf{x}-\mathbf{a}) = m \quad \text{and} \quad |(\mathbf{x}-\mathbf{a})_m| < \delta_f$$

imply that

$$k(\mathbf{f}(\mathbf{x})-\mathbf{f}(\mathbf{a})) \geq n \quad \text{and} \quad |(\mathbf{f}(\mathbf{x})-\mathbf{f}(\mathbf{a}))_n| < \varepsilon/2$$

and that for any $\mathbf{x} \in \mathbf{GNS}$,

$$k(\mathbf{x}-\mathbf{a}) = m \quad \text{and} \quad |(\mathbf{x}-\mathbf{a})_m| < \delta_g$$

imply that

$$k(\mathbf{g}(\mathbf{x})-\mathbf{g}(\mathbf{a})) \geq n \quad \text{and} \quad |(\mathbf{g}(\mathbf{x})-\mathbf{g}(\mathbf{a}))_n| < \varepsilon/2$$

Therefore, for any $\mathbf{x} \in \mathbf{GNS}$ satisfying

$$k(\mathbf{x}-\mathbf{a}) = m \quad \text{and} \quad |(\mathbf{x}-\mathbf{a})_m| < \min\{\delta_f, \delta_g\}$$

one has

$$k([\mathbf{f}(\mathbf{x}) \pm \mathbf{g}(\mathbf{x})] - [\mathbf{f}(\mathbf{a}) \pm \mathbf{g}(\mathbf{a})]) \geq n$$

and

$$|([\mathbf{f}(\mathbf{x}) \pm \mathbf{g}(\mathbf{x})] - [\mathbf{f}(\mathbf{a}) \pm \mathbf{g}(\mathbf{a})])_n| \leq |(\mathbf{f}(\mathbf{x}) - \mathbf{f}(\mathbf{a}))_n| + |(\mathbf{g}(\mathbf{x}) - \mathbf{g}(\mathbf{a}))_n|$$
$$< \varepsilon/2 + \varepsilon/2 = \varepsilon$$

That is, it has been shown that $\mathbf{h}(\mathbf{x})$ and $\mathbf{k}(\mathbf{x})$ are quasi-(m,n) continuous at the point $\mathbf{a}$. ■

Theorem 11.2.2. *Assume that the functions* $\mathbf{f}(\mathbf{x})$ *and* $\mathbf{g}(\mathbf{x})$ *are quasi-*(m,n_1) *and* (m, n_2) *continuous at a point* $\mathbf{a}$, *respectively. Then the product function* $\mathbf{h}(\mathbf{x}) = \mathbf{f}(\mathbf{x})\mathbf{g}(\mathbf{x})$ *is quasi-*(m,n) *continuous at point* $\mathbf{a}$, *where* $n = \min\{k(\mathbf{f}(\mathbf{a})) + n_2, n_1 + k(\mathbf{g}(\mathbf{a}))\}$.

Proof: Since $\mathbf{f}(\mathbf{x})$ and $\mathbf{g}(\mathbf{x})$ are quasi-(m,n_1) and quasi-(m,n_2) continuous at $\mathbf{a}$, respectively, from the definition of quasi-(m,n) continuity, it follows that for any real number $\varepsilon > 0$ there exist real numbers $\delta_f > 0$ and $\delta_g > 0$ such that, for any $\mathbf{x} \in \mathbf{GNS}$,

$$k(\mathbf{x}-\mathbf{a}) = m \quad \text{and} \quad |(\mathbf{x}-\mathbf{a})_m| < \delta_f$$

imply that

$$k(\mathbf{f}(\mathbf{x}) - \mathbf{f}(\mathbf{a})) \geq n_1$$

and

$$|(\mathbf{f}(\mathbf{x}) - \mathbf{f}(\mathbf{a}))_{n_1}| < \frac{\varepsilon}{2[|(\mathbf{g}(\mathbf{a}))_{k(\mathbf{g}(\mathbf{a}))}| + 1]}$$

$k(\mathbf{f}(\mathbf{x})) = k(\mathbf{f}(\mathbf{a})) \leq n_2$, and $|(\mathbf{f}(\mathbf{x}))_{k(\mathbf{f}(\mathbf{a}))}| \leq M$, for some real number M; and that for any $\mathbf{x} \in \mathbf{GNS}$,

$$k(\mathbf{x}-\mathbf{a}) = m \quad \text{and} \quad |(\mathbf{x}-\mathbf{a})_m| < \delta_g$$

imply that

$$k(\mathbf{g}(\mathbf{x}) - \mathbf{g}(\mathbf{a})) \geq n_2 \, |(\mathbf{g}(\mathbf{x}) - \mathbf{g}(\mathbf{a}))_{n_2}| < \varepsilon/2M$$

and

$$k(\mathbf{g}(\mathbf{x})) = k(\mathbf{g}(\mathbf{a})) \leq n_2$$

Therefore, for any $\mathbf{x} \in \mathbf{GNS}$ satisfying

$$k(\mathbf{x}-\mathbf{a}) = m \quad \text{and} \quad |(\mathbf{x}-\mathbf{a})_m| < \min\{\delta_f, \delta_g\}$$

one has

$$\begin{aligned}
k(\mathbf{f}(\mathbf{x})\mathbf{g}(\mathbf{x})-\mathbf{f}(\mathbf{a})\mathbf{g}(\mathbf{a})) &= k(\mathbf{f}(\mathbf{x})\mathbf{g}(\mathbf{x})-\mathbf{f}(\mathbf{x})\mathbf{g}(\mathbf{a})+\mathbf{f}(\mathbf{x})\mathbf{g}(\mathbf{a})-\mathbf{f}(\mathbf{a})\mathbf{g}(\mathbf{a}))\\
&= k(\mathbf{f}(\mathbf{x})[\mathbf{g}(\mathbf{x})-\mathbf{g}(\mathbf{a})]+\mathbf{g}(\mathbf{a})[\mathbf{f}(\mathbf{x})-\mathbf{f}(\mathbf{a})])\\
&\geq n\\
&= \min\{k(\mathbf{f}(\mathbf{x}))+n_2, k(\mathbf{g}(\mathbf{a}))+n_1\}\\
&= \min\{k(\mathbf{f}(\mathbf{a}))+n_2, k(\mathbf{g}(\mathbf{a}))+n_1\}
\end{aligned}$$

and

$$\begin{aligned}
&|(\mathbf{f}(\mathbf{x})\mathbf{g}(\mathbf{x})-\mathbf{f}(\mathbf{a})\mathbf{g}(\mathbf{a}))_n|\\
&\leq |(\mathbf{f}(\mathbf{x})[\mathbf{g}(\mathbf{x})-\mathbf{g}(\mathbf{a})])_n| + |(\mathbf{g}(\mathbf{a})[\mathbf{f}(\mathbf{x})-\mathbf{f}(\mathbf{a})])_n|\\
&\leq \left|(\mathbf{f}(\mathbf{x})[\mathbf{g}(\mathbf{x})-\mathbf{g}(\mathbf{a})])_{k(\mathbf{f}(\mathbf{x}))+n_2}\right| + \left|(\mathbf{g}(\mathbf{a})[\mathbf{f}(\mathbf{x})-\mathbf{f}(\mathbf{a})])_{k(g(a))+n_1}\right|\\
&= \left|(\mathbf{f}(\mathbf{x}))_{k(\mathbf{f}(\mathbf{x}))}\right| \times |(\mathbf{g}(\mathbf{x})-\mathbf{g}(\mathbf{a}))_{n_2}| + \left|(\mathbf{g}(\mathbf{a}))_{k(\mathbf{g}(\mathbf{a}))}\right| \times |(\mathbf{f}(\mathbf{x})-\mathbf{f}(\mathbf{a}))_{n_1}|\\
&\leq M \times \frac{\varepsilon}{2M} + \left|(\mathbf{g}(\mathbf{a}))_{k(\mathbf{g}(\mathbf{a}))}\right| \times \frac{\varepsilon}{2[\left|(\mathbf{g}(\mathbf{a}))_{k(\mathbf{g}(\mathbf{a}))}\right|+1]}\\
&< \frac{\varepsilon}{2}+\frac{\varepsilon}{2}\\
&= \varepsilon
\end{aligned}$$

It has thus been shown that $\mathbf{h}(\mathbf{x}) = \mathbf{f}(\mathbf{x})\mathbf{g}(\mathbf{x})$ is quasi-(m,n) continuous at $\mathbf{a}$. ■

Theorem 11.2.3. *Assume that* $\mathbf{f}(\mathbf{x})$ *is quasi-*(m,n_f) *continuous at a point* $\mathbf{a}$ *and that* $\mathbf{g}(\mathbf{x})$ *is quasi-*$(m,k(\mathbf{g}(\mathbf{a})))$ *continuous at* $\mathbf{a}$ *such that* $\mathbf{g}(\mathbf{a}) \neq \mathbf{0}$. *Then, the quotient function* $\mathbf{f}(\mathbf{x})/\mathbf{g}(\mathbf{x})$ *is quasi-*(m,n) *continuous at* $\mathbf{a}$, *where* $n = \min\{n_f - k(\mathbf{g}(\mathbf{a})), k(\mathbf{f}(\mathbf{a})) - k(\mathbf{g}(\mathbf{a}))\}$.

Proof: From the assumption that $\mathbf{g}(\mathbf{x})$ is quasi-$(m,k(\mathbf{g}(\mathbf{a})))$ continuous at $\mathbf{a}$, it follows that for any real number $\varepsilon > 0$ there exists a real number $\delta_g > 0$ such that, for any $\mathbf{x} \in \mathbf{GNS}$,

$$k(\mathbf{x}-\mathbf{a}) = m \quad \text{and} \quad |(\mathbf{x}-\mathbf{a})_m| < \delta_g$$

imply that

$$k(\mathbf{g}(\mathbf{x})-\mathbf{g}(\mathbf{a})) \geq k(\mathbf{g}(\mathbf{a}))$$

$$\left|(\mathbf{g}(\mathbf{x})-\mathbf{g}(\mathbf{a}))_{k(\mathbf{g}(\mathbf{a}))}\right| < \frac{\varepsilon \times m \times \left|(\mathbf{g}(\mathbf{a}))_{k(\mathbf{g}(\mathbf{a}))}\right|}{2 \times [\left|(\mathbf{f}(\mathbf{a}))_{k(\mathbf{f}(\mathbf{a}))}\right|+1]}$$

and $k(\mathbf{g}(\mathbf{x})) = k(\mathbf{g}(\mathbf{a}))$, where $0 < m \leq \left|(\mathbf{g}(\mathbf{x}))_{k(\mathbf{g}(\mathbf{a}))}\right| \leq M$ for some fixed positive numbers $m, M \in \mathbb{R}$. The quasi-$(m,k(\mathbf{g}(\mathbf{a})))$ continuity of $\mathbf{g}(\mathbf{x})$ at $\mathbf{a}$ guarantees the existence of m and M.

Now from the assumption that $\mathbf{f}(\mathbf{x})$ is quasi-(m, n_f) continuous at $\mathbf{a}$, it follows that for any real number $\varepsilon > 0$, there exists a real number $\delta_f > 0$ such that for any $\mathbf{x} \in \mathbf{GNS}$,

$$k(\mathbf{x}-\mathbf{a}) = m \quad \text{and} \quad |(\mathbf{x}-\mathbf{a})_m| < \delta_f$$

imply that

$$k(\mathbf{f}(\mathbf{x}) - \mathbf{f}(\mathbf{a})) \geq n_f$$

and

$$\left|(\mathbf{f}(\mathbf{x}) - \mathbf{f}(\mathbf{a}))_{n_f}\right| < \frac{\varepsilon m \left|(\mathbf{g}(\mathbf{a}))_{k(\mathbf{g}(\mathbf{a}))}\right|}{2M}$$

Now let $\mathbf{x} \in \mathbb{R}$ satisfy

$$k(\mathbf{x}-\mathbf{a}) = m \quad \text{and} \quad |(\mathbf{x}-\mathbf{a})_m| < \min\{\delta_f, \delta_g\}$$

Then from Theorem 11.1.3 it follows that

$$\begin{aligned}
k\left(\frac{\mathbf{f}(\mathbf{x})}{\mathbf{g}(\mathbf{x})} - \frac{\mathbf{f}(\mathbf{a})}{\mathbf{g}(\mathbf{a})}\right) &= k\left(\frac{\mathbf{f}(\mathbf{x})\mathbf{g}(\mathbf{a}) - \mathbf{g}(\mathbf{x})\mathbf{f}(\mathbf{a})}{\mathbf{g}(\mathbf{x})\mathbf{g}(\mathbf{a})}\right) \\
&= k(\mathbf{f}(\mathbf{x})\mathbf{g}(\mathbf{a}) - \mathbf{g}(\mathbf{x})\mathbf{f}(\mathbf{a})) - k(\mathbf{g}(\mathbf{x})\mathbf{g}(\mathbf{a})) \\
&= k(\mathbf{f}(\mathbf{x})\mathbf{g}(\mathbf{a}) - \mathbf{f}(\mathbf{a})\mathbf{g}(\mathbf{a}) + \mathbf{f}(\mathbf{a})\mathbf{g}(\mathbf{a}) - \mathbf{f}(\mathbf{a})\mathbf{g}(\mathbf{x})) - 2k(\mathbf{g}(\mathbf{a})) \\
&= k([\mathbf{f}(\mathbf{x}) - \mathbf{f}(\mathbf{a})]\mathbf{g}(\mathbf{a}) + \mathbf{f}(\mathbf{a})[\mathbf{g}(\mathbf{a}) - \mathbf{g}(\mathbf{x})]) - 2k(\mathbf{g}(\mathbf{a})) \\
&\geq \min\{k(\mathbf{g}(\mathbf{a})) + n_f - 2k(\mathbf{g}(\mathbf{a})), k(\mathbf{f}(\mathbf{a})) + k(\mathbf{g}(\mathbf{a})) - 2k(\mathbf{g}(\mathbf{a}))\} \\
&= n
\end{aligned}$$

and that

$$\begin{aligned}
&\left|\left(\frac{\mathbf{f}(\mathbf{x})}{\mathbf{g}(\mathbf{x})} - \frac{\mathbf{f}(\mathbf{a})}{\mathbf{g}(\mathbf{a})}\right)_n\right| = \left|\left(\frac{\mathbf{f}(\mathbf{x})\mathbf{g}(\mathbf{a}) - \mathbf{g}(\mathbf{x})\mathbf{f}(\mathbf{a})}{\mathbf{g}(\mathbf{x})\mathbf{g}(\mathbf{a})}\right)_n\right| \\
&\leq \left|\left(\frac{[\mathbf{f}(\mathbf{x}) - \mathbf{f}(\mathbf{a})]\mathbf{g}(\mathbf{a})}{\mathbf{g}(\mathbf{x})\mathbf{g}(\mathbf{a})}\right)_n\right| + \left|\left(\frac{[\mathbf{g}(\mathbf{x}) - \mathbf{g}(\mathbf{a})]\mathbf{f}(\mathbf{a})}{\mathbf{g}(\mathbf{x})\mathbf{g}(\mathbf{a})}\right)_n\right| \\
&= \left|\frac{([\mathbf{f}(\mathbf{x}) - \mathbf{f}(\mathbf{a})]\mathbf{g}(\mathbf{a}))_{n+2k(\mathbf{g}(\mathbf{a}))}}{(\mathbf{g}(\mathbf{x})\mathbf{g}(\mathbf{a}))_{2k(\mathbf{g}(\mathbf{a}))}}\right| + \left|\frac{([\mathbf{g}(\mathbf{x}) - \mathbf{g}(\mathbf{a})]\mathbf{f}(\mathbf{a}))_{n+2k(\mathbf{g}(\mathbf{a}))}}{(\mathbf{g}(\mathbf{x})\mathbf{g}(\mathbf{a}))_{2k(\mathbf{g}(\mathbf{a}))}}\right| \\
&\leq \frac{M}{m\left|(\mathbf{g}(\mathbf{a}))_{k(\mathbf{g}(\mathbf{a}))}\right|}\left|(\mathbf{f}(\mathbf{x}) - \mathbf{f}(\mathbf{a}))_{n_f}\right| + \frac{\left|(\mathbf{f}(\mathbf{a}))_{k(\mathbf{f}(\mathbf{a}))}\right|}{m\left|(\mathbf{g}(\mathbf{a}))_{k(\mathbf{g}(\mathbf{a}))}\right|}\left|(\mathbf{g}(\mathbf{x}) - \mathbf{g}(\mathbf{a}))_{k(\mathbf{g}(\mathbf{a}))}\right| \\
&\leq \frac{\varepsilon}{2} + \frac{\varepsilon}{2} \\
&= \varepsilon
\end{aligned}$$

Therefore, the quotient function is quasi-(m, n) continuous at $\mathbf{a}$. ∎

11.3. Differential Calculus

In order to simplify the discussion, let us look at the concept of $\langle \varepsilon$–$\delta \rangle$ limits in **GNS** first. Assume that $\mathbf{f} : \mathbf{GNS} \to \mathbf{GNS}$ is a function and $\mathbf{a} \in \mathbf{GNS}$ a fixed generalized number. For a fixed $L \in \mathbb{R}$,

$$\lim_{(m,n)\mathbf{x}\to\mathbf{a}} \mathbf{f}(\mathbf{x}) = L \tag{11.11}$$

stands for the fact that for fixed integers m and n, for any real number $\varepsilon > 0$ there exists a real number $\delta > 0$ such that for any $\mathbf{x} \in \mathbf{GNS}$,

$$k(\mathbf{x}-\mathbf{a}) = m \quad \text{and} \quad |(\mathbf{x}-\mathbf{a})_m| < \delta$$

imply that

$$k(\mathbf{f}(\mathbf{x}) - \mathbf{f}(\mathbf{a})) \geq n \quad \text{and} \quad |(\mathbf{f}(\mathbf{x}) - \mathbf{f}(\mathbf{a}))_n| < \varepsilon$$

In this case, Eq. (11.11) is read as the quasi-(m,n) limit of $f(x)$ at $\mathbf{a}$ is L.

Let $S \in \mathbb{R}$ be fixed such that for fixed $m, n \in \mathbb{Z}$, for each real number $\varepsilon > 0$ there exists a real number $\delta > 0$ such that, for any $\mathbf{x} \in \mathbf{GNS}$,

$$k(\mathbf{x}-\mathbf{a}) = m \quad \text{and} \quad |(\mathbf{x}-\mathbf{a})_m| < \delta \tag{11.12}$$

imply that

$$k(\mathbf{f}(\mathbf{x}) - \mathbf{f}(\mathbf{a})) \geq n \tag{11.13}$$

and

$$\left| \frac{(\mathbf{f}(\mathbf{x}) - \mathbf{f}(\mathbf{a}))_n}{(\mathbf{x}-\mathbf{a})_m} - S \right| < \varepsilon \tag{11.14}$$

Then the generalized number $S \times \mathbf{1}_{(n-m)}$ is called the (m,n)-derivative of the function $\mathbf{f}$ at $\mathbf{a}$ (Wang, 1985). In this case, $\mathbf{f}$ is (m,n)-differentiable at $\mathbf{a}$, and the (m,n)-derivative will be written as

$$\begin{aligned} \mathbf{f}'_{(m,n)}(\mathbf{a}) &= \frac{d_n \mathbf{f}(\mathbf{a})}{d_m \mathbf{x}} \\ &= \lim_{(m,n)\mathbf{x}\to\mathbf{a}} \frac{\mathbf{f}(\mathbf{x}) - \mathbf{f}(\mathbf{a})}{\mathbf{x}-\mathbf{a}} \\ &= S \times \mathbf{1}_{(n-m)} \end{aligned} \tag{11.15}$$

It is obvious that the $(0,0)$-derivative of $\mathbf{f}$ is the ordinary derivative of real-valued functions.

Example 11.3.1. Let $g:\mathbb{R}\to\mathbb{R}$ be an infinitely differentiable real-valued function such that $\{x\in\mathbb{R}: g(x)\neq 0\}$ is bounded and $\int_{-\infty}^{\infty} g(t)\,dt = 1$. Define

$$\mathbf{g}(\mathbf{x}) = \begin{cases} g(x_1)\times \mathbf{1}_{(-1)} & \text{if } \mathbf{x} = (\dots,0,x_1,x_2,\dots) \\ \mathbf{0} & \text{otherwise} \end{cases}$$

It can be shown that the $(1,-1)$-derivative of $\mathbf{g}:\mathbf{GNS}\to\mathbf{GNS}$ is

$$\frac{d_{-1}\mathbf{g}(\mathbf{x})}{d_1\mathbf{x}} = \begin{cases} g'(x_1)\times \mathbf{1}_{(-2)} & \text{if } \mathbf{x} = (\dots,0,x_1,x_2,\dots) \\ \mathbf{0} & \text{otherwise} \end{cases}$$

Theorem 11.3.1. *If a function* $\mathbf{f}:\mathbf{GNS}\to\mathbf{GNS}$ *is* (m,n)*-differentiable at a point* $\mathbf{a}\in\mathbf{GNS}$, *then* $\mathbf{f}$ *is quasi-*(m,n) *continuous at* $\mathbf{a}$.

Proof: To show that $\mathbf{f}$ is quasi-(m,n) continuous at $\mathbf{a}$, let ε be an arbitrary real number such that $0<\varepsilon<1$. It suffices to show that there exists a real number $\delta>0$ such that, for any $\mathbf{x}\in\mathbf{GNS}$,

$$k(\mathbf{x}-\mathbf{a}) = m \quad \text{and} \quad |(\mathbf{x}-\mathbf{a})_m| < \delta$$

imply that

$$k(\mathbf{f}(\mathbf{x})-\mathbf{f}(\mathbf{a})) \geq n \quad \text{and} \quad |(\mathbf{f}(\mathbf{x})-\mathbf{f}(\mathbf{a}))_n| < \varepsilon$$

Since $\mathbf{f}$ is (m,n)-differentiable at $\mathbf{a}$, there exists an $S\in\mathbb{R}$ such that for the chosen $\varepsilon>0$, there exists a real number $\delta>0$ such that, for any $\mathbf{x}\in\mathbf{GNS}$,

$$k(\mathbf{x}-\mathbf{a}) = m \quad \text{and} \quad |(\mathbf{x}-\mathbf{a})_m| < \delta \leq \min\left\{\frac{\varepsilon}{2(|S|+1)},\frac{\varepsilon}{2}\right\}$$

imply that

$$|k(\mathbf{f}(\mathbf{x})-\mathbf{f}(\mathbf{a}))| \geq n$$

and

$$\left|\frac{(\mathbf{f}(\mathbf{x})-\mathbf{f}(\mathbf{a}))_n}{(\mathbf{x}-\mathbf{a})_m} - S\right| < \varepsilon$$

So

$$\left|\frac{(\mathbf{f}(\mathbf{x})-\mathbf{f}(\mathbf{a}))_n}{(\mathbf{x}-\mathbf{a})_m}\right| - |S| < \varepsilon$$

and

$$\begin{aligned}|(\mathbf{f}(\mathbf{x})-\mathbf{f}(\mathbf{a}))_n| &< (\varepsilon+|S|)\,|(\mathbf{x}-\mathbf{a})_m| \\ &= \varepsilon\,|(\mathbf{x}-\mathbf{a})_m| + |S|\times|(\mathbf{x}-\mathbf{a})_m| \\ &< \frac{\varepsilon^2}{2} + |S|\,\frac{\varepsilon}{2(|S|+1)} \\ &< \frac{\varepsilon}{2}+\frac{\varepsilon}{2} \\ &= \varepsilon\end{aligned}$$

That is, $\mathbf{f}$ is quasi-(m,n) continuous at $\mathbf{a}$. ■

Theorem 11.3.2.

(1) If $\mathbf{f}(\mathbf{x}) = \mathbf{c} \in$ **GNS** *is a constant function, then* $d_n\mathbf{f}(\mathbf{x})/d_m\mathbf{x} = \mathbf{0}$.

(2) If s *is a positive integer and*

$$\mathbf{f}(\mathbf{x}) = \mathbf{x}^s = \underbrace{\mathbf{x}\times\mathbf{x}\times\cdots\times\mathbf{x}}_{s\ times}$$

then

$$\frac{d_n\mathbf{f}(\mathbf{x})}{d_m\mathbf{x}} = s\mathbf{x}_m^{s-1}$$

where $m = k(\mathbf{x})$ *and* $n = sm$.

Proof:

(1) From Eq. (11.15), it follows that for any fixed $\mathbf{a} \in$ **GNS**,

$$\begin{aligned}\frac{d_n\mathbf{f}(\mathbf{a})}{d_m\mathbf{x}} &= \lim_{(m,n)\mathbf{x}\to\mathbf{a}} \frac{\mathbf{f}(\mathbf{x})-\mathbf{f}(\mathbf{a})}{\mathbf{x}-\mathbf{a}} \\ &= \lim_{(m,n)\mathbf{x}\to\mathbf{a}} \frac{\mathbf{c}-\mathbf{c}}{\mathbf{x}-\mathbf{a}} \\ &= \mathbf{0}\end{aligned}$$

(2) For any $\mathbf{x},\mathbf{a} \in$ **GNS**, the axioms of a non-Archimedean ordered field (see the proof of Theorem 11.1.1) can be used to prove that

$$\mathbf{x}^s - \mathbf{a}^s = (\mathbf{x}-\mathbf{a})(\mathbf{x}^{s-1} + \mathbf{x}^{s-2}\mathbf{a} + \cdots + \mathbf{x}\mathbf{a}^{s-2} + \mathbf{a}^{s-1})$$

Therefore, for $m = k(\mathbf{a})$ and $n = sm$,

$$\begin{aligned}\frac{d_n\mathbf{f}(\mathbf{a})}{d_m\mathbf{x}} &= \lim_{(m,n)\mathbf{x}\to\mathbf{a}} \frac{\mathbf{f}(\mathbf{x})-\mathbf{f}(\mathbf{a})}{\mathbf{x}-\mathbf{a}} \\ &= \lim_{(m,n)\mathbf{x}\to\mathbf{a}} \frac{\mathbf{x}^s-\mathbf{a}^s}{\mathbf{x}-\mathbf{a}} \\ &= \lim_{(m,n)\mathbf{x}\to\mathbf{a}} (\mathbf{x}^{s-1}+\mathbf{x}^{s-2}\mathbf{a}+\cdots+\mathbf{x}\mathbf{a}^{s-2}+\mathbf{a}^{s-1}) \\ &= s\mathbf{a}_m^{s-1}\end{aligned}$$

■

Theorem 11.3.3. *Suppose that $c \in \mathbb{R}$ is a constant and $\mathbf{f}'_{(m,n)}(\mathbf{x})$ and $\mathbf{g}'_{(m,n)}(\mathbf{x})$ exist. Then*

(a) If $\mathbf{F}(\mathbf{x}) = c\mathbf{f}(\mathbf{x})$, then $\mathbf{F}'_{(m,n)}(\mathbf{x}) = c\mathbf{f}'_{(m,n)}(\mathbf{x})$.

(b) If $\mathbf{G}(\mathbf{x}) = \mathbf{f}(\mathbf{x})+\mathbf{g}(\mathbf{x})$, then $\mathbf{G}'_{(m,n)}(\mathbf{x}) = \mathbf{f}'_{(m,n)}(\mathbf{x})+\mathbf{g}'_{(m,n)}(\mathbf{x})$.

(c) If $\mathbf{H}(\mathbf{x}) = \mathbf{f}(\mathbf{x})-\mathbf{g}(\mathbf{x})$, then $\mathbf{H}'_{(m,n)}(\mathbf{x}) = \mathbf{f}'_{(m,n)}(\mathbf{x})-\mathbf{g}'_{(m,n)}(\mathbf{x})$.

The proof is straightforward and is omitted.

One problem with the concept of (m,n)-derivative is that the derivative is a generalized number in $I_{(n-m)}$. In (Wang, 1991), this problem was corrected by the following methods.

Suppose $\mathbf{y} = \mathbf{f}(\mathbf{x}) = (\ldots, y_{-m}, \ldots, y_0, \ldots, y_n, \ldots)$ is a function from **GNS** into **GNS**. Then, for each fixed index $k \in \mathbb{Z}$, y_k is a function of $\mathbf{x}$. That is, in general y_k depends on an infinite number of variables $x_0, x_1, x_{-1}, x_2, x_{-2}, \ldots$ As in the calculus of real-valued functions, let $\Delta\mathbf{x}$ and $\Delta\mathbf{y}$ represent increments of $\mathbf{x}$ and $\mathbf{y}$ such that

$$\Delta\mathbf{x} = (\ldots, (\Delta\mathbf{x})_i, \ldots), \qquad \Delta\mathbf{y} = (\ldots, (\Delta\mathbf{y})_i, \ldots)$$

where $(\Delta\mathbf{y})_i = (\mathbf{f}(\mathbf{x}+\Delta\mathbf{x})-\mathbf{f}(\mathbf{x}))_i$ for each $i \in \mathbb{Z}$.

Let $\mathbf{a} \in$ **GNS** be fixed. Let

$$\Delta\mathbf{y} = \mathbf{f}(\mathbf{a}+\Delta\mathbf{x})-\mathbf{f}(\mathbf{a}) \tag{11.16}$$

If there exists a generalized number, denoted $\mathbf{f}'(\mathbf{a})$, such that for each generalized number $\boldsymbol{\varepsilon} > \mathbf{0}$, there exists a generalized number $\boldsymbol{\delta} > \mathbf{0}$ satisfying, for all $\Delta\mathbf{x} \in$ **GNS** with $|\Delta\mathbf{x}| = \max\{\Delta\mathbf{x}, -\Delta\mathbf{x}\} < \boldsymbol{\delta}$,

$$\left|\frac{\Delta\mathbf{y}}{\Delta\mathbf{x}} - \mathbf{f}'(\mathbf{a})\right| < \boldsymbol{\varepsilon} \tag{11.17}$$

Then $\mathbf{f}'(\mathbf{a})$ is called the derivative of the function $\mathbf{y} = \mathbf{f}(\mathbf{x})$ at $\mathbf{a}$. Symbolically,

$$\begin{aligned} \mathbf{f}'(\mathbf{a}) &= \lim_{\Delta\mathbf{x}\to\mathbf{0}} \frac{\Delta\mathbf{y}}{\Delta\mathbf{x}} \\ &= \lim_{\Delta\mathbf{x}\to\mathbf{0}} \frac{\mathbf{f}(\mathbf{a}+\Delta\mathbf{x}) - \mathbf{f}(\mathbf{a})}{\Delta\mathbf{x}} \end{aligned} \tag{11.18}$$

Remark. Topologically speaking, the concept of derivative of $\mathbf{f}(\mathbf{x})$ at $\mathbf{a} \in \mathbf{GNS}$ is introduced so that the topological space $\mathbf{GNS} \subseteq \mathbb{R}^{\mathbb{Z}}$ has the box product topology. For more about this topology, see (Kuratowski and Mostowski, 1976).

For $\mathbf{x}, \mathbf{y} \in \mathbf{GNS}$, a different ordering relation, denoted by $\ll$, on $\mathbf{GNS}$ can be defined as follows: $\mathbf{x} \ll \mathbf{y}$ iff for each $i \in \mathbb{Z}$, $x_i < y_i$. With this ordering relation on $\mathbf{GNS}$, the weak derivative of $\mathbf{y} = \mathbf{f}(\mathbf{x})$, a function from $\mathbf{GNS}$ into $\mathbf{GNS}$, can be introduced as follows (Wang, 1991): Suppose $\mathbf{K} \in \mathbf{GNS}$ is fixed. If for each real number $\varepsilon > 0$, there exists $\boldsymbol{\delta} = (\ldots, 0, \delta_{k(\boldsymbol{\delta})}, \delta_{k(\boldsymbol{\delta})+1}, \ldots)$ with the property that $\delta_j > 0$, for each $j \geq k(\boldsymbol{\delta})$, such that whenever $|\Delta\mathbf{x}| \ll \boldsymbol{\delta}$, one has

$$\left| \left(\frac{\mathbf{f}(\mathbf{a}+\Delta\mathbf{x}) - \mathbf{f}(\mathbf{a})}{\Delta\mathbf{x}} \right)_s - \mathbf{K}_s \right| < \varepsilon$$

for each $s \in \mathbb{Z}$, then, $\mathbf{K}$ is called the weak derivative of $\mathbf{f}(\mathbf{x})$ at $\mathbf{a}$, denoted $\mathbf{f}'_w(\mathbf{a}) = \mathbf{K}$.

11.4. Integral Calculus

11.4.1. Lebesgue Measures

For convenience and completeness of this chapter, we briefly look at some elementary definitions and properties of Lebesgue integration, which is the center of the theory of real analysis.

In classical mathematical analysis, functions which are continuous everywhere are basically the center of the discussion of interest. However, as the theory grew more mature and more applications were found, it seemed that it was no longer sufficient to consider the everywhere continuous functions only. At first, it was found that based on the consideration of continuous functions, the theory established was often incomplete and also its applications were not flexible, at the same time, it also seriously affected the development of the theory itself. For example, the theory of integration of continuous functions was developed with the theory of linear algebra as its background. This background has had great impact on the establishment and development of the theory. However, even though it was restricted to the study of continuous functions, under certain circumstances the harmony between the two theories was destroyed. For example, the concept of limits introduced in the space L^2, denoted by $\lim_{n\to\infty} f_n = f$, is different from the concept of $\lim_{n\to\infty} f_n(x) = f(x)$ in mathematical analysis, where L^2 stands for the space of all the functions which are Lebesgue square-integrable on $\mathbb{R}$. This new concept of convergence has widespread applications. There are many good reasons for introducing new concepts. For example, if it has been shown that the boundary problems of partial differential equations of some problems in physics do not have differentiable solutions, can one claim, based on this fact, that these possess no practical meaning? Obviously not! since boundary problems are often approximations of some realistic situations, they are idealized mathematical models, and the coefficients of the differential equations in the models are not absolutely accurate. Therefore, one should look at boundary problems from the viewpoint of approximation. It is unreasonable to restrict ourselves to the range of continuous and differentiable functions. Because of this, the concept of generalized solutions was introduced.

In the theory of Riemann integrals, to integrate each term of a given series, the series must be uniformly convergent. Without uniform convergence, the limit of a sequence of integrable functions may not be integrable at all, and therefore there is no way for one to talk about integrating each terms and summing the results. On the other hand, in applied situations, the requirement of uniform convergence is often not satisfied or is so complicated that it only causes more trouble. Similar problems exist in different areas of the old theory of integration. Most arise because in the definition of definite integrals, too much emphasis is put on the requirement of continuity. The essence of Riemann integrations involves (1) cutting the interval into finitely many subintervals; (2) on each subinterval, looking at the function value $f(x)$ as a constant; (3) improving accuracy of the approximation by increasing the number of subintervals; (4) taking the limit to achieve the desired accuracy. For this process to work, $f(x)$ cannot vary "too much" on each of these subintervals. That is, $f(x)$ must be "fundamentally" continuous. If $f(x)$ is "extremely discontinuous," no matter how small each subinterval is, there is no way that one can look at $f(x)$ as approximately a constant. The function is then considered to be not integrable.

If $f(x)$ is discontinuous, then we should try some new methods. For instance, partition the interval into small subsets, on each of which $f(x)$ does not vary "too much." This is how the theory of Lebesgue integration was born.

The open interval A in the space $\mathbb{R}^n$ is assumed to be the set of points $\{x \in \mathbb{R}^n : a_i < x_i < b_i, \quad i = 1,2,\dots,n\}$, where a_i and b_i are constants. If A is an interval, it means that some of the inequalities in its definition can contain the equals sign. For any interval A, $|A|$ stands for $\prod_{i=1}^{n}(b_i - a_i)$, called the volume of A. Let E be a set of points in $\mathbb{R}^n$, and $A_1, A_2, \dots, A_n, \dots$ a sequence of open intervals such that $\bigcup_{i=1}^{\infty} A_i \supseteq E$. Then $\sum_{i=1}^{\infty} |A_i| = u > 0$. The exterior measure of E, denoted $m * E$, is defined by

$$m * E = \inf\left\{ \sum_{i=1}^{\infty} |A_i| : A_i\text{'s are open intervals and } \bigcup_{i=1}^{\infty} A_i \supseteq E \right\}$$

Let A be an interval such that $A \supseteq E$, and $A_1, A_2, \ldots, A_n, \ldots$ a sequence of open intervals such that $A - E \subseteq \bigcup_{i=1}^{\infty} A_i$. Hence, $A - \cup_{i=1}^{\infty} A_i \subseteq E$. Define the interior measure of E, denoted as $m_* E$, by

$$m_* E = \sup\{|A| - \sum_{i=1}^{\infty} |A_i|\} = |A| - m * (A - E)$$

If $m * E = m_* E$, the set E is Lebesgue measurable or simply measurable. Therefore, the collection of all bounded subsets of $\mathbb{R}^n$ is classified as measurable or not measurable. An unbounded set $B \subseteq \mathbb{R}^n$ is measurable if $B \cap A$ is measurable for each chosen interval A. One equivalent condition for a subset $A \subseteq \mathbb{R}^n$ to be measurable is given in the next theorem.

Theorem 11.4.1. *A subset $A \subseteq \mathbb{R}^n$ is measurable iff, for any subset $T \subseteq \mathbb{R}^n$,*

$$m^* T = m^*(T \cap E) + m^*(T \cap (\mathbb{R}^n - E))$$

In this case, the exterior measure of A is called the measure of A and is denoted by mA.

Theorem 11.4.2. *Assume that A and B are measurable subsets in $\mathbb{R}^p$ and $\mathbb{R}^q$, respectively. Then $C = A \times B$ is a measurable subset in $\mathbb{R}^{p+q} = \mathbb{R}^r \times \mathbb{R}^q$, and $m(C) = mA \times mB$.*

Let $f(x)$ be a real-valued function defined on a measurable set $E \subseteq \mathbb{R}^n$. If there is a partition of the subset $E = \bigcup_{i=1}^{n} E_i$ such that on each E_i the function $f(x)$ is a constant, $f(x)$ is a simple function. It is obvious that the sum and product of simple functions are still simple functions. However, the limit of a sequence of simple functions may not be a simple function at all, since a simple function can only have a finite number of different function values. A function $f(x)$ is measurable on E if there is a sequence $\{\phi_i(x)\}_{i=1}^{\infty}$ of simple functions, defined on E, such that $f(x) = \lim_{n \to \infty} \phi_n(x)$.

Let $f(x) = f(x_1, x_2, \ldots, x_n)$ be a function defined on a subset $E \subseteq \mathbb{R}^n$. Define

$$G^+(E; f) = \{(x_1, x_2, \ldots, x_n, z) \in \mathbb{R}^{n+1} : \\ x = (x_1, x_2, \ldots, x_n) \in E \text{ and } 0 \leq z < f(x)\}$$

and

$$G^-(E; f) = \{(x_1, x_2, \ldots, x_n, z) \in \mathbb{R}^{n+1} : \\ x = (x_1, x_2, \ldots, x_n) \in E \text{ and } f(x) < z \leq 0\}$$

The set

$$G(E; f) = G^+(E; f) \cup G^-(E; f) \subseteq \mathbb{R}^{n+1}$$

is called the graph of $f(x)$.

Theorem 11.4.3. *A function $f(x)$ defined on a measurable subset $E \subseteq \mathbb{R}^n$ is measurable iff the graph $G(E;f)$ of the function is measurable in the $(n+1)$-dimensional space $\mathbb{R}^{n+1}$.*

Now it is time to look at the concept of Lebesgue integrals.

Case 1: Let $E \subseteq \mathbb{R}^n$ be a measurable subset with $mE < +\infty$, and $f(x)$ a bounded function defined on E. Let us see how the Lebesgue integral of $f(x)$ is defined.

Assume that $E = \bigcup_{i=1}^n E_i^*$ and $E = \bigcup_{j=1}^m E_j^{**}$ are two partitions such that each E_i^* and E_j^{**} are measurable. It is obvious that

$$E = \bigcup_{i=1}^n \bigcup_{j=1}^m E_i^* E_j^{**}$$

is also a partition, which is a finer partition than either of the previous two. For $E = \bigcup_{i=1}^n E_i$, let b_i and B_i be the greatest lower bound and the least upper bound of $f(x)$ on E_i; i.e., $b_i \le f(x) \le B_i$ for each $x \in E_i$. Define

$$s = \sum_{i=1}^n b_i m E_i \quad \text{and} \quad S = \sum_{i=1}^n B_i m E_i$$

which are called the small sum and the big sum of the function with respect to the partition $D : E = \bigcup_{i=1}^n E_i$. Obviously, one has

$$bmE \le s \le S \le BmE$$

where b and B are the greatest lower and the least upper bounds of $f(x)$ on E, respectively. It can be shown that if D^* is a finer partition than D and if s^* and S^* are the small sum and the big sum of $f(x)$ with respect to D^*, then

$$s \le s^* \le S^* \le S$$

No matter what kind of partition D is, the small sum and the big sum are always between bmE and BmE one can thus define

$$\underline{\int}_E f(x)\,dx = \sup_D\{s\} \quad \text{and} \quad \overline{\int}_E f(x)\,dx = \inf_D\{S\}$$

Then

$$bmE \le \underline{\int}_E f(x)\,dx \le \overline{\int}_E f(x)\,dx \le BmE$$

where $\underline{\int}_E f(x)\,dx$ and $\overline{\int}_E f(x)\,dx$ called the lower integral and the upper integral of $f(x)$ on E. If $\underline{\int}_E f(x)\,dx = \overline{\int}_E f(x)\,dx$, then $f(x)$ is integrable and written as

$$\int_E f(x)\,dx = \underline{\int}_E f(x)\,dx = \overline{\int}_E f(x)\,dx$$

Case 2: Let $E \subseteq \mathbb{R}^n$ be a measurable subset with $mE < +\infty$, and $f(x) \geq 0$ a nonnegative function defined on E. To define the integral of $f(x)$ on E, let

$$\{f(x)\}_n = \begin{cases} f(x) & \text{if } f(x) \leq n \\ n & \text{if } f(x) > n \end{cases}$$

for any fixed $n \in \mathbb{N}$. Then $\{f(x)\}_n$ is a bounded function on E and satisfies $\{f(x)\}_n \leq \{f(x)\}_{n+1}$. If for each $n \in \mathbb{N}$, $\{f(x)\}_n$ is integrable on E, then $f(x)$ is integrable on E, and we write

$$\int_E f(x)\,dx = \lim_{n\to\infty} \int_E \{f(x)\}_n dx$$

For a general function $f(x)$ defined on E, define

$$f^+(x) = \begin{cases} f(x) & \text{when } f(x) \geq 0 \\ 0 & \text{when } f(x) < 0 \end{cases}$$

and

$$f^-(x) = \begin{cases} -f(x) & \text{when } f(x) \leq 0 \\ 0 & \text{when } f(x) > 0 \end{cases}$$

If $f^+(x)$ and $f^-(x)$ have integral values on E which are not both infinities, then $f(x)$ has integral value on E, which is denoted by

$$\int_E f(x)\,dx = \int_E f^+(x)\,dx - \int_E f^-(x)\,dx$$

If the integral values of $f^+(x)$ and $f^-(x)$ are finite, $f(x)$ is integrable on E.

Case 3: Let E be a measurable subset of $\mathbb{R}^n$ and $f(x)$ a function defined on E. Define

$$F(x) = \begin{cases} f(x) & \text{if } x \in E \\ 0 & \text{otherwise} \end{cases}$$

Then $F(x)$ is a function defined on the entire space $\mathbb{R}^n$. Without loss of generality, let $f(x)$ be a function defined on $\mathbb{R}^n$. First, assume $f(x) \geq 0$ for all $x \in \mathbb{R}^n$. For each $k \in \mathbb{N}$, define

$$\mathbb{R}_k = \{x \in \mathbb{R}^n : -k \leq x_i \leq k, \quad i = 1, 2, \ldots, n\}$$

Then

$$\mathbb{R}_1 \subseteq \mathbb{R}_2 \subseteq \mathbb{R}_3 \subseteq \cdots \subseteq \mathbb{R}_k \subseteq \cdots$$

If $f(x)$ is integrable on each $\mathbb{R}_k$, define

$$\int_{\mathbb{R}^n} f(x)\,dx = \lim_{k\to\infty} \int_{\mathbb{R}_k} f(x)\,dx$$

as the integral value of $f(x)$ on $\mathbb{R}^n$. When this is value is finite, $f(x)$ is integrable on $\mathbb{R}^n$.

Second, let $f(x)$ be a general function defined on $\mathbb{R}^n$. Define

$$f^+(x) = \begin{cases} f(x) & \text{when } f(x) \geq 0 \\ 0 & \text{when } f(x) < 0 \end{cases}$$

and

$$f^-(x) = \begin{cases} -f(x) & \text{when } f(x) \leq 0 \\ 0 & \text{when } f(x) > 0 \end{cases}$$

If $\int_{\mathbb{R}^n} f^+(x)\,dx$ and $\int_{\mathbb{R}^n} f^-(x)\,dx$ are meaningful, define the integral value of $f(x)$ as

$$\int_{\mathbb{R}^n} f(x)\,dx = \int_{\mathbb{R}^n} f^+(x)\,dx - \int_{\mathbb{R}^n} f^-(x)\,dx$$

If these two integrals are finite, $f(x)$ is integrable on $\mathbb{R}^n$.

11.4.2. Integrals on GNS

Let us now turn our attention to the concept of integrals of functions defined on **GNS**. Let $\mathbf{f}: E \to F$ be a function from a subset E of **GNS** into a subset F of **GNS**. Define a new function $\mathbf{f}^*$: **GNS** $\to$ **GNS** as follows:

$$\mathbf{f}^*(\mathbf{x}) = \begin{cases} \mathbf{f}(\mathbf{x}) & \text{if } \mathbf{x} \in E \\ 0 & \text{otherwise} \end{cases}$$

Without loss of generality, it can be assumed that each function **f** from a subset of **GNS** into a subset of **GNS** is a function from **GNS** into **GNS**.

To make this chapter self-contained, the definition of integrals of functions defined on **GNS**, studied in Chapter 6 will be reintroduced. After that, new concepts of definite integrals and indefinite integrals will be studied. We define the integral, called a GNL-integral, with GNL standing for generalized number and Lebesgue, of functions defined on **GNS** inductively as follows. For a given function $\mathbf{f}$: **GNS** $\to$ **GNS**, let $E = \{\mathbf{x} \in \mathbf{GNS} : \mathbf{f}(\mathbf{x}) \neq \mathbf{0}\}$.

Case 1: Suppose that for each $\mathbf{x} \in E$, $x_{-m} = 0$, for all $m \in \mathbb{N}$ and that $\mathbf{y} = \mathbf{f}(\mathbf{x}) = (\ldots, y_{-m}, \ldots, y_0, \ldots, y_n, \ldots)$, where, in general, $y_k = y_k(\mathbf{x})$ is a function of all x_m, $m \in \mathbb{N} \cup \{0\}$.

Step 1: Define a subset $H^{(0)} \subseteq \mathbb{R}$ as follows: $a_0 \in H^{(0)}$ iff

$$y_{-m} = 0 \quad \text{if } m > 0 \qquad \text{and} \qquad y_0 = y_0(a_0) \quad \text{a function of } a_0 \tag{11.19}$$

where

$$\mathbf{y} = \mathbf{f}(\mathbf{x}) = (\ldots, 0, \ldots, y_{-m}, \ldots, y_0, \ldots, y_n, \ldots)$$

whenever $\mathbf{x} = (\ldots, 0, a_0, x_1, x_2, \ldots)$, with $\mathbf{x}(k(\mathbf{x})) = a_0$; let

$$f^{(0)}(a_0) = \begin{cases} y_0(a_0) & \text{if } a_0 \in H^{(0)} \\ \infty & \text{otherwise} \end{cases} \tag{11.20}$$

If the real-valued function $f^{(0)}$ is Lebesgue integrable on $\mathbb{R}$, then $\mathbb{R} - H^{(0)}$ has Lebesgue measure zero, and set $\int_{\mathbb{R}} f^{(0)}(x_0)\,dx_0 = a^{(0)}$.

Step n: Assume the real numbers $a^{(0)}, a^{(1)}, \ldots, a^{(n-1)}$ have all been defined. For any fixed $x_0, \ldots, x_{n-1}$, define a subset $H^{(n)} \subseteq \mathbb{R}$ as follows: $a_n \in H^{(n)}$ iff

$$y_{-m} = 0 \quad \text{if } m > n \qquad \text{and} \qquad y_{-n} = y_{-n}(a_n) \quad \text{a function of } a_n \tag{11.21}$$

where

$$\mathbf{y} = \mathbf{f}(\mathbf{x}) = (\ldots, y_{-m}, \ldots, y_0, \ldots, y_n, \ldots)$$

whenever $\mathbf{x} = (\ldots, 0, x_0, \ldots, x_{n-1}, a_n, x_{n+1}, \ldots)$; let

$$f^{(n)}_{(x_0, \ldots, x_{n-1})}(a_n) = \begin{cases} y_{-n}(a_n) & \text{if } a_n \in H^{(n)} \\ \infty & \text{otherwise} \end{cases} \tag{11.22}$$

If the real-valued function $f^{(n)}_{(x_0, \ldots, x_{n-1})}$ is Lebesgue integrable on $\mathbb{R}$, denote the integral on $\mathbb{R}$ by $a^{(n)}(x_0, \ldots, x_{n-1})$. If $\sum_{x_0, \ldots, x_{n-1}} a^{(n)}(x_0, \ldots, x_{n-1})$ is meaningful, that is — if there are only countably many nonzero summands and the summation is absolutely convergent, then the sum is denoted by $a^{(n)}$.

Definition 11.4.1 [Wang (1985)]. If $\sum_{n=1}^{\infty} a^{(n)}$ is convergent and equals a, then the function $\mathbf{f}(\mathbf{x})$ is GNL-integrable, written

$$(\text{GNL}) \int_{\mathbf{GNS}} \mathbf{f}(\mathbf{x})\,d\mathbf{x} = a$$

Case 2: Partition the set $E = \{\mathbf{x} \in \mathbf{GNS} : \mathbf{f}(\mathbf{x}) \neq \mathbf{0}\}$ as follows: $E = \bigcup\{E_k : k \in \mathbb{N} \cup \{0\}\}$, where $E_k = \{\mathbf{x} \in E : k(\mathbf{x}) = -k\}$. For each fixed $k \in \mathbb{N}$, consider the right-shifting of $E_k : E_k^* = \{\mathbf{x} = \mathbf{x}^* \times \mathbf{1}_{(k)} : \mathbf{x}^* \in E_k\}$ and the left-shifting of $\mathbf{f}(\mathbf{x})$:

$$\mathbf{f}^{(k)}(\mathbf{x}) = \begin{cases} \mathbf{f}(\mathbf{x} \times \mathbf{1}_{(-k)}) \times \mathbf{1}_{(-k)} & \text{if } \mathbf{x} \in E_k^* \\ 0 & \text{otherwise} \end{cases}$$

First, define a new function $\mathbf{f}^{(0)} : \mathbf{GNS} \to \mathbf{GNS}$ by

$$\mathbf{f}^{(0)}(\mathbf{x}) = \begin{cases} \mathbf{f}(\mathbf{x}) & \text{for } \mathbf{x} \in E_0 \\ \mathbf{0} & \text{otherwise} \end{cases}$$

If $\mathbf{f}^{(0)}$ is GNL-integrable, write (**GNS**) $\int_{\mathbf{GNS}} \mathbf{f}^{(0)}(\mathbf{x})\,d\mathbf{x} = \mathbf{a}_0$. Second, define

$$E_1^* = E_1 \times \mathbf{1}_{(1)} = \{\mathbf{x} \in \mathbf{GNS} : \mathbf{x} = \mathbf{z} \times \mathbf{1}_{(1)}, \mathbf{z} \in E_1\}$$

That is, for any $\mathbf{x} \in E_1^*$, $k(\mathbf{x}) = 0$. Define a new function $\mathbf{f}^{(1)} : \mathbf{GNS} \to \mathbf{GNS}$ by

$$\mathbf{f}^{(1)}(\mathbf{x}) = \begin{cases} \mathbf{f}(\mathbf{x} \times \mathbf{1}_{(-1)}) \times \mathbf{1}_{(-1)} & \text{for } \mathbf{x} \in E_1^* \\ \mathbf{0} & \text{otherwise} \end{cases}$$

If $\mathbf{f}^{(1)}$ is GNL-integrable, write (GNL) $\int_{\mathbf{GNS}} \mathbf{f}^{(1)}(\mathbf{x})\,d\mathbf{x} = \mathbf{a}_1$. In general, write (GNL) $\int_{\mathbf{GNS}} \mathbf{f}^{(k)}(\mathbf{x})\,d\mathbf{x} = \mathbf{a}_k$ if $\mathbf{f}^{(k)}$ is GNL-integrable.

Definition 11.4.2 [Wang (1985)]. If $\sum_k \mathbf{a}_k$ is convergent, then the function $\mathbf{f}(\mathbf{x})$ is GNL-integrable and written

$$(\text{GNL}) \int_{\mathbf{GNS}} \mathbf{f}(\mathbf{x})\,d\mathbf{x} = \sum_k \mathbf{a}_k$$

A concept called (G) integrals was introduced by Wang (1985). Let $\mathbf{f} : \mathbf{GNS} \to \mathbf{GNS}$ be a function and $E = \{\mathbf{x} \in \mathbf{GNS} : \mathbf{f}(\mathbf{x}) \neq \mathbf{0}\}$. Let

$$\mathbf{y} = \mathbf{f}(\mathbf{x}) = (\ldots, y_{-m}, \ldots, y_0, \ldots, y_n, \ldots)$$

As before, the concept of (G) integrals will be introduced in two parts. The first part is a special case, while the second part generalizes the first.

Case 1: Suppose that for any $\mathbf{x} \in E$ the generalized number $\mathbf{x}$ takes the form $\mathbf{x} = (\ldots, 0, x_0, x_1, \ldots, x_n, \ldots)$. That is, $x_{-m} = 0$, for each $m \in \mathbb{N}$.

Step 0: Define a subset $H^{(0)} \subseteq \mathbb{R}$ such that $a_0 \in H^{(0)}$ iff $y_{-m} = y_{-m}(a_0)$ is a function of a_0 only for all $m \geq 0$. Let $\mathbf{f}^{(0)} : \mathbb{R} \to \mathbb{R}$ be a function defined by

$$\mathbf{f}^{(0)}(a_0) = \begin{cases} \sum_{s \geq 0} y_{-(s)}(a_0) \mathbf{1}_{(-s)} & \text{for } a_0 \in H^{(0)} \\ \infty & \text{otherwise} \end{cases}$$

If each $y_{-(s)}(a_0)$ is Lebesgue integrable, $s = 0, 1, 2, \ldots$, write

$$a_{-s}^{(0)} = \int_{\mathbb{R}} y_{-(s)}(x_0)\,dx_0 \times \mathbf{1}_{(-s)}$$

Step n: Let $x_0, x_1, \ldots, x_{n-1}$ be fixed. Define a subset $H^{(n)} \subseteq \mathbb{R}$ such that $a_n \in H^{(n)}$, iff $y_{-m} = y_{-m}(a_0)$ is a function of a_0 only for all $m \geq n$. Let $\mathbf{f}^{(n)}(x_0, x_1, \ldots, x_{n-1}) : \mathbb{R} \to \mathbb{R}$ be a function defined as follows: if $\mathbf{x} = (\ldots, 0, \ldots, 0, x_0, x_1, \ldots, x_{n-1}, a_n, x_{n+1}, \ldots)$,

$$\mathbf{f}^{(n)}(x_0, x_1, \ldots, x_{n-1})(a_n) = \begin{cases} \sum_{s \geq 0} y_{-(n+s)}(a_n) \mathbf{1}_{(-s)} & \text{for } a_n \in H^{(n)} \\ \infty & \text{otherwise} \end{cases}$$

If each $y_{-(n+s)}(a_n)$ is Lebesgue integrable, $s = 0, 1, 2, \ldots$, write

$$a_{-s}^{(n)}(x_0, x_1, \ldots, x_{n-1}) = \int_{\mathbb{R}} y_{-(n+s)}(x_n)\, dx_n$$

and if $\sum_{x_0, x_1, \ldots, x_{n-1}} a_{-s}^{(n)}(x_0, x_1, \ldots, x_{n-1}) \times \mathbf{1}_{(-s)}$ exists, denote it by

$$a_{-s}^{(n)} = \sum_{x_0, x_1, \ldots, x_{n-1}} a_{-s}^{(n)}(x_0, x_1, \ldots, x_{n-1}) \times \mathbf{1}_{(-s)}$$

for $s = 0, 1, 2, \ldots$.

Definition 11.4.3 [Wang (1985)]. For each $s \geq 0$, if the sequence $\{a_{-s}^{(n)}\}_{n=0}^{\infty}$ exists and $\sum_{n=0}^{\infty} a_{-s}^{(n)}$ is convergent, define

$$a_{-s} = \sum_{n=0}^{\infty} a_{-s}^{(n)}$$

If there are only a finite number of natural numbers s such that $a_{-s} \neq 0$, then the function $f(x)$ is (G)-integrable, denoted

$$(\mathrm{G}) \int_{\mathbf{GNS}} \mathbf{f}(\mathbf{x})\, d\mathbf{x} = \sum_{s \geq 0} a_{-s} \times \mathbf{1}_{(s)}$$

Case 2: Suppose that for some $\mathbf{x} \in E$ there exists $m \in \mathbb{N}$ such that $x_{-m} \neq 0$. Partition the set $E = \{\mathbf{x} \in \mathbf{GNS} : \mathbf{f}(\mathbf{x}) \neq \mathbf{0}\}$ as

$$E = \bigcup \{E_k : k \in \mathbb{N} \cup \{0\}\}$$

where for each $k \in \mathbb{N} \cup \{0\}$, $E_k = \{\mathbf{x} \in E : k(\mathbf{x}) = -k\}$. First, define

$$\mathbf{f}^{(0)}(\mathbf{x}) = \begin{cases} \mathbf{f}(\mathbf{x}) & \text{for } \mathbf{x} \in E_0 \\ \mathbf{0} & \text{otherwise} \end{cases}$$

If $\mathbf{f}^{(0)}$ is (G)-integrable, let $(\mathrm{G}) \int_{\mathbf{GNS}} \mathbf{f}^{(0)}(\mathbf{x})\, d\mathbf{x} = \mathbf{a}_0$. Second, define the right-shifting E_1^* of E_1 and a function $\mathbf{f}^{(1)} : \mathbf{GNS} \to \mathbf{GNS}$ by

$$E_1^* = E_1 \times \mathbf{1}_{(1)} = \{\mathbf{x} \in \mathbf{GNS} : \mathbf{x} = \mathbf{z} \times \mathbf{1}_{(1)}, \mathbf{z} \in E_1\}$$

and

$$\mathbf{f}^{(1)}(\mathbf{x}) = \begin{cases} \mathbf{f}(\mathbf{x} \times \mathbf{1}_{(-1)}) \times \mathbf{1}_{(-1)} & \text{for } x \in E_1^* \\ \mathbf{0} & \text{otherwise} \end{cases}$$

If $\mathbf{f}^{(1)}$ is (G)-integrable, set $(\mathrm{G}) \int_{\mathbf{GNS}} \mathbf{f}^{(1)}(\mathbf{x})\, d\mathbf{x} = \mathbf{a}_1$.

Inductively, assume that for $n \in \mathbb{N}$, the generalized numbers $\mathbf{a}_0, \mathbf{a}_1, \mathbf{a}_2, \ldots, \mathbf{a}_n$ have been defined. Now a generalized number $\mathbf{a}_{n+1}$ can be defined as follows: First, define the right-shifting E_n^* of E_n and a function $\mathbf{f}^{(n)}$: **GNS** $\to$ **GNS** by

$$E_n^* = E_n \times \mathbf{1}_{(n)} = \{\mathbf{x} \in \mathbf{GNS} : \mathbf{x} = \mathbf{z} \times \mathbf{1}_{(n)}, \mathbf{z} \in E_n\}$$

and

$$\mathbf{f}^{(n)}(\mathbf{x}) = \begin{cases} \mathbf{f}(\mathbf{x} \times \mathbf{1}_{(-n)}) \times \mathbf{1}_{(-n)} & \text{for } x \in E_n^* \\ \mathbf{0} & \text{otherwise} \end{cases}$$

If $\mathbf{f}^{(n)}$ is (G)-integrable, let $(\mathrm{G}) \int_{\mathbf{GNS}} \mathbf{f}^{(n)}(\mathbf{x})\, d\mathbf{x} = \mathbf{a}_n$.

By mathematical induction it follows that if all the generalized numbers $\mathbf{a}_0, \mathbf{a}_1, \mathbf{a}_2, \ldots, \mathbf{a}_n, \ldots$ have been defined as before, then we can make the following definition.

Definition 11.4.4 [Wang (1985)]. If $\sum_{n=0}^{\infty} \mathbf{a}_n$ is convergent to a generalized number $\mathbf{a}$, then the function $\mathbf{f}$ is (G)-integrable,

$$(\mathrm{G}) \int_{\mathbf{GNS}} \mathbf{f}(\mathbf{x})\, d\mathbf{x} = \mathbf{a}$$

The following result is clear.

Theorem 11.4.4. *If* $\mathbf{f}$: **GNS** $\to$ **GNS** *is (*GNL*)-integrable, it is then* (G)*-integrable and*

$$(\mathrm{G}) \int_{\mathbf{GNS}} \mathbf{f}(\mathbf{x})\, d\mathbf{x} = (\mathrm{GNL}) \int_{\mathbf{GNS}} \mathbf{f}(\mathbf{x})\, d\mathbf{x} \tag{11.23}$$

Now let $\mathbf{F}$ be a family of (GNL)-integrable (resp., (G)-integrable) functions. Suppose that a convergence relation among functions defined on **GNS** has been chosen such that there exist a sequence $\{\mathbf{f}_n\}_{n=1}^{\infty} \subseteq \mathbf{F}$ and a function $\mathbf{f}$: **GNS** $\to$ **GNS** such that

$$\mathbf{f}_n \to \mathbf{f} \tag{11.24}$$

The $(\mathbf{GNS})_{\mathbf{F}}$-integral (resp., $(\mathrm{G})_{\mathbf{F}}$-integral) of $\mathbf{f}$ is defined as follows:

$$(\mathrm{GNL})_{\mathbf{F}} \int_{\mathbf{GNS}} \mathbf{f}(\mathbf{x})\, d\mathbf{x} = (\mathrm{GNL}) \int_{\mathbf{GNS}} \mathbf{f}(\mathbf{x})\, d\mathbf{x} \quad \text{if } \mathbf{f} \in \mathbf{F} \tag{11.25}$$

$$(\mathrm{G})_{\mathbf{F}} \int_{\mathbf{GNS}} \mathbf{f}(\mathbf{x})\, d\mathbf{x} = (\mathrm{G}) \int_{\mathbf{GNS}} \mathbf{f}(\mathbf{x})\, d\mathbf{x} \quad \text{if } \mathbf{f} \in \mathbf{F} \tag{11.26}$$

$$(\mathrm{GNL})_{\mathbf{F}} \int_{\mathbf{GNS}} \mathbf{f}(\mathbf{x})\, d\mathbf{x} = \lim_{n\to\infty} (\mathrm{GNL})_{\mathbf{F}} \int_{\mathbf{GNS}} \mathbf{f}_n(\mathbf{x})\, d\mathbf{x} \quad \text{if } \mathbf{f} \notin \mathbf{F} \tag{11.27}$$

and

$$(\mathrm{G})_{\mathbf{F}}\int_{\mathbf{GNS}}\mathbf{f}(\mathbf{x})\,d\mathbf{x}=\lim_{n\to\infty}(\mathrm{G})_{\mathbf{F}}\int_{\mathbf{GNS}}\mathbf{f}_n(\mathbf{x})\,d\mathbf{x}\quad \text{if } \mathbf{f}\notin\mathbf{F} \tag{11.28}$$

where it is assumed that the limits exist.

The concept of indefinite integral of a function $\mathbf{f}:\mathbf{GNS}\to\mathbf{GNS}$ was studied by Wang (1985) as follows. Assume that (GNL) $\int_{\mathbf{GNS}}\mathbf{f}(\mathbf{x})\,d\mathbf{x}$ exists and that the function value of $\mathbf{f}(\mathbf{x})$ depends only on the real variable $x_{k(\mathbf{x})}$. Then the indefinite integral of $\mathbf{y}=\mathbf{f}(\mathbf{x})$ is defined as

$$\int_{\mathbf{a}}^{\mathbf{x}}\mathbf{f}(\mathbf{z})\,d\mathbf{z}=$$

$$\begin{cases} (\mathrm{GNL})\int_{\mathbf{GNS}}\widetilde{\mathbf{f}}(\mathbf{z})\,d\mathbf{z}+ & \sum_{i=1}^{\infty}[\mathbf{f}(x_{k(\mathbf{x})})\times x_{k(\mathbf{x})+i}]\times\mathbf{1}_{(k(\mathbf{x})+1)} \\ & \text{for } \mathbf{x}=(\dots,0,x_{k(\mathbf{x})},x_{k(\mathbf{x})+1},\dots) \\ (\mathrm{GNL})\int_{\mathbf{GNS}}\mathbf{f}(\mathbf{z})\,d\mathbf{z}, & \text{for } \mathbf{x} \text{ with } x_{k(\mathbf{x})-1}>0 \\ 0 & \text{otherwise} \end{cases}$$

where $\mathbf{a}=(\dots,a_{-m},\dots,a_0,\dots,a_n,\dots)$ is fixed, and $\widetilde{\mathbf{f}}:\mathbf{GNS}\to\mathbf{GNS}$ is defined by

$$\widetilde{\mathbf{f}}(\mathbf{z})=\begin{cases}\mathbf{f}(\mathbf{z}) & \text{for } \mathbf{a}\le\mathbf{z}\le\mathbf{x} \\ \mathbf{0} & \text{otherwise}\end{cases}$$

11.5. Schwartz Distribution and GNS

11.5.1. A Nontechnical Introduction to Schwartz Distribution

The theory of distributions was developed to correspond to situations presented to the world of learning by physical experiences which are not adequately covered by the classical $y=f(x)$ notion of a function in mathematical analysis. Here, the word "physical" really means "phenomenological" — i.e., pertaining to the phenomena of nature. The concept of distributions, as introduced and codified by Laurent Schwartz (1950; 1951), has been widely used to tie many important and different instances, occurring in different parts of the world of learning, together.

A "distribution" is a kind of "generalized function." The class of all distributions or generalized functions includes many objects which are not functions at all in the sense of classical mathematical analysis. The reason why it is necessary to study the theory of distributions is not mere generality; rather, the theory has a coherence and a power that the theory of mathematical analysis lacks. There are many aspects of this conceptual power. For example, the operation of taking a derivative applies without restriction to distributions. That is, the derivative of a

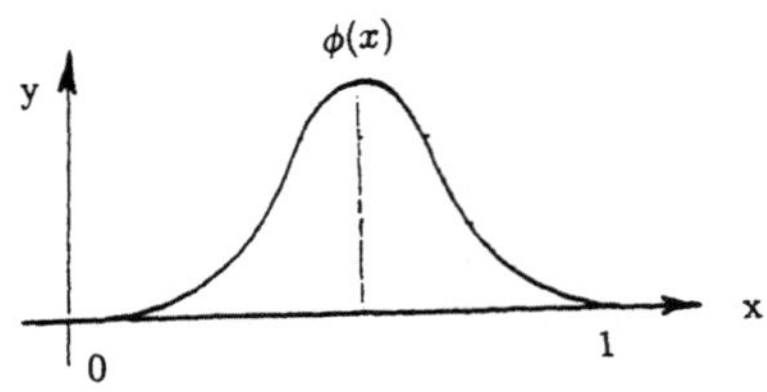

Figure 11.1. A pulse function supported on $[0,1]$.

distribution always exists and is another distribution. Since continuous functions are distributions, the operation of taking a derivative in the theory of distributions eliminates the defect of mathematical analysis, where many continuous functions have no derivatives.

A function $\phi : \mathbb{R} \to \mathbb{R}$ is a test function if

(a) ϕ is infinitely differentiable.

(b) ϕ has compact support; i.e., $\phi(x)$ vanishes outside of some interval $[a,b]$.

The motivations underlying the term "test function" are that test functions serve as tools in studying other functions, and that curves with certain crude geometrical shapes, like pulses or mesas (Figs. 11.1 and 11.2), can be constructed in an infinitely differentiable manner.

Let us now see how the idea of test functions can be used to evaluate an unknown function of interest. Assume that a chemist needs to test the properties of a certain substance at a fixed temperature t_0. So, he gathers a batch of the stuff with temperatures distributed in a pulse $\phi(T)$ near $T = t_0$. His objective is to find some law. Mathematically speaking, the law will normally be written as a function, say $f(T)$. Let us see how the test function ϕ is used to determine the unknown function f.

There is no doubt that in reality some temperatures appear more often than others. Hence, the temperatures have a density $\phi(T)$ supported on $[a,b]$, and a

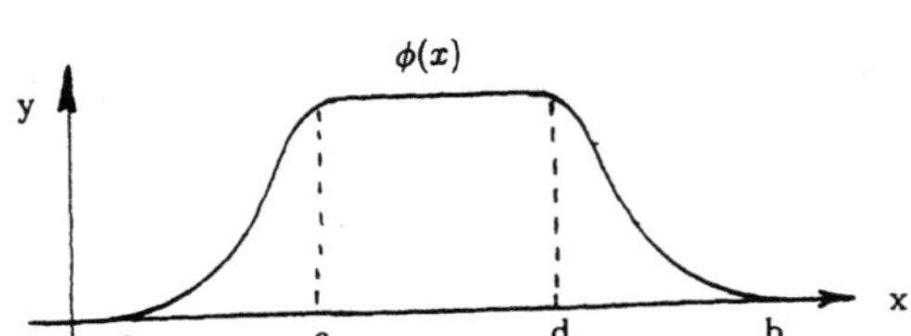

Figure 11.2. A mesa function supported on $[a,b]$.

"weighted average" is natural to consider:

$$\int_a^b f(T)\phi(T)\,dT = \int_{-\infty}^{\infty} f(T)\phi(T)\,dT$$

Now ϕ is used to measure f. If ϕ is chosen differently with different temperature ranges, eventually the structure of f will be known. In the language of the chemist, he might say that by enough experiments ϕ we learn the "law" described by f.

Let $f : \mathbb{R} \to \mathbb{R}$ be such that $\int_a^b |f(x)|\,dx$ exists and is finite for any finite interval $[a,b]$. Then the action of a test function ϕ on f is defined to be $\int_{-\infty}^{\infty} f(x)\phi(x)\,dx$. The following result supports this example.

Theorem 11.5.1 [Richards and Youn (1990)]. *Let $f : \mathbb{R} \to \mathbb{R}$ and $g : \mathbb{R} \to \mathbb{R}$ be continuous and satisfy that every test function ϕ has the same action on f and g. Then $f = g$.*

A sequence $\{\phi_n\}_{n=1}^{\infty}$ of test functions converges to zero, written $\phi_n \to 0$ if

(a) For each $k \in \mathbb{N}$, the sequence of the kth derivatives $\{\phi_n^{(k)}\}_{n=1}^{\infty}$ converges uniformly to zero.

(b) The ϕ_n have uniformly bounded supports; i.e., there is an interval $[a,b]$, independent of n, such that every $\phi_n(x)$ vanishes outside of $[a,b]$. Similarly, $\phi_n \to \phi$ means $(\phi - \phi_n) \to 0$. The set $\mathcal{D}$ of all test functions together with the convergence relation, defined earlier, is called the elementary space.

Definition 11.5.1 [Richards and Youn (1990)]. A distribution T is a mapping from the set $\mathcal{D}$ of all test functions into the real or complex numbers, such that the following conditions hold, where $T(\phi)$ is written as $\langle T, \phi\rangle$:

(a) (Linearity) $\langle T, a\phi(x) + b\psi(x)\rangle = a \cdot \langle T, \phi(x)\rangle + b \cdot \langle T, \psi(x)\rangle$, for all $\phi, \psi \in \mathcal{D}$ and constants a and b.

(b) (Continuity) If $\phi_n(x) \to 0$, then $\langle T, \phi_n(x)\rangle \to 0$.

Example 11.5.1. Let $f : \mathbb{R} \to \mathbb{R}$ be piecewise continuous. For any $\phi \in \mathcal{D}$, define

$$\langle T, \phi\rangle = \int_{-\infty}^{\infty} f(x)\phi(x)\,dx$$

It can be shown readily that T is a distribution.

Example 11.5.2. Dirac's δ function δ is a symbol which satisfies the following conditions:

$$\delta(x) = 0 \quad \text{whenever } x \neq 0$$
$$\delta(0) = \infty$$
$$\int_{-\infty}^{\infty} \phi(x)\delta(x)\,dx = \phi(0) \tag{11.29}$$

where $\phi \in \mathcal{D}$ is any test function. For a different discussion on the δ function, see Chapter 6.

In terms of distributions, the δ function is defined by the equation

$$\langle \delta, \phi \rangle = \phi(0) \quad \text{for any } \phi \in \mathcal{D}$$

Now shift the δ function through a distance $+a$. This should give the function $\delta(x-a)$, and

$$\begin{aligned}\langle \delta(x-a), \phi \rangle &= \int_{-\infty}^{\infty} \delta(x-a)\phi(x)\,dx \\ &= \int_{-\infty}^{\infty} \delta(y)\phi(y+a)\,dy \\ &= \langle \delta, \phi(x+a) \rangle \\ &= \phi(a)\end{aligned}$$

That is, the shifting $\delta(x-a)$ evaluates the test function ϕ at $x = a$. The following may be a little surprising:

$$\delta(2x) = \frac{1}{2}\delta(x)$$

and in general, $\delta(ax) = |a|^{-1}\,\delta(x)$.

Let $\{T_n\}_{n=0}^{\infty}$ be a sequence of distributions and T a distribution. $\{T_n\}_{n=0}^{\infty}$ is convergent to T, written as $T_n \to T$ if

$$\langle T_n, \phi \rangle \to \langle T, \phi \rangle \quad \text{for each } \phi \in \mathcal{D}$$

Based on this concept, the following theorem is interesting.

Theorem 11.5.2 [Richards and Youn (1990)]. *Let $f(x)$ be a piecewise continuous function such that*

$$\int_{-\infty}^{\infty} |f(x)|\,dx < \infty \quad \textit{and} \quad \int_{-\infty}^{\infty} f(x)\,dx = 1$$

Denote $f_a(x) = af(ax)$. Then

$$f_a(x) = af(ax) \to \delta(x) \quad \textit{as } a \to \infty$$

where the convergence is in the sense of distribution.

For a more comprehensive nontechnical introduction to the theory of Schwartz distributions, refer to the splendid work by Richards and Youn (1990).

11.5.2. GNS *Representation of Schwartz Distributions*

Let $f_1(x), f_2(x), \ldots, f_k(x)$ be functions from $\mathbb{R}$ into $\mathbb{R}$. These functions are linearly independent if for any constants $a_1, a_2, \ldots, a_k$, the equation

$$a_1 f_1(x) + a_2 f_2(x) + \cdots + a_k f_k(x) = 0$$

implies that $a_1 = a_2 = \cdots = a_k = 0$. Otherwise, the functions are linearly dependent. For a set F of functions from $\mathbb{R}$ into $\mathbb{R}$, F is linearly independent if each finite subset of F contains linearly independent functions.

Lemma 11.5.1. *For linearly independent continuous functions $f_i : \mathbb{R} \to \mathbb{R}$, $i = 1,2,\ldots$, there are sequences $\{x_k^{(n)}\}_{n=k}^{\infty}$ of real numbers, $k = 1,2,3,\ldots$, such that for each $k \in \mathbb{N}$,*

$$\det \begin{bmatrix} f_1(x_1^{(n)}) & f_2(x_1^{(n)}) & \cdots & f_n(x_1^{(n)}) \\ f_1(x_2^{(n)}) & f_2(x_2^{(n)}) & \cdots & f_n(x_2^{(n)}) \\ \vdots & \vdots & & \vdots \\ f_1(x_n^{(n)}) & f_2(x_n^{(n)}) & \cdots & f_n(x_n^{(n)}) \end{bmatrix} \neq 0 \qquad (11.30)$$

and, if $(n,i) \neq (m,j)$, then $x_i^{(n)} \neq x_j^{(n)}$.

Proof: Since the set $F = \{f_i : \mathbb{R} \to \mathbb{R} : i \in \mathbb{N}\}$ is assumed to be linearly independent for any constants $a_1, a_2, \ldots, a_n$, the equation $a_1 f_1(x) + a_2 f_2(x) + \cdots + a_n f_n(x) = 0$ implies that $a_1 = a_2 = \cdots = a_n = 0$. Therefore, none of the functions in F is the zero function. Thus, there exists a real number $x_1^{(1)}$ such that

$$\det[f_1(x_1^{(1)})] = f_1(x_1^{(1)}) \neq 0$$

Assume that for a natural number n, the real numbers

$$x_1^{(1)}, x_1^{(2)}, x_2^{(2)}, x_1^{(3)}, x_2^{(3)}, x_3^{(3)}, \ldots, x_1^{(n)}, x_2^{(n)}, \ldots, x_n^{(n)}$$

have been defined such that Eq. (11.30) holds and that $(n,i) \neq (m,j)$ implies $x_i^{(n)} \neq x_j^{(n)}$. Now we define the real numbers $x_1^{(n+1)}, x_2^{(n+1)}, \ldots, x_{n+1}^{(n+1)}$ such that the desired conditions are satisfied.

Since f_{n+1} is not the zero function, pick a real number $x_{n+1}^{(n+1)}$ such that $x_{n+1}^{(n+1)}$ is different from all previously chosen $x_k^{(n)}$ and that

$$\det[f_{n+1}(x_{n+1}^{(n+1)})] = f_{n+1}(x_{n+1}^{(n+1)}) \neq 0$$

The existence of $x_{n+1}^{(n+1)}$ comes from the fact that f_{n+1} is continuous and that all previously chosen $x_k^{(n)}$ are a finite number. Now pick a real number $x_n^{(n+1)}$ such that

$$0 \neq \det \begin{bmatrix} f_n(x_n^{(n+1)}) & f_{n+1}(x_n^{(n+1)}) \\ f_n(x_{n+1}^{(n+1)}) & f_{n+1}(x_{n+1}^{(n+1)}) \end{bmatrix}$$
$$= f_n(x_n^{(n+1)}) f_{n+1}(x_{n+1}^{(n+1)}) - f_n(x_{n+1}^{(n+1)}) f_{n+1}(x_n^{(n+1)})$$

The existence of such a number $x_n^{(n+1)}$ comes from the fact that f_n and f_{n+1} are linearly independent, which implies that

$$f_{n+1}(x_{n+1}^{(n+1)}) f_n(x) - f_n(x_{n+1}^{(n+1)}) f_{n+1}(x) \neq 0$$

The continuity of f_n and f_{n+1} guarantees that the $x_n^{(n+1)}$ can be different from all previous $x_k^{(m)}$'s.

By mathematical induction it can be shown that the real numbers $x_{n+1}^{(n+1)}, x_n^{(n+1)}, x_{n-1}^{(n+1)}, \ldots, x_1^{(n+1)}$ can be chosen one after another so that each one is different from all previous $x_k^{(m)}$'s and that for each i satisfying $n+1 \geq i \geq 1$,

$$\det \begin{bmatrix} f_i(x_i^{(n+1)}) & f_{i+1}(x_i^{(n+1)}) & \cdots & f_{n+1}(x_i^{(n+1)}) \\ f_i(x_{i+1}^{(n+1)}) & f_{i+1}(x_{i+1}^{(n+1)}) & \cdots & f_{n+1}(x_{i+1}^{(n+1)}) \\ \vdots & \vdots & & \vdots \\ f_i(x_{n+1}^{(n+1)}) & f_{i+1}(x_{n+1}^{(n+1)}) & \cdots & f_{n+1}(x_{n+1}^{(n+1)}) \end{bmatrix} \neq 0$$

Therefore, induction guarantees that there are sequences $\{x_k^{(n)}\}_{n=k}^{\infty}$ of real numbers, $k = 1, 2, \ldots$, such that for each $n \in \mathbb{N}$, Eq. (11.30) holds true and if $(n,i) \neq (m,j)$, then $x_i^{(n)} \neq x_j^{(m)}$. ■

Lemma 11.5.2. *The elementary space $\mathcal{D}$ of all test functions is separable; i.e., there exists a countable subset $F \subseteq \mathcal{D}$ such that for any $\phi \in \mathcal{D}$, there exists a sequence $\{\phi_n\}_{n=1}^{\infty} \subseteq F$ such that $\phi_n \to \phi$.*

Proof: Let $h : \mathbb{R} \to \mathbb{R}$ be defined by

$$h(x) = \begin{cases} e^{-1/x} & \text{if } x > 0 \\ 0 & \text{otherwise} \end{cases} \tag{11.31}$$

Then $\phi(x) = h(x)h(1-x)$ is a test function with support $[0,1]$. For any $a, b \in \mathbb{R}$, define

$$\phi_{a,b}(x) = \phi\left(\frac{x-a}{b-a}\right) = h\left(\frac{x-a}{b-a}\right) h\left(\frac{b-x}{b-a}\right)$$

Then $\phi_{a,b}$ is a test function with support $[a,b]$.

Let $p_n(x)$ be a polynomial of degree n with all rational number coefficients. Then $p_n(x)\phi_{a,b}(x)$ is a test function. Let $F = \{p_n(x)\phi_{a,b}(x) : p_n(x)$ is a polynomial of degree n with all rational number coefficients, $n \in \mathbb{N}$, and a and b are rational$\}$. Then $F \subseteq \mathcal{D}$ is countable. It is left to the reader to show that for any $\phi \in \mathcal{D}$ there exists a sequence $\{\phi_n\}_{n=1}^{\infty} \subseteq F$ such that $\phi_n \to \phi$. ■

For each infinitely differentiable function $f : \mathbb{R} \to \mathbb{R}$, define a function $\mathbf{f}$: **GNS** → **GNS** as follows:

$$\begin{aligned}\mathbf{f}(\mathbf{x}) &= f(x_0) + f'(x_0)\left[\sum_{m=1}^{\infty} x_m \mathbf{1}_{(m)}\right] + \cdots + \frac{1}{n!}f^{(n)}(x_0)[\sum_{m=1}^{\infty} x_m \mathbf{1}_{(m)}]^n + \cdots \\ &= \sum_{n=0}^{\infty} \frac{1}{n!} f^{(n)}(x_0)\left[\sum_{m=1}^{\infty} x_m \mathbf{1}_{(m)}\right]^n \end{aligned} \tag{11.32}$$

Here $\mathbf{f}$ is called the **GNS** function induced by f.

Theorem 11.5.3 [Wang (1985)]. *Let T be a distribution and F as in Lemma 11.5.2. Then there exists* $\mathbf{K}$: **GNS** → **GNS** *such that, for each $f \in F$,*

$$(\text{GNL}) \int_{\mathbf{GNS}} \mathbf{f}(\mathbf{x})\mathbf{K}(\mathbf{x})\, d\mathbf{x} = \langle T, f\rangle \tag{11.33}$$

where $\mathbf{f}$: **GNS** → **GNS** *is induced by f.*

Proof: Let

$$H(x) = \begin{cases} e^{-1/x} & \text{if } x \neq 0 \\ 0 & \text{otherwise} \end{cases}$$

For any rational numbers a and b, define

$$\begin{aligned}\Phi_{a,b}(x) &= H\left(\frac{x-a}{b-a}\right) H\left(1 - \frac{x-a}{b-a}\right) \\ &= H\left(\frac{x-a}{b-a}\right) H\left(\frac{b-x}{b-a}\right) \\ &= \begin{cases} \exp(-\frac{b-a}{x-a} - \frac{b-a}{b-x}) & \text{if } x \neq a,\ x \neq b \\ 0 & \text{otherwise} \end{cases}\end{aligned}$$

Let $G = \{p_n(x)\Phi_{a,b}(x) : p_n(x)$ is a polynomial of degree n with all rational coefficients, for $n \in \mathbb{N}$, and a and b are rational numbers$\}$. Then the families F and G can be put into 1–1 correspondence by matching up $p_n(x)\phi_{a,b}(x) \longleftrightarrow p_n(x)\Phi_{a,b}(x)$.

Now we show that the family G is linearly independent. Pick an arbitrary finite subset $H \subseteq G$, say

$$H = \{f_1, f_2, \ldots, f_k\}$$

Let $d_1, d_2, \ldots, d_k$ be constants satisfying

$$d_1 f_1(x) + d_2 f_2(x) + \cdots + d_k f_k(x) = 0 \tag{11.34}$$

It suffices to show that $d_1 = d_2 = \cdots = d_k = 0$.

Case 1: For each i with $1 \le i \le k$, assume

$$\begin{aligned} f_i(x) &= \Phi_{a_i,b_i}(x) \\ &= \begin{cases} \exp\left[-\frac{(a_i-b_i)^2}{(x-a_i)(b_i-x)}\right] & \text{if } x \neq a_i,\ x \neq b_i \\ 0 & \text{otherwise} \end{cases} \end{aligned}$$

and that $(a_i, b_i) \neq (a_j, b_j)$, whenever $i \neq j$. Thus, Eq. (11.34) becomes

$$\begin{aligned} d_1 \exp\left[-\frac{(a_1-b_1)^2}{(x-a_1)(b_1-x)}\right] &+ d_2 \exp\left[-\frac{(a_2-b_2)^2}{(x-a_2)(b_2-x)}\right] + \cdots \\ &+ d_k \exp\left[-\frac{(a_k-b_k)^2}{(x-a_k)(b_k-x)}\right] = 0 \end{aligned} \tag{11.35}$$

Now choose k different values of x for which it is easy to evaluate the e-functions. We obtain S, k linear equations in k variables $d_1, d_2, \ldots, d_k$, from which it can be shown that the system has zero solution only. That is, $d_1 = d_2 = \cdots = d_k = 0$. That completes the proof that $f_1, f_2, \ldots, f_k$ are linearly independent.

Case 2: For each i with $1 \le i \le k$, assume

$$f_i(x) = p_{n_i}(x)\Phi_{a_i,b_i}(x)$$

where $\Phi_{a_i,b_i}(x)$ is defined by Eq. (11.35) satisfying $(a_i, b_i) \neq (a_j, b_j)$, whenever $i \neq j$, and the polynomial

$$p_{n_i}(x) = c_{in_i}x^{n_i} + c_{in_i-1}x^{n_i-1} + \cdots + c_{i1}x + c_{i0} \tag{11.36}$$

where the coefficients c_{ij}'s are rational. Equation (11.34) then becomes

$$\begin{aligned} d_1 p_{n_1}(x) \exp\left[-\frac{(a_1-b_1)^2}{(x-a_1)(b_1-x)}\right] &+ d_2 p_{n_2}(x) \exp\left[-\frac{(a_2-b_2)^2}{(x-a_2)(b_2-x)}\right] \\ + \cdots &+ d_k p_{n_k}(x) \exp\left[-\frac{(a_k-b_k)^2}{(x-a_k)(b_k-x)}\right] = 0 \end{aligned} \tag{11.37}$$

Now group the terms of Eq. (11.37) according to the powers of x appearing in the polynomials $p_{n_i}(x)$, $i = 1, 2, \ldots, k$, so that each power of x appears at most once.

Equation (11.37) is true iff the coefficient of each power of x equals zero. Hence, Eq. (11.37) is equivalent to $n = \max\{n_1, n_2, \ldots, n_k\} + 1$ many equations, each of which looks similar to Eq. (11.35). The only difference which might occur is that some constant coefficients may show up in front of some d_i's in Eq. (11.35). Therefore, the argument in Case 1 implies that $d_1 = d_2 = \cdots = d_k = 0$. That is, $f_1, f_2, \cdots, f_k$ are linearly independent. This completes the proof that the family G is linearly independent.

Since G is countable, enumerate the elements in G as

$$f_1, f_2, \cdots, f_n, \cdots, \quad n \in \mathbb{N} \tag{11.38}$$

From Lemma 11.5.1 it follows that there are sequences $\{x_k^{(n)}\}_{n=k}^{\infty}$ of real numbers, for $k = 1, 2, 3, \ldots$, satisfying Eq. (11.30).

(1) Since $f_1(x_1^{(1)}) \neq 0$, there must be a real number $c_1^{(0)}$ such that

$$c_1^{(0)} f_1(x_1^{(1)}) = \langle T, f_1 \rangle \tag{11.39}$$

(2) When $n = 2$, Eq. (11.30) is true. Hence the system

$$\begin{aligned} c_1^{(1)} f_1(x_1^{(2)}) + c_2^{(1)} f_2(x_1^{(2)}) &= 0 \\ c_1^{(1)} f_1(x_2^{(2)}) + c_2^{(1)} f_2(x_2^{(2)}) &= \langle T, f_2 \rangle - c_1^{(0)} f_2(x_1^{(1)}) \end{aligned} \tag{11.40}$$

has a unique solution which is denoted $c_1^{(1)}$ and $c_2^{(1)}$.

(3) In general, assume that for a fixed $m_0 \in \mathbb{N}$, the $c_k^{(m)}$'s have been defined for $k \leq m+1 \leq m_0 + 1$. Since Eq. (11.30) holds for $n = m_0 + 1$, the following system with unknowns $c_k^{(m_0+1)}$, $k = 1, 2, \ldots, m_0 + 1$ has A unique solution:

$$\begin{bmatrix} f_1(x_1^{(n)}) & f_2(x_1^{(n)}) & \cdots & f_n(x_1^{(n)}) \\ f_1(x_2^{(n)}) & f_2(x_2^{(n)}) & \cdots & f_n(x_2^{(n)}) \\ \vdots & \vdots & & \vdots \\ f_1(x_n^{(n)}) & f_2(x_n^{(n)}) & \cdots & f_n(x_n^{(n)}) \end{bmatrix} \begin{bmatrix} c_1^{(n)} \\ c_2^{(n)} \\ \vdots \\ c_n^{(n)} \end{bmatrix} = \begin{bmatrix} 0 \\ 0 \\ \vdots \\ 0 \\ K_n \end{bmatrix} \tag{11.41}$$

where $K_n = \langle T, f_n \rangle - \sum_{n=1}^{m_0} \sum_{k=1}^{m+1} c_k^{(m)} f_n(x_k^{(m+1)})$.

Choose a real-valued infinitely differentiable function $g(x)$ satisfying $\{x \in \mathbb{R} : g(x) \neq 0\}$ is bounded and $\int_{-\infty}^{\infty} g(x)\,dx = 1$. Define a function $\mathbf{K} : \mathbf{GNS} \to \mathbf{GNS}$ as follows: for $m \geq 0$,

$$\mathbf{K}(\mathbf{x}) = \begin{cases} c_k^{(m)} g(x_{m+1}) \mathbf{1}_{(-m-1)}, & \text{for } \mathbf{x} \text{ with } k(\mathbf{x}) = 0, \\ & x_i = x_0 = c_k^{(m+1)},\ i = 1, \ldots, m \\ 0 & \text{otherwise} \end{cases}$$

It then follows from the definition of (GNL)-integrals and Eq. (11.41) that $\mathbf{K}$: $\mathbf{GNS} \to \mathbf{GNS}$ satisfies Eq. (11.33). This completes the proof of the theorem. ■

For the countable subset $F \subseteq \mathcal{D}$ given in Lemma 11.5.2 and for the function $\mathbf{K} : \mathbf{GNS} \to \mathbf{GNS}$ introduced in Theorem 11.5.3, let $K \cdot F = \{\mathbf{K} \cdot \mathbf{f} : f \in F\}$, where for each $f \in F$, $\mathbf{f} : \mathbf{GNS} \to \mathbf{GNS}$ is the **GNS** function induced by f; see Eq. (11.32). Then the following result is not difficult to see.

Theorem 11.5.4 [Wang (1985)]. *For each arbitrary test function $f \in \mathcal{D}$ of elementary space,*

$$(\mathrm{GNL})_{\mathbf{K}\cdot F} \int_{\mathbf{GNS}} \mathbf{K}(\mathbf{x})\mathbf{f}(\mathbf{x})\, d\mathbf{x} = \langle T, f \rangle$$

where T is the distribution used to determine $\mathbf{K}$.

11.6. A Bit of History

Research in non-Archimedean fields has been closely related to those of non-Archimedean topology and non-Archimedean analysis. The study of non-Archimedean topological structures has been valuable in the study of dimension theory and in research in algebra and algebraic geometry. To see this, one need only to consult the p-adic topology and Stone duality theorem in Boolean algebra (Nyikos and Reichel, 1975). The concept of linear uniformity, introduced by Frechet (1945), and the theory of ω_μ-additive topological spaces, studied by many, including Hausdorff, Sikorski (1950), Shu-Tang Wang (1964), Stevenson and Thron (1971), Hodel (1975), Juhasz (1965), Husek and Reichel (1983), Yasui (1975), and Nyikos and Reichel (1976), are related to the general study of non-Archimedean topology. Also, non-Archimedean topology has been applied in research of abstract distances, introduced and studied by Frechet (1946), Kurepa (1934; 1936a; 1936b; 1956), Doss (1947), Colmez (1947), etc.

As for the study of non-Archimedean analysis, Italian differential geometer Levi-Civita (1892) introduced and studied the concept of formal power series. In particular, he discussed some algebraic problems of his non-Archimedean field and applications in Voronese's non-Archimedean geometry (Voronese, 1896; Voronese, 1891), at the time criticized and not trusted by many. The purpose of Levi-Civita's work was to establish an extremely rigorous "arithmetic" foundation for Voronese's geometry. Later, Hilbert employed Levi-Civita's work on formal power series in his work on foundations of geometry. Hans Hahn has been commonly recognized as the first person to study ordered algebraic structures. In fact, his work on ordered algebraic structures is a continuation of Levi-Civita's formal power series. The generalized number system, introduced by

Wang (1985), and studied in some detail in this chapter, was first studied with the definition of multiplication operation (by Wang) in 1963. His goal was to study strange functions, which appear in modern physics. Unfortunately, the paper was not published and got lost. Until 1977, Wang did not have a chance to reorganize his work, due to the Cultural Revolution in China. In 1978, Wang learned of Laugwitz's work (1968), through whom Levi-Civita's formal power series was introduced to Wang. In Laugwitz's research, he discussed formal power series from the angle of classical mathematical analysis. The major difference between the study of formal power series and that of generalized number system is that even though algebraically the generalized number system is a subfield of the formal power series, Wang, on **GNS**, introduced the concepts of various integrals and applied the theory of **GNS** to the study of various problems in modern physics. Particularly, Dirac's δ function in quantum mechanics and the Heaviside function in quantum field theory have been given concrete representations in **GNS**. Also, the concept of Taylor expansion, and all elementary functions were extended to the **GNS** field.

As shown in Chapter 6, one can be sure that the theory of generalized numbers will be developed a great deal, along with the successes which have been or will be accomplished in the realm of systems science and its applications.

11.7. Some Leads for Further Research

Comparing the study of the generalized number system with those contained in mathematical analysis, real analysis, and complex analysis, it can be seen that understanding the **GNS** is still at the germinating stage. More work needs to be carried out in order for us to gain a deeper understanding of nature and the world in which we live.

For instance, it is necessary to

(1) Study the ordering relation, defined in Section 11.1. It is obvious that the **GNS** field is not complete in the sense of Dedekind cuts.

(2) Find the relation between the ordering relations defined in Sections 11.1 and 11.3.

(3) Develop workable and concrete formulas for the operations of multiplication, division, and taking roots.

(4) Introduce the concept of continuity based on the ordering relation defined in Section 11.3.

(5) Establish a law, similar to the transfer principle in nonstandard analysis (Davis, 1977), governing the relation between a set of properties of real numbers and a set of properties of generalized numbers.

(6) Develop the results of generalized numbers corresponding to theorems in mathematical analysis, such as intermediate value theorem, maximum and minimum theorem, etc.

(7) Find detailed formulas to compute various derivatives, definite and indefinite integrals of given functions from **GNS** into **GNS**.

(8) Generalize the Fundamental Theorem of Calculus to the case of **GNS**, linking **GNS** differential calculus to **GNS** integral calculus.

(9) Develop an integral calculus of **GNS** based on Lebesgue integrals.

(10) Use **GNS** representations of Schwartz's distributions to explore new explanations of equations, appearing in the theory of distributions, which do not make sense in classical mathematical analysis.

(11) Employ **GNS** representations of Schwartz's distributions to solve open problems in the theory of distributions, such as the problem of defining products of distributions. Can the theory of **GNS** be used to explain why, in general, the products of distributions might not exist?

(12) Redevelop quantum mechanics based on **GNS** so that, hopefully, more puzzles in current quantum mechanics can be explained more rationally.

(13) Explore applications of the theory of **GNS** in other branches of human knowledge. The future success of the theory of **GNS** may well depend on various successful applications.

CHAPTER 12

Some Unsolved Problems in General Systems Theory

In this chapter some open problems in set-theoretic general systems theory are listed. A common feature of these problems is that their studies will lead to some new understanding in several research areas and the introduction of new concepts. The main part of this chapter is also contained in (Lin et al., 1997).

12.1. Introduction

Von Bertalanffy introduced the concept of systems in biology in the late 1920s (von Bertalanffy, 1934), and it has been developed a great deal in the last 70 years or so. Starting in the early 1960s, scholars began to establish a theory of general systems in order to lay down a theoretical foundation for all approaches of systems analysis developed in different disciplines. Led by von Bertalanffy and supported by several of the most powerful minds of our time, a global systems movement has been going on for several decades.

Along the development path of the systems movement, many scholars from different disciplines have tried to first define the concept of general systems and then apply the concept and the theory, developed on the concept of general systems, to solving problems in various research areas. Technical difficulties of introducing an ideal definition for general systems, and notably unsuccessful applications of general systems theory [for details see (Berlinski, 1976; Lilienfeld, 1978)], have taught many painful lessons. Among them are: (1) There might not exist an ideal definition for general systems, upon which a general systems theory could be developed so that this theory would serve as the theoretical foundation for all approaches of systems analysis, developed in various disciplines. (2) Without an ideal definition of general systems in place, "general systems theory" would be meaningless. In this circumstance, under the name of systems analysis, senseless transfer of results from one discipline to another have really made many scholars doubt the value of "general systems theory." (3) Based on historical experience and the "unreasonable effectiveness" of mathematics (Wigner, 1960), several fruitful definitions of general systems have been introduced since the early 1960s. Even though abstract general systems theory has been established through the efforts of many scholars, including Mesarovic and Takahara, the real challenge from scholars — namely to show the need and power of the concept of general systems and its theory it is necessary to obtain results which have not been nor could be, achieved without using the concept and its theory — has not been very well met.

To meet the challenge of showing the need and power of the concept and theory of general systems, we list some the following problems: infinite or finite divisibility of the material world; why the concept of systems was not studied more intensively until now; meaning of a system with contradictory relations; existence of absolute truths unreasonable effectiveness of mathematics; philosophy of mathematical modeling; the need to study an applied set theory; multilevel structure of the world; the Law of Conservation of Matter–Energy; the material foundation of human thought; the origin of the universe; the need to develop a mathematics of multilevels and its application in Einstein's relativity theory.

From this brief list, one can see that the study of general systems theory will surely impact the entire spectrum of human knowledge. How the systems movement is regarded is each scholar's opinion and position. No matter how suspicious some scholars are, history will still proceed at its own pace. To this end, the development histories of Cantor's set theory, Einstein's relativity theory, etc., have taught us that no new theory would be born without suffering from suspicions, prejudices, and calumnies. In fact, meeting challenges and self-renewing are among the most important motivations for a newborn theory to grow, to mature, and to gain a visible position in scientific history.

12.2. The Concept of Systems

The fundamental characteristics of a system are the unification of "isolated" objects, relations between the objects, and the structure of layers.

The idea of systems appeared at least as early as the one-element Ionians (624–500 B.C.). For example, in the search for order, Ionians had a naturalistic and materialistic bent. They sought causes and explanations in terms of the eternal working of things themselves rather than in any divine, mythological, or supernatural intervention. Looking for a single basic reality, each Ionian believed that all things have their origin in a single knowable element: water, air, fire, or some indeterminate, nebulous substance (Perlman, 1970).

In history the idea of systems has been used by many great thinkers to study different problems. For example, Aristotle's statement "the whole is greater than the sum of its parts" could be an early definition of a basic systems problem. Nicholas of Cusa, a profound thinker of the fifteenth century, linking medieval mysticism with the first beginnings of modern science, introduced the notion of the *coincidentia oppositorum*. Leibniz's hierarchy of monads looks quite like that of modern systems. Gustav Fechev, known as the author of the psychophysical law, elaborated, in the way of the naive philosophers of the nineteenth century, supraindividual organizations of higher order than the usual objects of observation, thus romantically anticipating the ecosystems of modern parlance (von Bertalanffy, 1972). We list only a few of the great thinkers.

In the second decade of this century, von Bertalanffy (1934) wrote

> Since the fundamental character of the living thing is its organization, the customary investigation of the single parts and processes cannot provide a complete explanation of the vital phenomena. This investigation gives us no information about the coordination of parts and processes. Thus the chief task of biology must be to discover the laws of biological systems (at all levels of organization). We believe that the attempts to find a foundation for theoretical biology point at a fundamental change in the world picture. This view, considered as a method of investigation, we shall call "organismic biology" and, as an attempt at an explanation, "the system theory of the organism."

Here, a new concept, "system" was introduced. After 70 years of testing, this concept has been widely accepted by scientists in all disciplines (Blauberg et al., 1977). To establish a theoretical foundation for all approaches of systems analysis, developed in different disciplines, Mesarovic, in the early 1960s introduced the mathematical definition of (general) systems, based on Cantor's set theory, as follows (Mesarovic and Takahara, 1975): A (general) system S is a relation on nonempty sets V_i:

$$S \subseteq \prod\{V_i : i \in I\}$$

where I is an index set. This structure of general systems reflects the unification of "isolated" objects, relations between the objects, and the structure of layers. Here, elements in the nonempty sets V_i, $i \in I$, are the objects of the system, S represents a relation between objects, and elements of the sets V_i, $i \in I$, could be systems too. Using this idea of layers inductively, we can see that an element S_1 of S could be a system, an element S_2 of the system S_1 could be a system, ... Can this process go on forever? If the answer is "yes," then "the world would be infinitely divisible"! If the answer to the question is "No," does that mean the world is made up of fundamental elements?

The idea that the world is infinitely divisible can be found in Anaxagoras' "seeds" philosophy (510–428 B.C.). He was unwilling to submerge the tremendous varieties of things into any common denominator, and preferred to accept the immediate diversity of things as is. With his philosophy, every object is infinitely divisible. No matter how far an object is divided, what is left would have characteristics of the original substance. The opposite idea that the world is made up of fundamental particles appeared during 500–55 B.C. The Leucippus–Democritus atom combined features of the Ionians' single element, Anaxagoras' seeds, and some thoughts of other schools, and yet was an improvement over all of them. "Atom" means "not divisible" in Greek. This term was intentionally chosen by Democritus to emphasize a particle so small that it could no longer be divided. To Leucippus and Democritus, the universe originally and basically consisted entirely of atoms and a "void" in which atoms move.

Upon consideration of the interrelationship between the systems under concern and some environments of the systems, Bunge (1979) gave a model of systems as follows: For nonempty set T, the ordered triple $W = (C, E, S)$ is a system over T if and only if C and E are mutually disjoint subsets of T and S is a nonempty set of relations on $C \cup E$; C and E are called the composition and an environment of the system W, respectively. Klir (1985) introduced a philosophical concept of general systems, which reads "a system is what is distinguished as a system." Theoretically, Klir's definition contains the most general meaning of the concept of systems originally posed by von Bertalanffy. Starting in 1976, Xuemou Wu and his followers have studied many different theories, under the name "pansystems." Pansystems analysis is a new research of multilevels across all disciplines. The theory deals with general systems, relations, symmetry, transformation, generalized calculus, and *shengke* (means survival and vanquishing), called the emphases of pansystems. Based on research of these emphases, the theory blends philosophical reasoning, mathematical logic, and mechanical structures into one solid body of knowledge.

The discussion here gives rise naturally to the following question: Why was the concept of systems not studied more before now? One answer is that the development of modern technology, e.g., computer technology, designs of satellites, climate control of giant buildings, etc., reveals one fact: History is in a special moment, where the accumulation of knowledge has reached such a level that each discovery of a relation between different areas of knowledge will materially produce a useful or consumable product.

Closely related to the concept of systems are the following questions: (1) What is the meaning of a system with two contradictory relations, e.g., $\{x > y, x \leq y\}$ as the relation set? (2) How can we know whether there exist contradictory relations in a given realistic system? From Klir's definition of systems, which is one of the definitions with the widest meaning among all definitions of systems, it follows that generally these two questions cannot be answered. Concerning this, a theorem of Gödel shows that it is impossible to show whether the systems representation of mathematics, developed on the ZFC axioms is consistent. For a detailed discussion of the systems representation of mathematics, see (Lin, 1989c). This fact implies that there are systems, for example, the system of set theory on ZFC, so that we do not know if propositions with contradictory meanings exist or not. This means that maybe not every system is consistent or that not every system has no contradictory relations.

By rewriting Russell's famous paradox in the language of systems theory, the following is true: There does not exist a system whose object set consists of all systems, where S is a system, if and only if S is an ordered pair (M,R) of sets such that R is a set of some relations defined on the set M (Lin, 1987). The elements in M are called the objects of S. The sets M and R are called the object set and the relation set of S, respectively. Now, for any given system $S=(M,R)$, each relation $r \in R$ can be understood as an S-truth. That is, r is true among the objects in S. Hence, Russell's paradox implies that there does not exist any statement (or relation) which describes a fact among all systems. If any structure in the world can be studied as a system, as Klir's definition of systems says, does Russell's paradox say that there is no universal truth or absolute truth? It is well known in the mathematics community that no theorem is true in all axiom systems.

Descartes and Galileo developed individually the following methods about scientific research and administration: Divide the problem under consideration into as many small parts as possible, study each isolated part (Kline, 1972), and simplify the complicated phenomenon into basic parts and processes (Kuhn, 1962). In the history of science and technology, Descartes' and Galileo's methods have been very successfully applied. They guaranteed that physics would win great victories one by one (von Bertalanffy, 1972), and not only those, because their methods are still widely used for research of fundamental theories and in modern laboratories. At the same time, due to the tendency of modern science toward synthesis and the transverse development of modern technology, more and more scientists and administrators are forced to study problems with many cause–effect chains (i.e., systems). Some examples are the many-body problem in mechanics, the body structure of living things (but not the particles constituting the living things), etc. In the study of such problems, Descartes' and Galileo's methods need to be modified somehow, because their methods emphasize separating the problem and phenomenon under consideration into isolated parts and processes. But according to von Bertalanffy, it should be recognized that the world we live in is not a pile of innumerable isolated "parts"; and any practical problem and phenomenon cannot be described perfectly by only one cause–effect chain. The basic character of the world is its organization, and the connection between the interior and the exterior of different things. Thus, the chief task of modern science should be a systematic study of the world.

12.3. Mathematical Modeling and the Origin of the Universe

It is well known that almost all laws of nature can be studied in the language of mathematics. Countless examples have shown that mathematics is extremely effective when applied to the analysis of natural systems. However, mathematics seems to be an intellectual activity separate from the development of science and technology. Its postulates come from pure thought or experience. In either case, once formulated, they form a basis for which the structure of a mathematical theory can be developed independently of their genesis.

Wigner (1960) asserted that "the unreasonable effectiveness of mathematics" is a mystery whose understanding and solution have yet to come: The miracle of the appropriateness of the language of mathematics for the formulation of the laws of physics is a wonderful gift which we neither understand nor deserve. We should be grateful for it and hope that it will remain valid in future research and that it will extend, for better or for worse, for our pleasure, though perhaps also to our bafflement, to wide branches of learning.

In this section, we discuss some problems related to the structure of mathematics, applications of mathematics, and the origin of the universe.

By a (general) system, we mean an ordered pair $S = (M, R)$ of sets, where R is a set of some relations defined on M. The elements in M are called objects of S and M and R are called the object set and the relation set of S, respectively. It has been shown that each formal language can be described as a system (M, K), where M is a set containing all words in the language and K is the collection of all grammatical rules in a fixed grammar book (Lin, 1989c); and each theory can be described as a system $(M, T \cup K)$, where M and K are defined as before and T is a set of basic principles upon which the theory is developed (Lin, 1989c). As an example, mathematics can be described as a system $(M, T \cup K)$ with T the collection of all axioms in the ZFC axiom system (Kuratowski and Mostowski, 1976).

Now the application of mathematics is an activity of matching a subsystem of the systems representation $(M, T \cup K)$ of mathematics to the situation under consideration. For example, a hibiscus flower has five petals. The mathematical word "five" provides a certain description of the flower. This description serves to distinguish it from flowers with three, four, six, ..., petals.

The whole structure of mathematics might be said to be true by virtue of mere definitions (namely, of the nonprimitive mathematical terms) if the ZFC axioms are true. However, at this juncture, we cannot refer to ZFC axioms as propositions which are either true or false, for they contain free primitive terms, "set" and the relation "$\in$" of membership, which have not been assigned any specific meanings. All we can assert so far is that any specific interpretation of the primitives which satisfies the axioms — i.e., turns them into true statements — will also satisfy all the theorems deduced from them. Examples can be given to show that the free primitive terms "set" and $\in$ permit many different interpretations in everyday life as well as in the investigation of laws of nature and from each interpretation we understand something more about nature, even though these new understandings may not necessarily be correct. At the same time, we feel that mathematics is unreasonably effective as applied to the study of natural problems. The following example, called the vase puzzle, will show that different mathematical understandings could be contradictory.

12.3.1. Vase Puzzle

Suppose a vase and an infinite number of pieces of paper are available. The pieces of paper are labeled with natural numbers $1, 2, 3, \ldots$, such that each piece has exactly one label. The following recursive procedure is performed: *Step 1:* Put the pieces of paper, labeled from 1 through 10, into the vase, then remove the piece labeled 1. *Step n:* Put the pieces of paper, labeled from $10n - 9$ through $10n$, into the vase, then remove the piece labeled n. After the procedure is completed, how many pieces of paper are left in the vase?

First, define a function $f(n) = 9n$, which tells how many pieces of paper are left in the vase after step n, for $n = 1, 2, 3, \ldots$ Therefore, if the recursive procedure can be finished, the number of pieces of paper left in the vase should be equal to the limit of the function $f(n)$ as $n \to \infty$. Hence, the answer is that infinitely many pieces of paper are left in the vase. Second, for each natural number n, we define the set M_n of pieces of paper left in the vase after the nth step is finished. After the recursive procedure is finished, the set of pieces of paper left in the vase equals the intersection $\bigcap_{n=1}^{\infty} M_n = \emptyset$. That is, no piece of paper is left in the vase.

Lin, Ma, and Port (1990) pointed out theoretically that there must be an impassable chasm between pure and applied mathematics. The reason is that pure mathematics is established upon a set of axioms, say ZFC. The theory possesses a beauty analogous to painting, music, and poetry, and a harmony between numbers and figures, whereas in applied mathematics each object of interest is always first given some "mathematical meaning," and then conclusions are drawn based on the relations of these meanings to the objects involved. It is the assignment of a mathematical meaning to each object that causes problems, because different interpretations can be given to the same object. As described in (Lin et al., 1990), some interpretations can result in contradictory mathematical models.

If we go back to the vase puzzle, it can be seen that the contradiction has very fruitful implications. First, one could argue that since there is no way to finish the recursive procedure in the puzzle, the contradiction does not exist. However, if this argument is correct, a large portion of mathematics would be incorrect, including the concept of limits, existence of subsequences, and Cantor's diagonal method in set theory, since they all require completion of a similar recursive procedure. Second, if information about a phenomenon has not all been used, the phenomenon can either not be understand because of the lack of knowledge or be understood at a more global level and more information is needed to be more specific. However, the vase puzzle does not satisfy this general methodology of recognition. Information about the ordering really made understanding more controversial rather than more specific. Third, among the most common practices of modern scientific activities is the science and technology of data analysis. It is known in analysis that the more data that are collected, the more misleading conclusion could be. This is due to the accumulation of errors. An extraordinary amount of errors can easily conceal the actual state of a phenomenon. The vase puzzle is another example showing that the more facts we know, the more confused we become. At the same time, the vase puzzle also questions the accuracy of scientific predictions, because the ignorance of a fact will lead to a completely different prediction.

Now instead of looking at the concept of mathematical modeling exclusively, let us take a more general stand. General systems theory, derived from the work of von Bertalanffy (1968), has greatly influenced all current approaches of systems analysis. Because of its promise, many people have tried to apply general systems theory to solving practical problems (Klir, 1970). This attempt has been notably unsuccessful; for detailed discussion on this, see (Berlinski, 1976; Lilienfeld, 1978). Wood-Harper and Fitzgerald (1982) thought that the reason for this lack of success was that the very generality of the theory made it difficult to use and to develop a methodological solution. Lin (1988c) analyzed the points on which each application of general systems theory is based and pointed out that these are the most probable places from which unexpected troubles arise.

A well-known fact is that the basic concepts of general systems theory were introduced in the language of set theory. Therefore, in applications, an elementary question is, What is a set? For example, is the collection of all living organisms in a certain experimental "container" a set? This problem seems to be simple, since in applications no one ever doubts the reasonableness of the statement that the collection of living organisms is a set. It is also a fundamental problem, because in each application of general systems theory, determining if a collection of some objects is a set is always the first problem we encounter before we can draw any useful conclusions based on theoretical results. The second question is, What does the formula $x \in y$ mean? Here $\in$ is a primitive notation, so in applications of general systems theory $\in$ can be defined in different ways. This is apparently a source of confusion in applications of the theory. The third question is, what does the statement "x is a set" mean? It is well known that some collections of objects are not sets. This is another source of confusion in applications of general systems theory.

By using axioms in the ZFC system, we can see that if A is a collection of objects and there exists a one-to-one correspondence $h : A \to X$ for some set X, then A is a set. It is natural to consider the following: How can we establish a one-to-one correspondence between a set and a collection of objects that we are studying? For example, we are studying the collection of all products of a certain kind, which are continuously coming out of several assembly lines. The difficulty is that the collection of interest is not fixed and is dependent on time. Surely, the collection cannot be assumed to be finite; however, it cannot be assumed to be infinite either. Otherwise, we would run into the same situation as described in the vase puzzle.

It is a known fact that human eyes can offer the brain wrong information. Another problem we have to face before any mathematical modeling is established is, How do we know anything? Descartes had pondered this question for a long time; for details see (Kline, 1972, p. 305). After comparing several definitions of systems, Lin (1989d) proposed the following axiom: The existence of any real matter is known by the existence of all particles in some level of the matter and by some relations between the particles in the matter and some particles in an environment of the matter.

As mentioned by Kunen (1980), a collection of cows cannot be considered a set. When can a collection of objects in an application be studied as a set so that the results in set theory and general systems theory can be used to make inferences? Can we introduce a procedure to check when a given collection of real matter is a "set"? Studying this question will lead to the creation of "applied set theory."

Since a system is what is distinguished as a system, according to Klir (1985), another difficulty we have to face to apply general systems theory is how to write a realistic relation, such as a human relation, as one or several relations in set theory, based upon the language of formal logic. After several systems models have been established, do we have a procedure to select additional models so that they produce the most desirable and most reasonable conclusions on the material system of interest?

When faced with practical problems, new abstract mathematical theories are developed to investigate them. At the same time, predictions based on the newly developed theories are so "accurate" that the following question appears naturally: Does the human way of thinking have the same structure as that of the material world?

As shown in (Lin, 1990d), mathematics can be built upon the "empty set, $\emptyset$." History gradually shows that almost all natural phenomena can be described and studied with mathematics. We thus ask: If all laws of nature could be written in the language of mathematics, could we conclude that the universe we live in rests on the empty set $\emptyset$, even as *Genesis* states that "in the beginning, God created the heaven and the earth. And the earth was without form, and void?"

In applications of general systems theory and mathematics, it has never been checked whether the collection of objects and the membership relation between the objects satisfy all the ZFC axioms. Only when $\in$ makes all ZFC axioms true propositions can general systems theory and mathematics be applied to a situation. Therefore, it is necessary to develop a method to check whether a collection of objects and a membership relation defined on the collection satisfy the ZFC axiom system.

Another interesting and important question, related to the concept of systems modeling and systems identification, is: How can we bring in the idea of "noise"? For instance, current achievements in modern quantum mechanics strongly support the fact that the laws of nature, if any, are written in terms of probability. Also, as pointed out by Masani (1994), each study of human relations is subject to the correction of the human noise. In some sense, human noises are different from those observed in any other systems containing no human beings.

12.4. Laws of Conservation and the Multilevel Structure of Nature

As shown in Section 12.1, the concept of systems characterizes the organizational structure of the world, including the structure of layers. One inference, based on the structure of layers, that can be made from the concept of systems is that the world is finitely divisible. Hence, a system $S_n = (M_n, R_n)$ is an nth-level object system of a system $S_0 = (M_0, R_0)$ if there exist systems $S_i = (M_i, R_i)$, for $i = 1, 2, \ldots, n-1$, such that $S_i \in M_{i-1}$, for $i = 1, 2, \ldots, n$, where n is a natural number (Lin and Ma, 1987). Under the assumption that the ZFC axiom system is consistent, the following has been shown: For any system S each chain of object systems of S must be finite, where a sequence $\{S_i\}_{i \in n}$ of systems, with n an ordinal number, is a chain of object systems of S if S_0 is an i_0th-level object system of S, and for any $i, j \in n$, if $i < j$, there then is a natural number $n(i,j)$ such that the system S_j is an $n(i,j)$th-level object system of S_i. An object x is an nth-level object of a system $S_0 = (M_0, R_0)$, where n is a positive integer, if there exist systems $S_i = (M_i, R_i)$, $i = 1, 2, \ldots, n$, such that $S_i \in M_{i-1}$, for $i = 1, 2, \ldots, n$, and $x \in M_n$. If the object x is no longer a system, then it is called a fundamental object of S_0. The previous result then says that each system must be finitely divisible.

Based on this discussion, it can be shown that for any system $S = (M, R)$ there exists a unique set $M(S)$ consisting of all fundamental objects in S. The set $M(S)$ is constructed as follows: For each object $x \in M(S)$, there exists a chain of object systems $\{S_i\}_{i \in n}$, for some natural number $n = n(x)$, such that $S_0 \in M$, $S_i \in M_{i-1}$, $i \in n$, and $x \in M_n$. Intuitively speaking, this result says that each system is built on fundamental objects. Does this result imply that the world is made up of fundamental particles, where a fundamental particle is a particle which cannot be divided into smaller particles? We know that the world consists of atoms. A fixed atom A can be thought of as a system S consisting of the set of all electrons and the nucleus in A and a set of some relations between the electrons and the nucleus. The collection of all electrons and the nucleus in A is a set because of the finiteness of the collection. Generally, if a particle X can be divided into smaller particles, it can be considered as an ordered pair (M, R), where M is the totality of the smaller particles in X, and R is a set of some relations between the particles in M. Therefore, if M is a set, X can be described as a system. Hence, the previous result implies that if M is a set, then X consists of fundamental particles. The last conclusion is subject to three possibilities:

(i) The world is made up of fundamental particles.

(ii) There exists a particle X such that X can be divided into smaller particles and the collection of all smaller particles in X is not a set.

(iii) The ZFC axiom system is not consistent.

Man has spent hundreds of years on the first possibility. When molecules were first found, they were thought to be the fundamental bricks of the natural world. Before long, smaller structures were found (i.e., atoms). At this time, people really believed that the smallest bricks of the world were found. But, once again, scientific achievement disproved that. Hence, people gradually began to believe that there did not exist any fundamental particles in the world at all. Here it is shown that each system is made up of fundamental objects. Can we conclude that the world is made up of fundamental particles? This depends on whether the second or third possibilities and the following reason are true:

There are some covert mistakes in the application of the result of systems theory.

Because all classical mathematics, which has been the "unreasonably" effective part of mathematics when applied to analyze practical problems, and consequently, all modern science and technology can be developed on ZFC (Kunen, 1980), we can be convinced that the ZFC axiom system is consistent. (Notice that there is no way to prove this yet!): Now the reasons are: (1) There are some mistakes in the application of systems theory. (2) There is a particle X such that X can be divided into smaller particles and the collection of all smaller particles in X is not a set.

On the other hand, if the foregoing discussion is correct and if the world is made up of fundamental particles, then each fundamental particle must be so small that it has no size. Otherwise, it can still be cut into smaller pieces. Thus, under the assumption that the world consists of fundamental particles, what the meaning of particles is becomes an important problem, because each fundamental particle cannot have a size.

As introduced by Klir (1985), a system is what is distinguished as a system. We can do the following systems modeling: For each chosen matter, real situation problem, or environment, such as a chemical reaction process, a system describing the matter of interest can always be defined. For example, let $S = (M, R)$ be a systems representation of a chemical reaction process, such that M stands for the set of all substances used in the reaction and R is the set of all relations between substances in M. It can be seen that S represents the chemical reaction of interest. Changes of the system represents changes of the objects in M and changes of relations in R. Based on this understanding and the result of systems theory that each chain of object systems of a given system must be finite, the following axiom, due to Lavoisier given in 1789, has been proven theoretically: In all the operations of art and nature, nothing is created! An equal quantity of matter exists both before and after the experiment ... and nothing takes place beyond changes and modifications in the combination of the elements. Upon this principle, the whole art of performing chemical experiments depends.

Based on this understanding, the conservation principle of ancient atomists, (Perlman, 1970, p. 414), which states that the total count of atoms in the universe is constant, so the total amount of matter remains the same regardless of changes, can be recast as follows:

12.4.1. Law of Conservation of Fundamental Particles (Lin, submitted)

The total count of fundamental particles in the universe is constant; therefore, the total amount of matter remains the same regardless of changes and modifications.

Let us discuss some impacts of what has been done on the Law of Conservation of Matter–Energy. The law states that the total amount of matter and energy is always the same. Historically, this law was developed based on the new development of science and technology: energy and matter are two different forms of the same thing. That is, matter can be transferred into energy, and energy into matter. All matter consists of fundamental particles, as already discussed and so does energy. Since energy can be in different forms, such as light, it confirms that each fundamental particle has no volume or size. Now, as a consequence of the Law of Conservation of Fundamental Particles, the Law of Conservation of Matter–Energy becomes clear and obvious.

One of the many important open problems is to develop conservation equations in terms of fundamental particles and to write them in modern symbolic form. Only with the indispensable "=" in the equations can the theory, developed on the Law of Conservation of Fundamental Particles, become an "exact science" with the capacity of prediction.

12.5. Set-Theoretic General Systems Theory

This section focuses on general systems theory developed on ZFC set theory. As mentioned in Section 12.2, this general theory will establish a theoretical foundation for all approaches of systems analysis developed in various disciplines.

Let S be a (general) system if S equals an ordered pair (M,R) of sets such that R is a set of some relations defined on the set M (Lin, 1987). The elements in M are called the objects of S and M and R are the object set and the relation set of S, respectively. As pointed out by Mesarovic and Takahara (1989), an object in M is recognized as a building block, a component of the system; it can be real matter or a conceptual entity. Since the existence of an object is recognized through its attributes or characteristics, a more appropriate definition for general systems should be an ordered pair of a set of attributes of some objects and a set of relations among the attributes of the objects. However, if one just simply uses the attributes here as the objects, then the definition of general systems would be reasonable.

For a relation $r \in R$, where $S = (M,R)$ is a system, there exists an ordinal number $n = n(r)$ such that $r \subseteq M^n$. The number n is called the length of r. It is assumed that the length of the empty relation $\emptyset$ is zero; that is, $n(\emptyset) = 0$.

One goal of systems theory, based on this definition, is to possess the same beauty as mathematics possesses. Beauty ensures that the theory will be handed down from generation to generation, since only beauty can make the theory assimilated into our daily thoughts process and brought again and again before the mind with ever-renewed encouragement. To achieve this goal, the general structure of a system, mappings from systems into systems, constructions of systems, hierarchies of systems, etc., need to be studied systematically. Along this line are several open problems.

It can be shown that for any system $S = (M,R)$, where for any $r \in R$, $n(r) = 2$, S is a subset of the set $p^2(M) \cup p(p(M) \cup p^4(M))$, where $p^1(X) = p(X)$ is the power set of X and $p^{i+1}(X) = p(p^i(X))$ for each $i = 1,2,3,\ldots$. Is it possible that a structural representation for general systems can be given similar to the one given earlier? The characteristic of this representation is that only ordinal numbers and power set operations are involved.

A system $S = (M,R)$ is called centralized system if each object in S is a system and there exists a system $C = (M_C, R_C)$ satisfying $M_C \neq \emptyset$ such that for any distinct elements $x,y \in M$, say $x = (M_x, R_x)$ and $y = (M_y, R_y)$, $M_C = M_x \cap M_y$ and $R_C \subseteq R_x|M_C \cap R_y|M_C$, where $R_x|M_C = \{r \cap M_C^{n(r)} : r \in R_x\}$ and $R_y|M_C = \{r \cap M_C^{n(r)} : r \in R_y\}$. The system C is called a center of S. If in, in addition, $R_C = R_x|M_C \cap R_y|M_C$, S is called strongly centralized. For centralized systems, it has been shown that under the assumption that ZFC axiom system is consistent, for κ an infinite cardinality and $\theta > \kappa$ a regular cardinality such that, for any $\alpha < \theta$,

$$\left|\alpha^{<\kappa}\right| < \theta$$

then for each system $S = (M,R)$ where $|M| \geq \theta$ and $|M_m| < \kappa$, for each object $m \in M$ a system with $m = (M_m, R_m)$, and where there exists an object contained in at least θ objects in M, there exists a partial system $S' = (M', R')$ of S such that S' forms a centralized system and $|M'| \geq \theta$. This result is a restatement of the general Δ-lemma in set theory (Kunen, 1980). Since this result has found several interesting applications in sociology and epidemiology (Lin, 1988a; Lin and Vierthaler, 1998; Lin and Forrest, 1995), it is worth asking: the question below: Under what conditions is the partial system S' a strongly centralized system?

It can be seen that the concept of systems is a higher-level abstraction of mathematical structures. For example, n-tuple relations, networks, abstract automatic machines, algebraic systems, topological spaces, vector spaces, algebras, fuzzy sets and fuzzy relations, manifolds, metric spaces, normed spaces, Frechet spaces, Banach spaces, Banach algebras, normed rings, Hilbert spaces, semigroups, Riesz spaces, semiordered spaces, and systems of axioms and formal languages are all systems. From these examples it is natural to consider the problem: Find properties of networks, automatic machines, topological spaces, vector spaces, algebras, fuzzy structures, systems of axioms, etc., such that in general systems theory they have the same appearance.

Let $S_i = (M_i, R_i)$, $i = 1,2$, be two systems and $h : M_1 \to M_2$ a mapping. By transfinite induction, two classes $\widehat{M}_i$, $i = 1,2$, and a class mapping $\widehat{h} : \widehat{M}_1 \to \widehat{M}_2$ can be defined with the following properties:

$$\widehat{M}_i = \bigcup_{n \in ON} M_i^n, \quad i = 1,2$$

and for each $x = (x_0, x_1, \ldots, x_\alpha, \ldots) \in \widehat{M}_1$

$$\widehat{h}(x) = (h(x_0), h(x_1), \ldots, h(x_\alpha), \ldots)$$

where ON stands for the class of all ordinal numbers. For each relation $r \in R$, $\widehat{h}(r) = \{\widehat{h}(x) : x \in r\}$ is a relation on the set M_2 with length $n(r)$. Without causing confusion, h will also be used to indicate the class mapping $\widehat{h}$, and h is a mapping from S_1 into S_2, denoted by $h : S_1 \to S_2$.

A mapping $h : S_1 = (M_1, R_1) \to S_2 = (M_2, R_2)$ is S-continuous (Ma and Lin, 1990e) if for any relation $r \in R_2$,

$$h^{-1}(r) = \{h^{-1}(x) : x \in r\} \in R_1$$

It then can be shown that if the systems S_i, $i = 1,2$, are topological spaces, h is S-continuous iff h is a continuous mapping from the topological space S_1 into the topological space S_2, and that if the systems S_i, $i = 1,2$, are rings in algebra, the S-continuity of h implies that h is a homomorphism from the ring S_1 into the ring S_2. This example shows the need to find the topological properties of topological spaces and algebraic properties of algebras, such as groups, rings, etc., such that in systems analysis they have the same appearance.

A mapping $f : S_1 \to S_2$ is a morphism from a system S_1 into a system S_2 if $f(R_1) \subseteq R$. Let $\mathrm{Hom}(S_1, S_2) = \{f : S_1 \to S_2 : f \text{ is a morphism}\}$. Suppose $\{S_i : i \in I\}$ is a set of systems. A system $S = (M, R)$ is a product (resp., Cartesian product) of S_i, if there is a family of morphisms (resp., S-continuous mappings) $\{\Phi_i \in \mathrm{Hom}(S, S_i) : i \in I\}$ (resp., $\{p_i : S \to S_i : i \in I\}$) such that for each system S' and a family $\{\Psi_i \in \mathrm{Hom}(S', S_i) : i \in I\}$ (resp., $\{q_i : S' \to S_i : i \in I\}$ of S-continuous mappings) there is a unique morphism (resp., S-continuous mapping) $\lambda \in \mathrm{Hom}(S', S)$ (resp., $\lambda : S' \to S$) so that

$$\Psi_i = \Phi_i \circ \lambda \text{ (resp., } q_i = p_i \circ \lambda) \quad \text{for each } i \in I$$

It can be shown that for each set $\{S_i : i \in I\}$ of systems, a product and a Cartesian product of S_i always exist, which are unique up to a similarity and denoted by $\prod_{i \in I} S_i$ and $\prod_{\mathrm{cp}} \{S_i : i \in I\}$, respectively. The reason why the system $\prod_{\mathrm{cp}} \{S_i : i \in I\}$ is called the Cartesian product of S_i is that when S_i are topological spaces, this system is the Cartesian product of the spaces S_i. In this case, the system $\prod_{i \in I} S_i$ is the box product of the spaces S_i. When the systems S_i are algebras, such as groups, rings, etc., the system $\prod_{i \in I} S_i$ is the direct sum of the algebras S_i. That is, the box product in topology and the direct sum in algebra have the same structure. This fact gives rise to the following problem: Find topological structures and algebraic structures such that in systems analysis they appear to be the same. Conversely, it is natural to ask: When can a system be studied as a topological space or a specific algebra?

Let $(T, \leq)$ be a partially ordered set with order type α. An α-type hierarchy S of systems [over the partially ordered set $(T, \leq)$] is a function defined on T such that for each $t \in T$, $S(t) = S_t = (M_t, R_t)$ is a system (Lin, 1989a). For an α-type hierarchy S of systems, let $\ell_{tr} : S_r \to S_t$ be a mapping from the system S_r into the system S_t, for any $r, t \in T$ with $r \geq t$, such that

$$\ell_{ts} = \ell_{tr} \circ \ell_{rs} \quad \text{and} \quad \ell_{tt} = \mathrm{id}_{S_t}$$

where r, s, t are arbitrary elements in T satisfying $s \geq r \geq t$ and $\mathrm{id}_{S_t} = \mathrm{id}_{M_t}$ is the identity mapping on the set M_t. The family $\{\ell_{ts} : t, s \in T, s \geq t\}$ is termed a family of linkage mappings of the α-type hierarchy S, and each ℓ_{ts} is termed a linkage mapping from the system S_s into the system S_t.

It has been shown (Lin, 1989a) that for each α-type hierarchy S of systems satisfying that for each $t \in T$, $M_t \neq \emptyset$, where $S_t = (M_t, R_t)$, there exists a family $\{\ell_{ts} : t, s \in T, s \geq t\}$ of linkage mappings of S. An interesting question is, how many families of linkage mappings are there for a given α-type hierarchy of systems? Or, how many families of linkage mappings with certain properties, such as S-continuous, are there for a given α-type hierarchy of systems? Another question, which is similar to the question of expanding a given function, is the following: Suppose S is an α-type hierarchy of systems over a partially ordered set $(T, \leq)$ and $\{\ell_{ts} : t, s \in T^*, s \geq t\}$ is a family of linkage mappings of the hierarchy of systems $\{S_t : t \in T^*\}$, where $T^* \subset T$. Under what conditions does there exist a family $\{\overline{\ell}_{ts} : t, s \in T, s \geq t\}$ of linkage mappings for S such that $\overline{\ell}_{ts} = \ell_{ts}$ for all s, $t \in T^*$ with $s \geq t$?

As shown by Omran (1971), there is strong evidence for the existence of the epidemiological transition, which describes the transition of health and disease patterns, namely that infectious diseases are gradually replaced by noncommunicable diseases as the economic and industrial development of a region takes place. Its main characteristic is that the transition history of the health and disease patterns of industrialized nations have been repeated, or are yet to be repeated, in developing regions or nations. To model this important concept, Lin and Lin (1995) employed the concepts of systems over groups and of periodic systems over groups. To be specific, suppose $(T, +, \leq)$ is an ordered Abelian group. A hierarchy S of systems over $(T, +, \leq)$ is called a system over a group. If, in addition, the hierarchy is linked, the system is termed a linked system over a group. For a fixed $t_0 \in T$, the right transformation of S, denoted S^{-t_0}, is defined to be $S^{-t_0}(t) = S(t - t_0)$ for each $t \in T$; and the left transformation of S, denoted S^{+t_0}, is defined by $S^{+t_0}(t) = S(t+t_0)$ for each $t \in T$. A system S over a group T is periodic if there exists $p \in T$ such that $p \neq 0$ and $S^{+p} = S$. The element p is called a period of S. As pointed out in (Lin and Forrest, 1995), there is a need to study multiperiodic systems where a system S over a group T is multiperiodic if there exists a subset $D \subseteq T$ such that $S^{+p} = S$ for each $p \in D$.

Feedback has been a hot topic in recent years since all living, and many non-living systems are feedback systems. How to study these systems is an important question. The concepts of feedback transformation and feedback decoupling have brought some light because they tell when the system under consideration can be divided into smaller systems. For recent research, consult Takahara and Asahi (1985), and Saito (1986; 1987). Let us take a brief look at the concepts of feedback transformation and feedback decoupling and several related open questions.

Let $\mathcal{A}$ be a field, X and Y linear spaces over the field $\mathcal{A}$, and S an (input–output) system satisfying the following conditions: (i) $\emptyset \neq S \subseteq X \times Y$; (ii) $s \in S$ and $s' \in S$ imply that $s + s' \in S$; (iii) $s \in S$ and $\alpha \in \mathcal{A}$ imply $\alpha \cdot s \in S$, where $+$ and $\cdot$ are addition and scalar multiplication in $X \times Y$, respectively, defined by: for any $(x_1,y_1),(x_2,y_2) \in X \times Y$ and any $\alpha \in \mathcal{A}$, $(x_1,y_1)+(x_2,y_2) = (x_1+x_2,y_1+y_2)$ and $\alpha \cdot (x_1,y_1) = (\alpha x_1, \alpha y_1)$. Then S is called an (input–output) linear system (Mesarovic and Takahara, 1975). An input–output system $S \subset X \times Y$ is a functional system if S is a function from the input set X into the output set Y. Let $S \subset X \times Y$ and $S_f : Y \to X$ be a linear system and a linear functional system, respectively. Then the feedback system of S by S_f is defined as the input–output system S' such that

$$(x,y) \in S' \leftrightarrow (\exists z \in X)((x+z,y) \in S \text{ and } (y,z) \in S_f)$$

The systems S and S_f are called an original system and a feedback component system, respectively. For details, see (Saito and Mesarovic, 1985).

Let $\mathcal{S} = \{S \subset X \times Y : S \text{ is a linear system}\}$, $\mathcal{S}_f = \{S_f : Y \to X : S_f \text{ is a linear functional system}\}$; and $\mathcal{S}' = \{S' \subset X \times Y : S' \text{ is a subset}\}$. Then a feedback transformation $F : \mathcal{S} \times \mathcal{S}_f \to \mathcal{S}'$ is defined by the last equivalence relation for each $(S,S_f) \in \mathcal{S} \times \mathcal{S}_f$. The transformation F is called the feedback transformation over the linear spaces X and Y.

Assume that the input space X and the output space Y can be written $X = \prod\{X_i : i \in I\}$ and $Y = \prod\{Y_i : i \in I\}$. For each $i \in I$, the projection $p_i = (p_{ix},p_{iy}) : X \times Y \to X_i \times Y_i$ is defined by $p_{ix} : X \to X_i$ and $p_{iy} : Y \to Y_i$ satisfying $p_{ix}(x) = x_i$ and $p_{iy}(y) = y_i$, for any $x \in X$ and $y \in Y$. Let $S_i = p_i(S)$ for $i \in I$. Then the system S is decomposed into a family of factor systems $\overline{S} = \{S_i : i \in I\}$. If $((x_i)_{i\in I},(y_i)_{i\in I}) \in S$ is identified with $((x_i,y_i)_{i\in I}) \in \prod\{S_i : i \in I\}$, then $S \subseteq \prod\{S_i : i \in I\}$. A linear system $S \in \mathcal{S}$ is decoupled by feedback (Saito and Mesarovic, 1985) if there is $S_f \in \mathcal{S}_f$ such that

$$F(S,S_f) = \prod\{p_i(F(S,S_f)) : i \in I\}$$

Under the assumption that ZFC is consistent, it has been shown (Lin, 1994) that if $S \in \mathcal{S}$ is decomposed into a family of factor systems $\overline{S} = \{S_i : i \in I\}$, then S is decoupled by feedback iff $R(S) = \prod\{R(S_i) : i \in I\}$ and $N(S) = \prod\{N(S_i) : i \in I\}$, where $R(S) = \{y \in Y : \exists x \in X((x,y) \in S)\}$ and $N(S) = \{x \in X : (x, 0_y) \in S\}$. Based on this result, (Lin, 1994) gives a concrete procedure to decouple the linear system

$$\begin{aligned} \dot{z} &= Az + Bx \\ y &= Cz + Dx \\ z(0) &= 0 \end{aligned}$$

where z is an $m \times 1$ variable vector, called the state of the system, A, B, C, and D are constant matrices of size $m \times m$, $m \times n$, $n \times m$, and $n \times n$, respectively, such that D is nonsingular, and the input space X and the output space Y are

$$X = Y = \{r : [0, \infty) \to \mathbb{R}^n : r \text{ is piecewise continuous}\}$$

where $\mathbb{R}$ is the set of all real numbers. The family of factor systems consists of n one-dimensional linear systems of the following form:

$$\begin{aligned} \dot{z} &= Az + B_i x_i \\ y_i &= C_i z + D_i x_i \\ z(0) &= 0 \end{aligned}$$

for $i = 1, 2, \ldots, n$, where $B = [B_1\, B_2 \cdots B_n]$, $C = [C_1\, C_2 \cdots C_n]^T$, D_i is a nonzero constant, and the input space X_i and the output space Y_i are given by $X_i = Y_i = \{r : [0, \infty) \to \mathbb{R} : r \text{ is piecewise continuous}\}$.

Many questions are left open. For example, do the input space X and the output space Y have to be the same dimension in order to apply the theory? Find examples of systems such that some of their important properties are not kept by feedback systems. If some properties of the original system are not kept by feedback systems, can the properties be studied by using the concept of pullback? For the concept of pullback, see (Negoita, 1992). Is it possible to construct the solution of the n-dimensional linear system described earlier, based on the solutions of the one-dimensional factor systems?

12.6. Analytic Foundation for an Applicable General Systems Theory

During a recent and fruitful conversation with Professor Pesi Masani, he pointed out a fact, which could also be considered a challenge to systems science, that to show the need for the concept of (general) systems it is absolutely necessary to obtain some results which have not or could not be achieved without using the concept of systems. To meet this challenge and to make up a deficiency of the current theory of general systems analysis, where no quantitative study based on general systems analysis can be carried out, it will be shown that the non-Archimedean number field, first introduced by Shu-tang Wang in the late 1970s and early 1980s, is an excellent starting point.

Historically speaking, nonstandard analysis, established in the early 1960s by Robinson (1977), introduced the concepts of infinitesimals and infinities formally and legally, based on mathematical logic. This theory broke through the limitation of classical concept of real numbers, where infinities are considered ideal objects. With nonstandard analysis, many classical theories, such as mathematical analysis, algebra, normed spaces, some classical mechanics problems, etc., can be greatly simplified and become more intuitive. However, the shortfall of Robinson's nonstandard analysis is that infinitesimals and infinities can only be shown to exist and have no concrete expressions of any tangible form. Based on the argument of the existence of infinitesimals and infinities given in Robinson's theory, the non-Archimedean number field, called the generalized number system (**GNS**) by Wang (1985), has a computational advantage, since each number, whether ordinary real number, infinitesimal, or infinity, has an exact expression in terms of the real numbers in the classical sense.

A generalized number $\mathbf{x}$ is a function from $\mathbb{Z}$ into $\mathbb{R}$, where $\mathbb{Z}$ is the set of all integers and $\mathbb{R}$ the set of all real numbers, such that there exists a number $k = k(\mathbf{x}) \in \mathbb{Z}$, called the level index of $\mathbf{x}$, satisfying $\mathbf{x}(t) = 0$, for $t < k$, and $\mathbf{x}(k(\mathbf{x})) \neq 0$. Let **GNS** stand for the field consisting of all generalized numbers and the zero function $\mathbf{0} : \mathbb{Z} \to \mathbb{R}$, with the following operations: (1) Addition and subtraction: for any $\mathbf{x}, \mathbf{y} \in \mathbf{GNS}$, $(\mathbf{x} \pm \mathbf{y})(t) = \mathbf{x}(t) \pm \mathbf{y}(t)$, for $t \in \mathbb{Z}$; (2) Ordering: for any $\mathbf{x}, \mathbf{y} \in \mathbf{GNS}$, $\mathbf{x} < \mathbf{y}$ iff there is a $k_0 \in I$ such that $\mathbf{x}(t) = \mathbf{y}(t)$, if $t < k_0$, and $\mathbf{x}(k_0) < \mathbf{y}(k_0)$; (3) Scalar multiplication: let $c \in \mathbb{R}$ and $\mathbf{x} \in \mathbf{GNS}$, and define $(c\mathbf{x})(t) = c\mathbf{x}(t)$ for $t \in \mathbb{Z}$; (4) Multiplication: for $\mathbf{x} \in \mathbf{GNS}$, write $\mathbf{x}$ as a generalized power series

$$\mathbf{x} = \sum_{t \geq k(\mathbf{x})} \mathbf{x}(t) \times \mathbf{1}_{(t)}$$

where $\mathbf{1}_{(t)} \in \mathbf{GNS}$ is defined by $\mathbf{1}_{(t)}(s) = 0$, if $s \neq t$, and $\mathbf{1}_{(t)}(t) = 1$. Let $\mathbf{x}, \mathbf{y} \in \mathbf{GNS}$. The product $\mathbf{x} \times \mathbf{y}$ or $\mathbf{xy}$ is defined by

$$\mathbf{x} \times \mathbf{y} = \sum_{t} \sum_{m+n=t} [\mathbf{x}(m) \times \mathbf{y}(n)] \times \mathbf{1}_{(t)}$$

(5) Division: For $\mathbf{x}, \mathbf{y} \in \mathbf{GNS}$, if there exists a $\mathbf{z} \in \mathbf{GNS}$ such that $\mathbf{y} \times \mathbf{z} = \mathbf{x}$, then the generalized number $\mathbf{z}$ is called the quotient of $\mathbf{x}$ divided by $\mathbf{y}$, denoted by $\mathbf{x} \div \mathbf{y}$ or $\mathbf{x}/\mathbf{y}$.

Throughout history, "laws" of nature have been frequently expressed and treated with the rigor and beauty of mathematical symbols. For the most part, this kind of application of mathematics has been very successful. However, in the past 40 years, theoretical physicists have often been faced with difficulties in mathematical calculation. According to Wang (1991), such difficulties arise mainly because the modern world of physics is multilevel in its character, while applicable mathematics still remains single level where the only infinities are denoted by $-\infty$ or $+\infty$, and are considered in our macroscopic world as ideal objects and are treated with a different set of rules, compared with those of ordinary real numbers. Now, in **GNS** the field $\mathbb{R}$ of all real numbers is isomorphic to $\mathbb{I}_{(0)} = \{\mathbf{x} \in \mathbf{GNS} : \mathbf{x}(t) = 0 \quad \text{if } t \neq 0\}$, which represents the quantities measurable within the macroscopic world. Furthermore, all numbers in **GNS**, whether infinities or infinitesimals, are treated with the same set of rules.

To give the reader a taste of how **GNS** can be applied to the study of physics, we look at the calculation of the rest mass of a photon. According to Einstein, the inertial mass m of a moving particle relates to its rest mass m_0 as follows:

$$m = \frac{m_0}{\sqrt{1 - v^2/c^2}} \tag{12.1}$$

where c stands for the speed of light and v is the particle speed. Since the photon's speed is $v = c$, the rest mass m_0 should be $m_0 = 0$. Otherwise, one would obtain from Eq. (12.1) that $m = \infty$, contradicting observations. However, even it is assumed that $m_0 = 0$, injustice involved in the calculation still remains. This is because in classical mathematics on the real number line, $m = 0 \div 0$ is undetermined.

This difficulty can be overcome by using generalized numbers. Suppose the rest mass $\mathbf{m}_0$ of a photon is

$$\mathbf{m}_0 = (\ldots, 0, \ldots, \underset{\substack{\uparrow \\ \text{0th level}}}{0}, m_1, m_2, \ldots)$$

and that the inertial mass m in Eq. (12.1) is a finite real number; that is, $\mathbf{m} \in \mathbb{I}_{(0)}$ has a representation

$$\mathbf{m} = (\ldots, 0, \underset{\substack{\uparrow \\ \text{0th level}}}{m}, 0, \ldots)$$

Now the velocity of a photon is

$$\mathbf{v} = (\ldots, 0, \ldots, 0, \underset{\substack{\uparrow \\ \text{0th level}}}{c}, 0, v_2, v_3, \ldots)$$

where $v_2 < 0$, since c is assumed by Einstein to be the maximum speed in the universe. So Eq. (12.1) can be written as

$$\mathbf{m} = \frac{\mathbf{m}_0}{\sqrt{\mathbf{1}_{(0)} - \mathbf{v}^2/\mathbf{c}^2}}$$

where **c** is understood in the same way as we did with $\mathbf{m} \in \mathbb{I}_{(0)}$. By substituting the **GNS** values of **m**, $\mathbf{m}_0$, $\mathbf{1}_{(0)}$, **v**, and **c** into this equation and comparing the leading nonzero terms, one obtains $(-2c)m^2 v_2 = m_1^2$. That gives the relation between v_2 and m_1. For technical details, see (Lin and Wang, 1998).

Comparing the generalized number system with those contained in mathematical analysis, real, and complex analysis, we see that our understanding of the **GNS** is still superficial. More work needs to be carried out to gain deeper understanding about nature and the world in which we live.

For instance, it is necessary to

(1) Study the ordering relation. It is obvious that the **GNS** field is not complete in the sense of Dedekind cuts.

(2) Explore applications of the theory of **GNS** in other branches of human knowledge. The future success of the theory of **GNS** may well depend on various successful applications.

(3) Develop workable and concrete formulas for the operations of multiplication, division, and taking roots.

(4) Redevelop quantum mechanics based on **GNS** so that, hopefully, more puzzles in the current quantum mechanics can be explained more rationally.

(5) Establish a law, similar to the transfer law in nonstandard analysis (Davis, 1977), governing the relation between a set of properties of real numbers and a set of properties of generalized numbers.

(6) Develop the results of generalized numbers, corresponding to theorems in mathematical analysis, such as the intermediate value theorem, maximum and minimum theorem, etc.

(7) Find detailed formulas to compute various derivatives, definite and indefinite integrals of given functions from **GNS** into **GNS**.

(8) Generalize the Fundamental Theorem of Calculus to the case of **GNS**, linking **GNS** differential calculus to **GNS** integral calculus.

(9) Develop an integral calculus of **GNS** based on the same mechanism of Lebesgue integrals.

(10) Use **GNS** representations of Schwartz's distributions to explore new explanations of equations, appearing in the theory of distributions, which do not make sense in classical mathematical analysis.

(11) Employ **GNS** representations of Schwartz's distributions to solve open problems in the theory of distributions, such as defining products of distributions. Can the theory of **GNS** be used to explain why in general the products of distributions might not exist?

12.7. A Few Final Words

To face challenges, the study of general systems theory must further the research of the concept of (general) systems and related topics so that the theory possesses more systems, more large-scale combination, more pragmatism, more accuracy, and more unification (blending philosophy, mathematics, physical science, technology, etc., into one body of knowledge).

Based on the history of general systems theory and its current development, among the keys for its future success are publicity and practicalities. Only with these two tasks being accomplished, will the scientific community be truly moved so that research of general systems theory and applications moves to a higher level.

Man is both a reflection reduced with the size of history and an enlarged image of history. He is the creature of interactions between himself and the physical world. Research and application of general systems theory are a natural extension of his intelligent and comprehensive pursuits for the past several thousand years. They are also creatures of interactions between them and the physical world and reduced reflections and enlarged images of the history of the world of learning. Any one with a prepared mind can achieve a great deal in the development and application of such a theory.

The general systems enterprise has been, and will continue to be, developing and self-completing. In the scientific garden with various theories which will stand the test of time and history, general systems theory has been set off from the grayer background, self-realizing, self-coquetting, self-dedicating, self-rebeling and self-surpassing. Even though the general systems enterprise has not been, and will not be, pursuing eternity, it will definitely become a powerful force sweeping through the world of learning.

References

Ackoff, R. L. (1959). Games, decisions and organizations. *General Systems*, **4**, 145–150.

Baldwin, J., and B. Pilsworth (1982). Dynamic programming for fuzzy systems with fuzzy environment. *J. Math. Anal. Appl.*, **85**, 1–23.

Bayliss, L. E. (1966). *Living Control Systems*. San Francisco.

Bellman, R. (1957). *Dynamic Programming*. Princeton University Press, Princeton, NJ.

Bellman, R., and L. A. Zadeh (1970). Decision making in fuzzy environment. *Management Sci.*, **17**, B 141–B 164.

Berlinski, D. (1976). *On Systems Analysis*. MIT Press, Cambridge, MA.

von Bertalanffy, L. (1924). *Einführung in Spengler's Werk*. Literaturblatt Kolnische Zeitung, May.

—— (1934). *Modern Theories of Development* (Transl. J. H. Woodge) Oxford University Press, Oxford; Harper Torch Books, New York (1962); German Original: *Kritische Theories der Formbildung*, Borntäger, Berlin (1928).

—— (1967). General systems theory: Application to psychology, social science. *Inform. Sci. Soc.*, **6**, 125–136.

—— (1968). *General Systems Theory*. George Braziller, New York.

—— (1972). The history and status of general systems theory. In: G. Klir (ed.), *Trends in General Systems Theory*. New York, pp. 21–41.

Billstein, R., S. Libeskind, and J. W. Lott (1990). *A Problem Solving Approach to Mathematics for Elementary School Teachers*. 4th ed., Benjamin/Cummings, Redwood City, CA.

Blauberg, I. V., V. N. Sadovsky, and E. G. Yudin (1977). *Systems Theory, Philosophical and Methodological Problems*. Progress Publishers, Moscow.

Box, G., W. Hunter, and J. S. Hunter (1978). *Statistics for Experimenters: An Introduction to Design, Data Analysis, and Model Building*. Wiley, New York.

Brown, T. A., and R. E. Strauch (1965). Dynamic programming in multiplicative lattices. *J. Math. Anal. Appl.*, **12**, 624–637.

Bunge, B. (1979). *Treatise on Basic Philosophy*. Vol. 4 : A World of Systems. Reidel, Dordrecht, Holland.

Cartan, H. (1980). Nicolas Bourbaki and contemporary mathematics. *Math. Intell.*, **2**, 175–180.

Casti, J. (1985). *Non-Linear Systems Theory*. Academic Press, Orlando.

Checkland, P. B. (1975). The development of systems thinking by systems practice — A methodology from action research programme. In: R. Trappl and F. de P. Hawika (eds.), *Progress in Cybernetics and System Research*, Vol. II. Hemisphere, Washington, DC, pp. 278–283.

Cohen, J., and I. Stewart (1994). *The Collapse of Chaos*. Viking Press, New York.

Colmez, J. (1947). Espaces a escart generalise regulier. *C. R. Acad. Sci. Paris*, **244**, 372–373.

Cornacchio, J. V. (1972). Topological concepts in the mathematical theory of general systems. In: G. Klir (ed.), *Trends in General Systems Theory*. Wiley-Interscience, New York.

Davis, M. (1977). *Applied Non-Standard Analysis*. Wiley, New York.

Denardo, E. V. (1967). *Contraction Mappings in the Theory Underlying Dynamic Programming*. SIAM Rev., No. 9, 1121–1129.

Deng, J.-L. (1985). *Grey Control Systems*. Hua-zhong Institute of Technology Press, Wuhan.

Department of Business Administration of Wuhan University (1983). *Fundamentals of Modern Administration*. Knowledge Press, Wuhan.

Department of Mathematics, Ji-Lin University (1978). *Mathematical Analysis* (in Chinese). People's Education Press.

Dieudonne, J. (1982). *A Panorama of Pure Mathematics: As Seen by N. Bourbaki*. Academic Press, New York.

Dirac, P. A. (1958). *The Principles of Quantum Mechanics*. Oxford University Press, London.

Doss, R. (1947). Sur les espaces ou la topologie peut etre definis a l' aid d'unecart abstrait symmetrique et regulier. *C. R. Acad. Sci. Paris*, **223**, 1087–1088.

Drake, F. R. (1974). *Set Theory: An Introduction to Cardinals*. North-Holland, Amsterdam.

Dreyfus, S., and A. Law (1977). *The Art and Theory of Dynamic Programming*. Academic Press, New York.

Einstein, A. (1922). *The World as I See It*. Covici, Friede, New York.

Ellis, H. C. (1978). Analyzing business information needs. In: R. N. Maddison (ed.), *Data Analysis for Information Systems Design*. BCS Conference Papers, British Computer Society, London.

Engelking, R. (1975). *General Topology*. Polish Scientific Publishers, Warszawa.

Fraenkel,, A. A., Y. Bar-Hillel, and A. Levy (1973). *Foundations of Set Theory*. North-Holland, Amsterdam.

Frechet, M. (1945). La notion d'uniformite et les ecart abstrait. *C. R. Acad. Sci. Paris*, **221**, 337–339.

—— (1946). De l'ecart numerique alecart abstrait. *Portugese Math.*, **5**, 121–131.

N. Furukawa (1980). Characterization of optimal policies in vector-valued Markovian decision process. *Math. Oper. Res.*, **5**, 271–279.

Gane, C., and T. Sarsons (1979). *Structured Analysis: Tools and Techniques*. Prentice-Hall, Englewood Cliffs, NJ.

Gleick, J. (1987). *Chaos: Making a New Science*. Viking Press, New York.

Goldberg, D. (1989). *Genetic Algorithms in Search, Optimization and Machine Learning*. Addison-Wesley, Reading, MA.

Graniner, J. V. (1988). *Math. Mag.*, **61**, 220–230.

Gratzer, G. (1978). *Universal Algebra*. Springer-Verlag, New York.

Hahn, E. (1967). Aktuelle entwicklungstendenzen der soziologischen theorie. *Deutsche Z. Phil.*, **15**, 178–191.

Hall, A. D., and R. E. Fagen (1956). Definitions of systems. *General Systems*, **1**, 18–28.

Halmos, P. R. (1957). Nicolas Bourbaki. *Scientific American*, May, 88–99.

—— (1960). *Naive Set Theory*. Van Nostrand, Princeton, NJ

M. I. Henig (1983). Vector-Valued Dynamic Programming. *SIAM J. Control Optim.*, No. 3, 237–245.

Hess, B. (1969). *Modelle Enzymatischer Prozesse*. Nova Acta Leopoldina (Halle, Germany).

Hodel, R. E. (1975). Extensions of metrization theorems to higher cardinality. *Fund. Math.*, **87**, 219–229.

Holland, J. (1975). *Adaptation in Natural and Artificial Systems*. University of Michigan Press.

Hu, T. C. (1982). *Combinatorial Algorithms*. Addison-Wesley, New York.

Husek, M., and H. C. Reichel (1983). Topological characterizations of linearly uniformizable spaces. *Topology Appl.*, **15**, 173–188.

Jech, T. J. (1973). *The Axiom of Choice*. American Elsevier, New York; North-Holland, Amsterdam.

—— (1978). *Set Theory*. Academic Press, New York.

I. Juhasz (1965). Untersuchungen über ω_μ-metrisierbare raume. *Ann. Univ. Sci. Sec. Mat., Budapest*, **8**, 129–145.

Kleene, S. C. (1952). *Introduction to Meta-Mathematics*. North-Holland, Amsterdam.

Kline, M. (1972). *Mathematical Thought from Ancient to Modern Times*. Oxford University Press, Oxford.

Klir, G. (1970). *An Approach to General Systems Theory*. Van Nostrand, Princeton, NJ.

—— (1985). *Architecture of Systems Problem Solving*. Plenum Press, New York.

J. Kreisel and J. L. Krivine (1967). *Elements of Mathematical Logic*. North-Holland, Amsterdam.

Krippendorff, K. (1980). *Content Analysis: An Introduction to Its Methodology*. Sage Publications, Beverly Hills, CA.

Kuhn, T. (1962). *The Structure of Scientific Revolutions*. University of Chicago Press, Chicago.

Kunen, K. (1980). *Set Theory: An Introduction to Independence Proofs*. North-Holland, Amsterdam.

Kuratowski, K. (1980). *A Half Century of Polish Mathematics: Remembrances and Reflections*. Pergamon, Oxford.

Kuratowski, K., and A. Mostowski (1976). *Set Theory: With an Introduction to Descriptive Set Theory*. North-Holland, Amsterdam.

Kurepa, D. (1934). Tableu ramifies densembles: Espaces pseudo-distancies. *C. R. Acad. Sci. Paris*, **198**, 1563–1565.

— (1936). Le probleme de Souslin et les espaces abstraits. *C. R. Acad. Sci. Paris*, **203**, 1049–1052.

— (1936). Sur les classes (E) et (D). *Publ. Math. Univ. Beograd.*, **5**, 124–132.

— (1936). L'Hypothese de Ramification. *C. R. Acad. Sci.*, Ser. 202, 185–187.

— (1956). Sur lecart abstrait. *Glasnik Mat. Fiz. Astr.*, **11**, 105–134.

R. Lauer (1976). Defining social problems: Public and professional perspectives. *Social Problems*, **24**, 122– 30.

Laugwitz, D. (1961). Anwendungen unendlichkleiner zahlen I. *J. Reine Angew. Math.*, **207**, 53–60.

— (1962). Anwendungen unendlichkleiner zahlen II. *J. Reine Angew. Math.*, **208**, 22–34.

— (1968). Eine nichtarchimidische erweiterung anordeneter korper. *Math. Nachr.*, **37**, 225–236.

Barry Lee (1978, 1979). *Introducing Systems Analysis and Design*, Vols. I, II. National Computer Center, Manchester.

Levi-Civita, T. (1892). Sugli infinitied infinitesimi attuali quali elementianalitici. *Atti. 1st Veneto Sc., Letted Art.*, Ser. 7^a, **4**, 1765–1815.

Lilienfeld, D. (1978). *The Rise of Systems Theory*. Wiley, New York.

Lin, Y. (submitted). *Centralized Systems and Their Existence*.

— (1987). A model of general systems. *Math. Modelling*, **9**, 95–104.

— (1988). An application of systems analysis in sociology. *Cybernetics and Systems*, **19**, 267–278.

— (1988). On periodic linked time systems. *Int. J. Systems Sci.*, **19**, 1299–1310.

— (1988). Can the world be studied in the viewpoint of systems? *Math. Computer Model.*, **11**, 738–742.

— (1989). Order structures of families of general systems. *Cybernetics and Systems*, **20**, 51– 66.

— (1989). Some properties of linear systems. *Int. J. Systems Sci.*, **20**, No. 6, 927–937.

— (1989). A multi-relation approach of general systems and tests of applications. *Synthese: Int. J. Epistem. Methodol. Phil. Sci.*, **79**, 473–488.

— (1989). Multi-level systems. *Int. J. Sys. Sci.*, **20**, 1875–1889.

— (1990). The concept of fuzzy systems. *Kybernetes: Int. J. Systems. Cybernet.*, textbf19, 45–51.

—— (1990). A few philosophical remarks on general systems theory. *Abstracts of 8th International Congress of Cybernetics and Systems*, June 11–15, New York.

—— (1990). Connectedness of general systems. *Systems Sci.*, **16**, 5–17.

—— (1990). A few systems-colored views of the world. In: R. E. Mickens (ed.), *Mathematics and Science*. World Scientific Press, Singapore.

—— (1994). Feedback transformation and its applications. *J. Systems Eng.*, **3**, 32–38.

Lin, Y., and K. Z. Forrest (1995). Existence of major diseases and modeling of epidemiological transition: A general systems theoretic approach. *Cybernetics and Systems*, **26**, 647–664.

Lin, Y. (1996). Developing a theoretical foundation for the laws of conservation. *Kybernetes: Int. J. Systems Cybernet.*, **24**, No. 5, 52–60.

Lin,, Y., Q. P. Hu, and D. Li (1997). Some unsolved problems in general systems theory I and II. *Cybernet. Systems*, **28**, 287–303, and 591–605.

Lin, Y., and Y.-H. Ma (1987). Remarks on analogy between systems. *Int. J. General Systems*, **13**, 135–141.

—— (1989). Remarks on the definition of systems. *Systems Anal. Model., Simu.*, **6**, 923–931.

—— (1990). General feedback systems. *Int. J. General Systems*, **18**, No. 2, 143–154.

—— (1993). System — a unified concept. *Cybernet. Systems*, **24**, 375–406.

Y. Lin, Y.-H. Ma, and R. Port (1998). Centralizability and its existence. *Math. Model. Sci. Comput.*

—— (1990). Several epistemological problems related to the concept of systems. *Math. Comput. Model.*, **14**, 52–57.

Y. Lin and Y.-P. Qiu (1987). Systems analysis of the morphology of polymers. *Math. Model.*, **9**, 493–498.

Lin, Y., and E. A. Vierthaler (1998). A systems identification of social problems and its application to public issues of contention during the twentieth century. *Math. Model. Sci. Comput.*

Lin, Y., and S.-T. Wang (1998). Developing a mathematical theory of computability which speaks the language of levels. *Math. Comput. Model.*

Ma, Y.-H., and Y. Lin (1987). Some properties of linked time systems. *Int. J. General Systems*, **13**, 125–134.

—— (1988). Limit properties of hierarchies of general systems. *Appl. Math. Lett.*, **1**, 157– 160.

—— (1990). Continuous mappings between general systems. *Math. Comput. Model.*, **14**, 58– 63.

T. de Marco (1980). *Structured Analysis: Systems Specifications*. Prentice-Hall, Englewood Cliffs, NJ.

Masani, P. R. (1994). The scientific methodology in the light of cybernetics. *Kybernetes: Int. J. Systems. Cybernet.*, **23**, No. 4, 5–132.

Mathematical sciences: a unifying and dynamical resource. (1985). *Notices of the American Mathematical Society*, **33**, 463–479.

Martin, D. A., and R. M. Solovay (1970). Internal Cohen extensions. *Ann. Math. Logic*, **2**, 143–178.

Mesarovic, M.,D. (1964). Views on general systems theory. In: M. D. Mesarovic (ed.), *Proceedings of the 2nd Systems Symposium at Case Institute of Technology* Wiley, New York.

Mesarovic, M. D., and Y. Takahara (1975). *General Systems Theory: Mathematical Foundations*. Academic Press, New York.

—— (1989). *Abstract Systems Theory*. Springer-Verlag, Berlin.

R. E. Mickens (1990). *Mathematics and Science*. World Scientific, Singapore.

Milsum, J. H. (1966). *Biological Control Systems Analysis*. New York, McGraw-Hill.

Minsky, M. L. (1967). *Computation, Finite and Infinite Machines*. Prentice Hall, Englewood Cliffs, NJ.

Mitten, L. G. (1974). Preference order dynamic programming. *Management Sci.*, **21**, 43–46.

Moore, A. W. (1991). In: T. Honderich (ed), *The Infinite: The Problems of Philosophy — Their Past and Present*. Routledge, London and New York.

Moritz, R. E. (1942). *On Mathematics and Mathematicians*. Dover, New York.

Mumford,, G., A. Land, and W. Hawgood (1978). A participative approach to computer systems. *Impact Sci. Soc.*, **28**, 235–253.

Negoita, C. V. (1992). Pullback versus feedback. In: C. V. Negoita (ed.), *Cybernetics and Applied Systems*. Marcel Dekker, New York, 239–247.

Nyikos, P., and H. C. Reichel (1975). On the structure of zero dimensional spaces. *Indag. Math.*, **37**, 120–136.

—— (1976). On uniform spaces with linearly ordered bases II. *Fund. Math.*, **93**, 1–10.

Omran, A. (1971). The epidemiologic transition: A theory of the epidemiology of population change. *Milband Memorial Fund Q.*, **49**, 509–538.

Perlman, J. S. (1970). *The Atom and the Universe*. Wadsworth, Belmont, CA.

Polya, G. (1973). *How to Solve It: A New Aspect of Mathematical Method*. 2nd ed. Princeton University Press, Princeton, NJ.

Popper, K. (1945). *The Open Society and Its Enemies*. Routledge, London.

Poston, T., and I. Stewart (1978). *Catastrophe Theory and Its Applications*. Pitman, London.

Qin, G.-G. (1988). *Pansystems Research about Dynamic Programming*. Guizhou Sci., No. 2, 73–79.

—— (1991). Dynamic programming and pansystems operation epitome principle. *J. Jiangsu Inst. Technol.*, **4**, No. 2, 62–68.

Quastler, H. (1955). *Information Theory in Biology*. The University of Illinois Press, Urbana, IL.

—— (1965). General principles of systems analysis. In: T. H. Waterman and H. J. Morowitz (eds.), *Theoretical and Mathematical Biology*. New York.

Quigley, F. D. (1970). *Manual of Axiomatic Set Theory*. Appleton-Century Crofts, New York.

Rapoport, A. (1949). Outline of a probabilistic approach to animal sociology, I–III. *Bull. Math. Biophys.*, **11**, 183–196, 273–281; **12** (1950), 7–17.

Rashevsky, N. (1956). Topology and life: In: search of general mathematical principles in biology and sociology. *General Systems*, **I**, 123–138.

Rescigno, A., and G. Segre (1966). *Drug and Tracer Kinetics*. Blaisdell, Waltham, MA.

Richards, I., and H. Youn (1990). *Theory of Distributions: A Non-Technical Introduction*. Cambridge University Press, Cambridge.

Rudin, M. E. (1977). Martin's axiom. In: J. Barwise (ed.), *Handbook of Mathematical Logic*. North-Holland, Amsterdam.

Rugh, W. J. (1981). *Nonlinear Systems Theory: The Volterra/Wiener Approach*. Johns Hopkins University Press, Baltimore.

Saito, T. (1986). Feedback transformation of linear systems. *Int. J. Systems Sci.*, **17**, No. 9, 1305–1315.

—— (1987). Structure of the class of linear functional time systems under output feedback transformation. *Int. J. Systems Sci.*, **18**, No. 3, 1017–1027.

Saito, T., and M. D. Mesarovic (1985). A meaning of the decoupling by feedback of linear functional time systems. *Int. J. General Systems*, **11**, no.1, 47–61.

Schetzen, M. (1980). *The Volterra and Wiener Theories of Non-Linear Systems*. Wiley, New York.

Schwartz, L. (1950). *Theorie des Distributions*, Vol. I. Hermann, Paris.

—— (1951). *Theorie des Distributions*, Vol. II. Hermann, Paris.

Shannon, C., and W. Weaver (1949). *The Mathematical Theory of Communication*. University of Illinois Press, Urbana, IL.

Shoenfield, J. R. (1975). Martin's axiom. *Am. Math. Monthly*, **82**, 610–617.

Shubik, M. (1983). *Game Theory in the Social Science: Concepts and Solutions*. MIT Press, Cambridge, MA.

Sikorski, R. (1950). Remarks on some topological spaces of higher power. *Fund. Math.*, **37**, 125–136.

Smith,, D., M. Eggen, and R. St. Andre (1990). *A Transition to Advanced Mathematics*. 3rd ed., Brooks/Cole, Pacific Grove, CA.

Spector, M., and J. I. Kitsuse (1977). *Constructing Social Problems*. Cummings, Menlo Park, CA.

Stevenson, F. W., and W. J. Thron (1971). Results on ω_μ-metric spaces and ω_μ- proximities. *Fund. Math.*, **73**, 171–178.

Stewart, J. (1987). *Calculus*. Brooks/Cole, Pacific Grove, CA.

Suslin, M. (1920). Problems 3. *Fund. Math.*, **1**, 223.

Systems Research, Yearbook (1969–1976). Vols. I–VIII, Blauberg, I. V., et al., (eds.), Moscow.

Takahara, Y. (1982). Decomposability conditions for general systems. *Large Scale Systems*, **3**, No. 1, 57–65.

Takahara, Y., and T. Asahi (1985). Causality invariance of general systems under output feedback transformations. *Int. J. Systems Sci.*, **16**, No. 10, 1185–1206.

Tamarkin, J. D. (1936). Twenty-five volumes of Fundamenta Mathematica. *Bull. Am. Math. Soc.*, **42**, 300.

Tarski, A. (1954, 1955). Contributions to the theory of models I, II, III. *Nederl. Akad. Wetensch. Proc. Ser. A.*, **57**, 572–581, 582–588; **58**, 56–64.

Tian, F., and X. C. Wang (1980). *The Course of Science of Sciences*. Scientific Press, Beijing.

Turing, A. M. (1936). On computable numbers, with an application to the Entscheidungs problem. *Proc. London Math. Soc.*, Ser. 2, **42**.

Vierthaler, E. A. (1993). Wholesale content analysis: Collecting long-term data to study large-scale societal changes of historic importance. *Math. Model. Sci. Comput.*, **2**, 310–19.

Voronese, G. (1891). *Fondamenti di Geometria a Piu Dimensioni*. Padova.

—— (1896). Intor noad alcune osservazioni sui segmeni infiniti e infinitesimi attuali. *Math. Ann.*, **47**, 423–432.

S.-T. Wang (1964). Remarks on ω_μ- additive spaces. *Fund. Math.*, **55**, 101–112.

—— (1985). Generalized number system and its applications I. *Tsukuba J. Math.*, **9**, No. 2, 203–15.

—— (1991). A non-Archimedean number field and its applications in modern physics. In: G. M. Rassias (ed.), *The Mathematical Heritage of C. F. Gauss*. World Scientific, Singapor, pp. 810–826.

Wigner, E. P. (1960). The unreasonable effectiveness of mathematics in the natural sciences. *Comm. Pure Appl. Math.*, **13**, 1–14.

Wonham, W. M. (1979). *Linear Multi- Variable Control: A Geometric Approach*. Springer-Verlag, New York.

Wood-Harper, A. T., and G. Fitzgerald (1982). A taxonomy of current approaches to systems analysis. *Comput. J.*, **25**, 12–16.

Wu, C. P. (1980). Multicriteria dynamic programming. *Sci. Sinica*, **23**, 814–822.

—— (1981). Optimal control problems with nonscalar valued performance criteria and dynamic programming. *Proceedings of the 8th Triennial World Congress of the IFAC*, pp. X72–X76.

S.-Z. Wu and X.-M. Wu (1986). Generalized fundamental equations in pansystems network analysis. *Kybernetes*, **15**, 181–184.

Wu, X.-M. (1981). Pansystems analysis — a new exploration of interdisciplinary investigation (with appendices on discussions on medicine, chemistry, logic and economics). *Sci. Exploration*, **1**, No. 1. 125–164.

—— (1982). Pansystems methodology: Concepts, theorems and applications I. *Sci. Exploration*, **2**, No. 1, 33–56.

—— (1982). Pansystems methodology: Concepts, theorems and applications II. *Sci. Exploration*, **2**, No. 2, 93–106.

—— (1982). Pansystems methodology: Concepts, theorems and applications III. *Sci. Exploration*, **2**, No. 4, 123–132.

—— (1983). Pansystems methodology: Concepts, theorems and applications IV. *Sci. Exploration*, **3**, No. 1, 125–137.

—— (1983). Pansystems methodology: Concepts, theorems and applications V. *Sci. Exploration*, **3**, No. 4 97–104.

—— (1984). Pansystems methodology: Concepts, theorems and applications VI. *Sci. Exploration*, **4**, No. 1, 107–116.

—— (1984). Pansystems methodology: Concepts, theorems and applications VII. *Sci. Exploration*, **4**, No. 4, 1–14.

—— (1984). Investigation and applications of pansystems recognition theory and pansystems-operations research of large-scale systems. *Appl. Math. Mech.*, **5**, No. 1, 995–1010.

—— (1984). *Approximation Transformation Theory and the Pansystems Concepts in Mathematics*. Hunan Science and Technology Press.

—— (1985). Pansystems methodology: Concepts, theorems and applications VIII. *Sci. Exploration*, **5**, No. 1, 9–23.

—— (1994). From pansystems philosophy to pansystems poetry. *Tianshui Shichuan Xue Bao*, **30**, Nos. 3–4, 91- -150.

S.-Z. Wu and X.-M. Wu (1984). Generalized principle of optimality in pansystems network analysis. *Kybernetes*, **13**, 231–235.

Yablonsky, A. I. (1984). The development of science as an open system. In: J. M. Gvishiani (ed.), *Systems Research: Methodological Problems*. Oxford University Press, New York, pp. 211–228.

Yasui, Y. (1975). On ω_μ- metrizable spaces. *Math. Japon.*, **20**, 159–180.

Zadeh, L. (1962). From circuit theory to systems theory. *Proc. IRE*, **50**, 856–865.

Zhu, X.-D., and X.-M. Wu (1987). An approach on fixed pansystems theorems: Panchaos and strange panattractor. *Appl. Math. Mech.*, **8**, No. 4, 339–344.

Index

MIX
Papier aus verantwortungsvollen Quellen
Paper from responsible sources
FSC® C105338

If you have any concerns about our products,
you can contact us on
ProductSafety@springernature.com

In case Publisher is established outside the EU,
the EU authorized representative is:
Springer Nature Customer Service Center GmbH
Europaplatz 3, 69115 Heidelberg, Germany

Printed by Libri Plureos GmbH
in Hamburg, Germany